T0298709

A BASIC COURSE IN REAL ANALYSIS

A BASIC COURSE IN REAL ANALYSIS

AJIT KUMAR

S. KUMARESAN

CRC Press
Taylor & Francis Group
Boca Raton   London   New York

CRC Press is an imprint of the
Taylor & Francis Group, an **informa** business

A CHAPMAN & HALL BOOK

CRC Press
Taylor & Francis Group
6000 Broken Sound Parkway NW, Suite 300
Boca Raton, FL 33487-2742

© 2014 by Taylor & Francis Group, LLC
CRC Press is an imprint of Taylor & Francis Group, an Informa business

No claim to original U.S. Government works

Version Date: 20130911

International Standard Book Number-13: 978-1-4822-1637-0 (Hardback)

Library of Congress Cataloging-in-Publication Data

Kumar, Ajit, 1972-
  A basic course in real analysis / Ajit Kumar and S. Kumaresan.
    pages cm
  "A CRC title."
  Includes bibliographical references and index.
  ISBN 978-1-4822-1637-0
  1. Functions of real variables--Textbooks. 2. Mathematical analysis--Textbooks. 3. Numbers, Real--Textbooks. I. Kumaresan, S. II. Title.

QA331.5.K86 2014
515'.8--dc23   2013035473

Visit the Taylor & Francis Web site at
http://www.taylorandfrancis.com

and the CRC Press Web site at
http://www.crcpress.com

Dedicated to

all who believe and take pride in the dignity of teaching

and especially to the

Mathematics Training and Talent Search Programme, India

Contents

Preface

This book is based on more than two decades of teaching Real Analysis in the famous Mathematics Training and Talent Search Programme across the country in India.

The unique features of our book are as follows:

1) We create an interest in Analysis by encouraging readers to think geometrically.

2) We encourage readers to investigate and explore pictures and guess the results.

3) We use pictures and leading questions to think of a possible strategy of proof. (There are more than a hundred pictures in the book.)

4) We preface all the major and difficult proofs with a strategy and explain how the strategy is translated into rigorous and precise (so-called) textbook proofs. (This will make sure that the reader does not miss the wood for the trees. This will also train readers to conceptualize and later write in such a way that it is acceptable to a professional.)

5) We explain the mystery and role of inequalities in analysis and train the students to arrive at estimates that will be useful for the proofs.

6) We emphasize the role of the least upper bound property of real numbers, which underlies all crucial results in real analysis. (The prevalent impression is that the Cauchy completeness of the real number system is the cornerstone of analysis.)

7) We keep a conversational tone so that the reader may feel that a teacher is with him all the way. (This will ensure that the book is eminently suitable for self-study, a much-felt need in a country like India where there is a paucity of teachers when compared to the large number of students.)

8) We attend to the needs of a conscientious teacher who would like to explain the hows and whys of the subject. (Typically, such teachers can explain the line-by-line proof or the logical steps, but are at a loss to explain how the results and concepts were arrived at and why the proof works or the central or crucial idea of the proof. In fact, these are often the features that help students develop a feel for the subject.)

9) We show both aspects of analysis, as a qualitative as well as quantitative study of functions. The prevalent practice is to introduce topological notions at a very early stage (which may be construed as qualitative) at the cost of traditional

ε-δ, ε-n_0 treatment (which may be considered as quantitative). Students who are exposed very early to the topological notions are invariably very uncomfortable while dealing with some proofs where the arguments fall under the name hard-analysis.

As some of the novelties, we would like to mention our proof of Cauchy completeness of $\mathbb{R}$ (Theorem 2.2.3), the sequential definition preceding the ε-δ definition or the limit definition of continuity, the repeated employment of the auxiliary function f_1 while dealing with differentiation (Theorem 4.1.3), the use of the curry leaf trick to give an understandable proof of many results such as $|f(x) - f_n(x)| = \lim_{m \to \infty} |f_m(x) - f_n(x)| \le \varepsilon$ (Theorem 7.3.12), and the emphasis on the LUB property of $\mathbb{R}$ throughout the course.

Experts may find that the present book lacks a few topics which are included in a standard first course in real analysis such as a rigorous treatment of real number system, basic notions of metric spaces, and perhaps, the Riemann-Stieltjes integral. We find that the students find a rigorous treatment of the real number system in a first course, especially in the beginning, too abstract and too abstruse, and unless they are extremely committed, they are driven away from analysis. We suggest that it may be dealt with after a first course in real analysis and the construction of $\mathbb{R}$ may be carried out in detail. As for the second topic, we believe that the introduction to topological and metric space concepts may be introduced in a second course in real analysis which may deal with several variable calculus and measure theory. It is our experience that a thorough introduction of Riemann integral via Darboux sums offers geometric insights to the concepts and proofs to the students while if we started with the Riemann-Stieltjes integral, it hides the very same insights. Students who have gone through the Riemann integral find it easier to master the Stieltjes integral within a couple of lectures.

Our original intention was to introduce these topics briefly in three appendices, but to keep the book in reasonable size we shelved the idea. We are open to include them in future editions if a large number of readers, especially the teachers, demand it.

The book has more than 100 pictures. All the pictures were drawn using the free software Sage, GeoGebra, and TikZ. The book, of course, is typeset in TEX/LATEX. We thank the creators of these wonderful tools.

We thank the participants of the MTTS Programme and the faculty who appreciated the way Analysis was taught in the MTTS camps and who urged us to write a book based on our experience in these camps. We shall be satisfied if students find that their confidence in Analysis is enhanced by the book.

We would like to receive comments, suggestions, and corrections from the readers of the book. They may be sent to our email addresses ajit72@gmail.com or to kumaresa@gmail.com. Please mark the subject as Real Analysis (Comments). We shall maintain a list of errata at http://main.mtts.org.in/downloads.

Ajit Kumar

S. Kumaresan

To the Students

We wrote this book with the aim of this being used for self-study.

There are many excellent books in real analysis. Many of them are student friendly. Their exposition is clear, precise, and exemplary more often than not. All of them, if we may say so, are aimed at students who are mathematically mature enough and who have an implicit faith that by repeated drill and rigor, they would absorb the tricks of the subject. Also, they assume that the readers have access to teachers who know the subject well and who can steer their students toward the understanding of the subject.

This book aims at students who are not sure of why they have to learn the subject. They are often puzzled about the mortal rigor of the subject. They start wondering how in the world the teacher knew how to prove the result. Most often, they do not even ask why those results were thought to be true or conceived of. The book addresses typically those students who do not have access to peers or teachers who are experts, but are willing to learn on their own. They also want to understand the whys and hows rather than simply believe it is good to learn these in an abstract and formal way and that things will work out.

Almost everything is learned in the early years of our life by observation and mimicking. We wrote the book keeping this principle in mind. In our writing, we often exhibit our raw thought process and the kind of questions we ask ourselves when we attempt to prove a result or solve a problem. Hopefully, you as the reader may pick up this process either consciously or unconsciously.

This books offers insights into the way a typical mathematician works. He observes a pattern, explores further or conducts experiments by means of looking at or creating examples, tries to understand the underlying principles and comes up with guesses or conjectures, and proves it rigorously based on his explorations. The proofs typically incorporate many of his nebulous ideas in a professional manner. In a typical exposition which tries to be concise and very precise, these aspects of the professional life are never brought out. Also, any subject has its own ethos, a particular way of looking at things and a few standard tricks which take care of almost nine cases out of ten, at least a large number of cases. This book makes a very serious and committed attempt to initiate the students in a friendly way to these aspects of Real Analysis.

All concepts, definitions, and result/theorems are motivated by a variety of means—by geometric thinking, by drawing analogies with real-world phenomena, or by drawing parallels within the realm of mathematics and so on. Insightful discussions and a plan of attack (called as strategy in the book) precede almost every proof. Then we carefully explain how these ideas translate into rigorous and precise proofs.

To master any subject, along with formalism, one needs to develop a feeling for it. This can be achieved by various means, by geometric thinking, by drawing analogies with real-world phenomena, or by drawing parallels within the realm of mathematics and so on. This book employs all these and more. It also aims to train the readers how the ideas gathered by these methods are translated into a formal and rigorous writing. Almost all proofs start with a strategy which captures the essential ideas of the proof and then we work out the details. This way the readers do not miss the forest for the trees. Many a time teachers go through the proofs in a very formal way. The students are convinced that the proof is logically correct but may be left with a feeling of inadequacy or overwhelmed with smothering details.

Let us warn you that some of our writing (especially the parts that aim to motivate or lay down strategy) may sound vague and you may not be able to appreciate it the first time. Please keep in mind that when trying to understand any phenomenon, human beings start with vague questions, nebulous explanations, which in turn give rise to more precise pointed questions. This process repeats and finally leads to a correct answer. Nothing, as a rule, was served on a platter while humans started wondering about various phenomena. After a few weeks of study, you will begin to understand their role in the development of the concepts and proofs.

There are quite a few exercises, and a few of them have hints. Some of the hints are again questions! Needless to say, we expect you to make serious attempts to solve them on your own and only after repeated failures should you look at the hints.

We are confident that this book fills a much felt need and empowers the students to think analysis on their own and take charge of their understanding.

We shall be amply rewarded if you appreciate our efforts and enjoy doing analysis. You may send your comments and suggestions to the email addresses mentioned in the preface.

Ajit Kumar
Mumbai, India

S. Kumaresan
Hyderabad, India

About the Authors

Dr. Ajit Kumar is a faculty member at the Institute of Chemical Technology, Mumbai, India. His main interests are differential geometry, optimization and the use of technology in teaching mathematics. He received his Ph.D. from University of Mumbai. He has initiated a lot of mathematicians into the use of open source mathematics software.

Dr. S Kumaresan is currently a professor at University of Hyderabad. His initial training was at Tata Institute of Fundamental Research, Mumbai where he earned his Ph.D. He then served as a professor at University of Mumbai. His main interests are harmonic analysis, differential geometry, analytical problems in geometry, and pedagogy. He has authored five books, ranging from undergraduate level to graduate level. He was the recipient of the C.L.C Chandna award for Excellence in Teaching and Research in Mathematics in the year 1998. He was selected for the Indian National Science Academy Teacher award for 2013. He was a member of the Executive Committee of International Commission on Mathematics Instruction during the period 2007–2009.

For the last several years, Dr. Ajit Kumar and Dr. Kumaresan have been associated with the Mathematics Training and Talent Search Programme, India, aimed at undergraduates.

List of Figures

Chapter 1

Real Number System

Contents

In this chapter, we shall acquaint ourselves with the real number system. We introduce the set of real numbers in an informal way. Once this is done, the most important property of the real number system, known as the least upper bound property, is introduced. We assume that you know the set $\mathbb{N}$ of natural numbers, the set $\mathbb{Z}$ of integers, and the set $\mathbb{Q}$ of rational numbers. You know that $\mathbb{N} \subset \mathbb{Z} \subset \mathbb{Q}$. You also know the arithmetic operations, addition and multiplication of two natural numbers, integers, and rational numbers. You also know the order relation $m < n$ between two integers, more generally between two rational numbers.

1.1 Algebra of the Real Number System

The set $\mathbb{R}$ of real numbers is a set which contains $\mathbb{Q}$ and on which we continue to have arithmetic operations and an order relation. We now list the properties of the real number system. The reader should not be overwhelmed by the list. All of them must be familiar to him.

The properties of the addition of real numbers are listed below.

(A1) $x + y = y + x$ for $x, y \in \mathbb{R}$. (Commutativity of Addition)

(A2) $(x + y) + z = x + (y + z)$ for all $x, y, z \in \mathbb{R}$. (Associativity of Addition)

(A3) There exists a unique element $0 \in \mathbb{R}$ such that $x + 0 = 0 + x = x$ for all $x \in \mathbb{R}$. (Existence of Additive Identity or Zero)

1

(A4) For every $x \in \mathbb{R}$, there exists a unique element $y \in \mathbb{R}$ such that $x + y = 0$. We denote this y by $-x$. (Existence of Additive Inverse)

The properties of the multiplication of real numbers are listed below.

(M1) $x \cdot y = y \cdot x$ for all $x, y \in \mathbb{R}$. (Commutativity of Multiplication)

(M2) $x \cdot (y \cdot z) = (x \cdot y) \cdot z$ for all $x, y, z \in \mathbb{R}$. (Associativity of Multiplication)

(M3) There exists a unique element $1 \in \mathbb{R}$ such that $x \cdot 1 = 1 \cdot x = x$ for all $x \in \mathbb{R}$. (Existence of Multiplicative Identity)

(M4) Given $x \neq 0$ in $\mathbb{R}$, there exists a unique y such that $xy = 1 = yx$. We denote this y by x^{-1} or by $\frac{1}{x}$. (Existence of Multiplicative Inverse or Reciprocal)

Finally we have the distributive law which says how the two operations interact with each other.

(D) For all $x, y, z \in \mathbb{R}$, we have $x \cdot (y + z) = x \cdot y + x \cdot z$. (Distributivity of multiplication $\cdot$ over addition $+$.)

There is an order relation on $\mathbb{R}$ which satisfies the Law of Trichotomy. Given any two real numbers x and y, one and exactly one of the following is true:

> Law of Trichotomy: $x = y$, $x < y$, or $y < x$.

We often write $y > x$ to denote $x < y$. Also, the symbol $x \leq y$ means either $x = y$ or $x < y$, and so on.

Remark 1.1.1. Let $x, y \in \mathbb{R}$. Assume that $x \leq y$ and $y \leq x$. We claim that $x = y$. If false, then either, $x < y$ or $y < x$ by law of trichotomy. Assume that we have $x < y$. Since $y \leq x$, either $x = y$ or $y < x$. Neither can be true, since we assumed $x \neq y$ and hence concluded $x < y$ from the first inequality $x \leq y$. Hence we conclude that the second inequality cannot be true, a contradiction. Thus our assumption that $x \neq y$ is not tenable.

Proposition 1.1.2. *We list some of the important facts about this order relation in* $\mathbb{R}$.

(1) *If $x < y$ and $y < z$, then $x < z$. (Transitivity)*

(2) *If $x < y$ and $z \in \mathbb{R}$, then $x + z < y + z$.*

(3) *If $x < y$ and $z > 0$, then $xz < yz$.*

(4) *If $x < y$, then $-y < -x$.*

(5) *For any $x \in \mathbb{R}$, $x^2 \geq 0$. In particular, $0 < 1$.*

(6) *If $x > 0$ and $y < 0$, then $xy < 0$.*

(7) *If $0 < x < y$, then $0 < 1/y < 1/x$.*

Many other well-known order relations can be derived from the above list. Just to test your understanding, do the following exercises.

Exercise Set 1.1.3.

(1) For $x < y$, we have $x < \frac{x+y}{2} < y$. The point $(x + y)/2$ is known as the midpoint between x and y.

(2) If $x \leq y + z$ for all $z > 0$, then $x \leq y$.

(3) For $0 < x < y$, we have $0 < x^2 < y^2$ and $0 < \sqrt{x} < \sqrt{y}$, assuming the existence of $\sqrt{x}$ and $\sqrt{y}$. More generally, if x and y are positive, then $x < y$ iff $x^n < y^n$ for all $n \in \mathbb{N}$.

(4) For $0 < x < y$, we have $\sqrt{xy} < \frac{x+y}{2}$.

An important note: We are certain that you would have been taught to associate real numbers as points on a number line. In this book we shall use this to understand many concepts and results and to arrive at ideas for a proof. But rest assured that all our proofs will be rigorous and in fact they will train you in translating geometric ideas into rigorous proofs that will be accepted by professional mathematicians.

1.2 Upper and Lower Bounds

Definition 1.2.1. Let $A \subset \mathbb{R}$ be nonempty. We say that a real number α is an *upper bound* of A if, for each $x \in A$, we have $x \leq \alpha$. Geometrically, this means that elements of A are to the left of α on the number line. Look at Figure 1.1.

In terms of quantifiers, α is an upper bound of A if

$$\forall x \in A(x \leq \alpha).$$

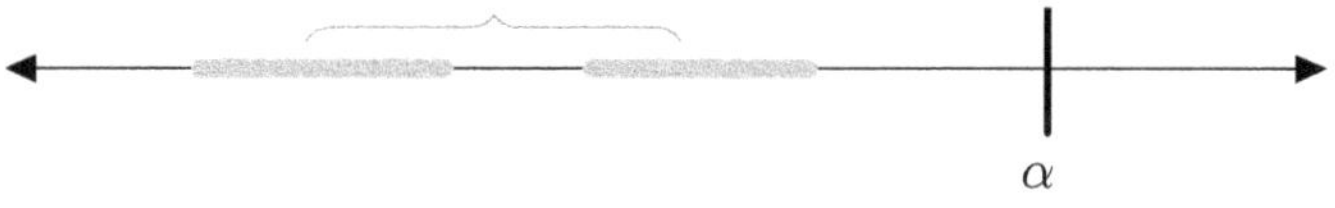

Figure 1.1: α is an upper bound of A.

A real number α is *not* an upper bound of A if there exists at least one $x \in A$ such that $x > \alpha$. This means that in the number line, we can find an element of A to the right of α. See Figure 1.2.

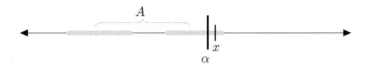

Figure 1.2: α is not an upper bound of A.

If α is an upper bound of A and $\alpha' > \alpha$, then α' is also an upper bound of A. See Figure 1.3.

Figure 1.3: Upper bound of a set is not unique.

Lower bounds of a nonempty subset of $\mathbb{R}$ are defined analogously.

If α is a lower bound of A, where can you find elements of A in the number line with reference to α? When do you say a real number is not a lower bound of A? Express these in terms of quantifiers.

There exists a lower bound for $\mathbb{N}$ in $\mathbb{R}$. Does there exist an upper bound for $\mathbb{N}$ in $\mathbb{R}$? The answer is No and it requires a proof which involves the LUB property of $\mathbb{R}$, the single most important property of $\mathbb{R}$. See Theorem 1.3.2.

Definition 1.2.2. $\emptyset \neq A \subset \mathbb{R}$ is said to be *bounded above* in $\mathbb{R}$ if there exists $\alpha \in \mathbb{R}$ which is an upper bound of A. That is, if there exists $\alpha \in \mathbb{R}$ such that for each $x \in A$, we have $x \leq \alpha$.

Note that in terms of quantifiers, we may write it as follows. A nonempty subset $A \subset \mathbb{R}$ is bounded above in $\mathbb{R}$ if

$$\exists\, \alpha \in \mathbb{R} \; (\forall x \in A \; (x \leq \alpha)).$$

Subsets of $\mathbb{R}$ *bounded below* in $\mathbb{R}$ are defined analogously. Readers are encouraged to write this definition.

We urge the readers to go through Appendix A to review the role of quantifiers in mathematics and the mathematical way of negating statements which involve quantifiers. A quick review will prepare you to understand what follows.

A is *not* bounded above in $\mathbb{R}$ if for each $\alpha \in \mathbb{R}$, there exists $x \in A$ (which depends on α) such that $x > \alpha$. Look at Figure 1.4.

In terms of quantifiers, we write it as follows. A subset $A \subset \mathbb{R}$ is not bounded above in $\mathbb{R}$ if

$$\forall \alpha \in \mathbb{R} \; (\exists x \in A \; (x > \alpha)).$$

Figure 1.4: Set which is not bounded above.

Can you visualize this in a number line?

When do you say $A \subset \mathbb{R}$ is not bounded below in $\mathbb{R}$?

An upper bound of a set need not be an element of the set. For example, if $A = (a, b)$ is an interval, then b is an upper bound of A but is not an element of A. If an upper bound of A is an element of A, then it is called a *maximum* of A. Can you show that if $M \in A$ is a maximum of A, then it is unique? That is, if M' is also a maximum of A, then you are required to prove $M = M'$.

When do you say an element of a set A is a minimum of the set? Is it unique?

Example 1.2.3.

(1) If $\emptyset \neq A \subset \mathbb{R}$ is finite, then an upper bound of A belongs to A.

Let us prove it by induction on the number n of elements in A. If $n = 1$, then the result is clear. Let $n = 2$ and $A = \{a, b\}$. Then by law of trichotomy, either $a < b$ or $a > b$. In the first case, b is the maximum while in the second a is the maximum. Assume the result for any subset of n elements. Let $A = \{a_1, \ldots, a_n, a_{n+1}\}$ be a set with $n + 1$ elements. Then by induction the set $B := \{a_1, \ldots, a_n\}$ has a maximum, say, $b = a_j \in B$. Now the two-elements set $C := \{a_j, a_{n+1}\}$ has a maximum, say $c \in C$. Let M be a maximum of the two–element set $\{b, c\}$. Note that $M \in \{a_1, \ldots, a_{n+1}\}$. Then for any $1 \leq i \leq n$, we have $a_i \leq b \leq M$ and $a_{n+1} \leq c \leq M$. Thus $M \in A$ is a maximum of A.

(2) Any lower bound of a nonempty subset A of $\mathbb{R}$ is less than or equal to an upper bound of A. Let α be a lower bound of A and β an upper bound of A. Fix an element $a \in A$. Then $\alpha \leq a$ and $a \leq \beta$. That is, $\alpha \leq a \leq \beta$ or $\alpha \leq \beta$. (Observe that we made use of the fact that A is nonempty.)

Let us look at $A = (0, 1)$. Clearly 1 is an upper bound of A. Intuitively, it is clear that any upper bound of A must be greater than or equal to 1. Thus, 1 is the least among all upper bounds of A. The next definition captures this idea.

Definition 1.2.4. Let $\emptyset \neq A \subset \mathbb{R}$ be bounded above. A real number $\alpha \in \mathbb{R}$ is said to be a *least upper bound* A if (i) α is an upper bound of A and (ii) if β is an upper bound of A, then $\alpha \leq \beta$.

A *greatest lower bound* of a subset of $\mathbb{R}$ bounded below in $\mathbb{R}$ is defined analogously.

Proposition 1.2.5. *Let $A \subset \mathbb{R}$ be a nonempty subset bounded above in $\mathbb{R}$. If α and β are least upper bounds of A, then $\alpha = \beta$, that is, the least upper bound of a nonempty subset bounded above in $\mathbb{R}$ is unique.*

Proof. Since α is a least upper bound of A and β is a least upper bound and hence an upper bound of A, we conclude that $\alpha \leq \beta$. Similarly, since β is a least upper bound of A and α is a least upper bound and hence an upper bound of A, we conclude that $\beta \leq \alpha$. It follows that $\alpha = \beta$, by Remark 1.1.1. $\square$

In view of the last proposition, it makes sense to say α is *the* least upper bound of A. We use the notation lub A to denote the least upper bound of A. The symbol LUB is a shorthand notation for least upper bound. The least upper bound α of a set A is also known as the *supremum* of A and is denoted by sup A.

Look at $B := (0, 1)$. Clearly 0 is a lower bound of B. It is intuitively clear that if β is a lower bound of A, then $\beta \leq 0$. Thus 0 is the greatest among all lower bounds of B. Therefore it makes sense to define a greatest lower bound of a nonempty subset of $\mathbb{R}$ which is bounded below in $\mathbb{R}$. The reader should attempt to formulate its definition before he reads it below.

We say that $\beta \in \mathbb{R}$ is a *greatest lower bound* of a nonempty subset $B \subset \mathbb{R}$ which is bounded below in $\mathbb{R}$ if (i) β is a lower bound of B and (ii) if γ is any lower bound of B, then $\gamma \leq \beta$. It is easy to show that if β and β' are greatest lower bounds of a set B, then $\beta = \beta'$. We therefore say *the* greatest lower bound of a set.

What do the symbols glb B and GLB stand for? The glb B is also called the *infimum* of B and is denoted by inf B.

The following is the most useful characterization of the LUB.

Proposition 1.2.6. *A real number $\alpha \in \mathbb{R}$ is the least upper bound of A iff (i) α is an upper bound of A and (ii) if $\beta < \alpha$, then β is not an upper bound of A, that is, if $\beta < \alpha$, then there exists $x \in A$ such that $x > \beta$.*

Proof. Let α be the LUB of A. Let $\beta < \alpha$ be given. Since $\beta < \alpha$, and α is the LUB of A, β cannot be an upper bound of A. Hence there exists $x \in A$ such that $x > \beta$. Look at Figure 1.5.

Conversely, let α be an upper bound of A with the property that for any $\beta < \alpha$, there exists $x \in A$ such that $x > \beta$. We need to prove that α is the LUB of A. Let β be an upper bound of A. We claim that $\beta \geq \alpha$. If false, then $\beta < \alpha$. Hence by hypothesis, there exists $x \in A$ such that $x > \beta$. That is, β is not an

Figure 1.5: α is the least upper bound of A.

upper bound, contradicting our assumption. Hence $\beta \geq \alpha$. $\square$

Remark 1.2.7. Most often the proposition above is used by taking $\beta = \alpha - \varepsilon$ for some $\varepsilon > 0$.

Example 1.2.8. If an upper bound α of A belongs to A, then lub $A = \alpha$. Thus the maximum of a set, if it exists, is the LUB of the set.

Exercise 1.2.9. State and prove the results for GLB analogous to Propositions 1.2.5 and 1.2.6.

Exercise 1.2.10. What is the analogue of Example 1.2.8 for GLB?

Example 1.2.11. Let $A = (0,1) := \{x \in \mathbb{R} : 0 < x < 1\}$. Then lub $A = 1$ and glb $A = 0$. To prove the first, observe that if $0 < \beta < 1$, then $(1 + \beta)/2 \in A$.

Clearly, 1 is an upper bound for A. Let b be an upper bound of A. Since $1/2 \in A$, we have $0 < 1/2 \leq b$. We claim $b \geq 1$. If not, $b < 1$. Hence $b \in (0,1)$. Now the midpoint $x := (b+1)/2 \in A$, but $x - b = (1-b)/2 > 0$ or $x > b$. That is, b is not an upper bound of A. Look at Figure 1.6.

Figure 1.6: LUB of $(0,1)$.

Can you adapt this argument to conclude lub $(a,b) = b$?

Example 1.2.12. Let I be a nonempty set. Assume that we are given two subsets A and B of $\mathbb{R}$ which are indexed by I. That is, $A := \{a_i : i \in I\} \subset \mathbb{R}$ and $B := \{b_i : i \in I\} \subset \mathbb{R}$. Assume further that $a_i \leq b_i$ for each $i \in I$. Let us assume that both the sets are bounded above and $\alpha := $ lub A and $\beta := $ lub B. Then we claim $\alpha \leq \beta$. For, β is an upper bound of A. If $a_i \in A$, then $a_i \leq b_i \leq \beta$ and hence the claim. Since $\alpha := $ lub A, and β is an upper bound of A, it follows that $\alpha \leq \beta$.

Now assume that $a_i \leq b_i$ for each $i \in I$. Let us assume that both the sets are bounded below and $\alpha := $ glb A and $\beta := $ glb B. Then α is a lower bound for B since if $b_j \in B$, then $\alpha \leq a_j \leq b_j$. Hence the lower bound α of B is less than or equal to the GLB of B, namely β.

1.3 LUB Property and Its Applications

Assume that $A \subset \mathbb{R}$ is bounded above in $\mathbb{R}$. Hence there is a real number which is an upper bound of A. The question now arises whether there exists a real number α which will be the least upper bound of A. The LUB property of $\mathbb{R}$ asserts the existence of such an α.

LUB Property of $\mathbb{R}$:

Given any nonempty subset of $\mathbb{R}$ which is bounded above in $\mathbb{R}$, there exists $\alpha \in \mathbb{R}$ such that $\alpha = \text{lub } A$.

Thus, any subset of $\mathbb{R}$ which has an upper bound in $\mathbb{R}$ has the lub **in** $\mathbb{R}$.

Note that lub A need not be in A. See Example 1.2.11.

The LUB property of $\mathbb{R}$ is the single most important property of the real number system and all key results in real analysis depend on it. It is also known as the order–completeness of $\mathbb{R}$.

Remark 1.3.1. The set $\mathbb{Q}$ of rational numbers satisfies all the properties listed in Section 1.1. However, it does not enjoy the LUB property. We shall see later (Remark 1.3.20) that the subset $\{x \in \mathbb{Q} : x^2 < 2\}$ is bounded above in $\mathbb{Q}$ and does not have an LUB *in* $\mathbb{Q}$.

As first two applications of the LUB property, we establish two versions of the *Archimedean property*.

Theorem 1.3.2 (Archimedean Property).
(AP1): $\mathbb{N}$ *is* not *bounded above in $\mathbb{R}$. That is, given any $x \in \mathbb{R}$, there exists $n \in \mathbb{N}$ such that $x > n$.*
(AP2): *Given $x, y \in \mathbb{R}$ with $x > 0$, there exists $n \in \mathbb{N}$ such that $nx > y$.*

Proof. (AP1): We prove this result by contradiction. Assume that $\mathbb{N}$ is bounded above in $\mathbb{R}$. By the LUB property of $\mathbb{R}$, there exists $\alpha \in \mathbb{R}$ such that $\alpha = \text{lub } \mathbb{N}$. Then for each $k \in \mathbb{N}$, we have $k \leq \alpha$. Since we wish to exploit the fact that α is the LUB of $\mathbb{N}$ and since we are dealing with integers, we consider $\alpha - 1 < \alpha$. Then $\alpha - 1$ is not an upper bound of $\mathbb{N}$ and hence there exists $N \in \mathbb{N}$ such that $N > \alpha - 1$. Adding 1 to both sides yields $N + 1 > \alpha$. Since $N + 1 \in \mathbb{N}$, we are forced to conclude that α is not an upper bound of $\mathbb{N}$. Hence our assumption that $\mathbb{N}$ is bounded above is wrong.

(AP2): Proof by contradiction. If false, then there exists $x > 0$, $y \in \mathbb{R}$ such that for each $n \in \mathbb{N}$, we must have $nx \leq y$. Since $x > 0$, we have $n \leq y/x$ for all $n \in \mathbb{N}$. That is, y/x is an upper bound for $\mathbb{N}$. This contradicts (AP1). $\qquad\square$

Remark 1.3.3. The version AP2 is the basis of all units and measurements! It says, any tiny quantity can be used as a unit against which others are measured.

Proposition 1.3.4. *Both the Archimedean principles are equivalent.*

Proof. Since we deduced AP2 from AP1, we need only show that AP2 implies AP1.

It is enough to show that no $\alpha \in \mathbb{R}$ is an upper bound of $\mathbb{N}$. Given $\alpha \in \mathbb{R}$, let $x = 1$ and $y = \alpha$. Then by AP2, there exists $n \in \mathbb{N}$ such that $nx > y$, that is, $n > \alpha$. This means α is not an upper bound of A. $\qquad\square$

The next couple of results are easy consequences of the Archimedean property. Let not the simplicity of their proofs deceive you. They are perhaps the most useful tools in analysis.

Theorem 1.3.5. (1) *Given $x > 0$, there exists $n \in \mathbb{N}$ such that $x > 1/n$.*
(2) *Let $x \geq 0$. Then $x = 0$ iff for each $n \in \mathbb{N}$, we have $x \leq 1/n$.*

Proof. We apply AP2 with $y = 1$ and x. Then there exists $n \in \mathbb{N}$ such that $nx > 1$. Hence $x > 1/n$. This proves (1).

(2). If $x = 0$, clearly, for each $n \in \mathbb{N}$, $x \leq 1/n$. We prove the converse by contradiction. Let, if possible, $x > 0$. Then by (1), there exists $N \in \mathbb{N}$ such that $x > 1/N$. This contradicts our hypothesis that for each $n \in \mathbb{N}$, $x \leq 1/n$. □

Exercise 1.3.6. Use the last result to give a proof of lub $(0, 1) = 1$ and glb $(0, 1) = 0$.

Remark 1.3.7. Typical use of Theorem 1.3.5 in Analysis: when we want to show two real numbers a, b are equal, we show that $|a - b| \leq 1/n$ for all $n \in \mathbb{N}$.

Exercise Set 1.3.8. Typical uses of the Archimedean property.

(1) Let $J_n := (0, \frac{1}{n})$. Show that $\cap_n J_n := \{x \in \mathbb{R} : \forall n \in \mathbb{N}, x \in J_n\} = \emptyset$.

(2) Let $J_n := [n, \infty)$. Show that $\cap_n J_n = \emptyset$.

(3) Let $J_n := (1/n, 1)$. Show that $\cup_n J_n := \{x \in \mathbb{R} : \exists n \in \mathbb{N}, x \in J_n\} = (0, 1)$.

(4) Write $[0, 1] = \cap_n J_n$ where J_n's are open intervals containing $[0, 1]$.

Exercise Set 1.3.9.

(1) Show that, for $a, b \in \mathbb{R}$, $a \leq b$ iff $a \leq b + \varepsilon$ for all $\varepsilon > 0$.

(2) Prove by induction that $2^n > n$ for all $n \in \mathbb{N}$. Hence conclude that for any given $\varepsilon > 0$, there exists $N \in \mathbb{N}$ such that if $n \geq N$, then $2^{-n} < \varepsilon$.

Proposition 1.3.10. *The set $\mathbb{Z}$ of integers is neither bounded above nor bounded below.*

Proof. Let us first prove that $\mathbb{Z}$ is not bounded above in $\mathbb{R}$. For, if $\alpha \in \mathbb{R}$ is an upper bound of $\mathbb{Z}$, then for each $n \in \mathbb{N} \subset \mathbb{Z}$, we have $n \leq \alpha$. Hence $\mathbb{N}$ is bounded above in $\mathbb{R}$, contradicting the Archimedean property (AP1).

Now we prove $\mathbb{Z}$ is not bounded below in $\mathbb{R}$. If $\beta \in \mathbb{R}$ is a lower bound of $\mathbb{Z}$, then for any $n \in \mathbb{N}$, we have $-n \in \mathbb{Z}$ and hence $-n \geq \beta$. That is, for each $n \in \mathbb{N}$, we have $n \leq -\beta$, again contradicting the Archimedean property (AP1). □

Proposition 1.3.11 (Greatest Integer Function). *Let $x \in \mathbb{R}$. Then there exists a unique $m \in \mathbb{Z}$ such that $m \leq x < m + 1$.*

Proof. Let $S := \{k \in \mathbb{Z} : k \le x\}$. We claim that $S \ne \emptyset$. For, otherwise, for each $k \in \mathbb{Z}$, we must have $k > x$. Let $n \in \mathbb{N}$ be arbitrary. Then $k = -n \in \mathbb{Z}$ and hence $-n = k > x$. It follows that $n < -x$. Hence $-x$ is an upper bound for $\mathbb{N}$, contradicting the Archimedean property. This proves our claim.

Furthermore, x is an upper bound of S, by the very definition of S.

Let $\alpha \in \mathbb{R}$ be its least upper bound. Since $\alpha - 1 < \alpha$, $\alpha - 1$ is not an upper bound of S. Then there exists $k \in S$ such that $k > \alpha - 1$. (See Figure 1.7.) Since $k \in S$, $k \le x$. We claim that $k + 1 > x$. For, if false, then $k + 1 \le x$. Therefore, $k + 1 \in S$. Since α is an upper bound for S, we must have $k + 1 < \alpha$ or $k < \alpha - 1$. This contradicts our choice of k. Hence we have $x < k + 1$. Thus, $k \le x < k + 1$.

The proposition follows if we take $m = k$.   □

Figure 1.7: Greatest integer function.

Next we prove that m is unique. Let n also satisfy $n \le x < n + 1$. If $m \ne n$, without loss of generality, assume that $m < n$ so that $n \ge m + 1$. (Why?) Since $m \le x < m + 1$ holds, we deduce that $m \le x < m + 1 \le n$. In particular, $n > x$, a contradiction to our assumption that $n \le x < n + 1$.   □

Observe that if $k + 1 < x$, the figure on the right of Figure 1.7 shows that the interval $[\alpha - 1, \alpha]$ of length 1 contains $[k, k + 1]$ of length 1 properly, as $k > \alpha - 1$. This apparent contradiction shows $k + 1 > x$, giving another proof.

Exercise 1.3.12. (i) Show that any nonempty subset of $\mathbb{Z}$ which is bounded above in $\mathbb{R}$ has a maximum.

(ii) Formulate an analogue for the case of subsets of $\mathbb{Z}$ bounded below in $\mathbb{R}$ and prove it.

The unique integer m such that $m \le x < m + 1$ is called the *greatest integer less than or equal to* x. It is often denoted by $[x]$. It is also called the *floor* of x and is denoted by $\lfloor x \rfloor$ in computer science. The number $x - \lfloor x \rfloor$ is called the *fractional part* of x. We observe that $0 \le x - \lfloor x \rfloor < 1$.

Theorem 1.3.13 (Density of $\mathbb{Q}$ in $\mathbb{R}$). *Given $a, b \in \mathbb{R}$ with $a < b$, there exists $r \in \mathbb{Q}$ such that $a < x < b$.*

Strategy: Assuming the existence of such an r, we write it as $r = \frac{m}{n}$ with $n > 0$. So, we have $a < \frac{m}{n} < b$, that is, $na < y < nb$. Thus we are claiming that the interval (nx, ny) contains an integer. It is geometrically obvious that a sufficient condition for an interval $J = (\alpha, \beta)$ to have an integer in it is that its length $\beta - \alpha$ is greater than 1. In our case, the length of (na, nb) is $nb - na = n(b - a) > 1$. Archimedean property assures of such n's. The m we are looking for is the one next to $[na]$.

Proof. Since $b - a > 0$, by AP2, there exists $n \in \mathbb{N}$ such that $n(b-a) > 1$. Let $k = [na]$ and $m := k + 1$. Then clearly, $na < m$. We claim $m < nb$. Look at Figure 1.8.

Figure 1.8: Density of $\mathbb{Q}$.

If $m > nb$, then the interval $(na, nb) \subset [m-1, m]$. The length of the interval $[na, nb]$ is > 1 while that of $[m-1, m] = 1$. This seems to be absurd. We turn this geometric reasoning into a proof using inequalities.

Consider

$$1 = (k+1) - k \geq nb - na = n(b-a) > 1, \text{ a contradiction.}$$

Hence we have $m < nb$. Thus, we obtain $na < m < nb$ or $a < m/n < b$.   □

Corollary 1.3.14. *Given $a, b \in \mathbb{R}$ with $a < b$, there exists $t \notin \mathbb{Q}$ such that $a < t < b$.*

Proof. Consider the real numbers $a - \sqrt{2} < b - \sqrt{2}$ and apply the last result. Let us work out the details.

By the density of rationals, there exists $r \in \mathbb{Q}$ such that $a - \sqrt{2} < r < b - \sqrt{2}$. We add $\sqrt{2}$ to each of the terms in the inequality to obtain $a < r + \sqrt{2} < b$. We claim $r + \sqrt{2}$ is irrational. For, otherwise, $s := r + \sqrt{2} \in \mathbb{Q}$. It follows that $s - r = \sqrt{2} \in \mathbb{Q}$. But we know that $\sqrt{2}$ is not rational. (See Proposition 1.3.16 below.)   □

Thus between any two real numbers there exists a rational number as well as an irrational number. How many such rational/irrational numbers exist between the given two distinct real numbers?

Exercise Set 1.3.15.

(1) Let $a \in \mathbb{R}$. Let $C_a := \{r \in \mathbb{Q} : r < a\}$. Show that lub $C_a = a$. Is the map $a \mapsto C_a$ of $\mathbb{R}$ into the power set $P(\mathbb{R})$ one to one?

(2) Given any open interval (a, b), $a < b$, show that the set $(a, b) \cap \mathbb{Q}$ is infinite.

(3) Let $t > 0$ and $a < b$ be real numbers. Show that there exists $r \in \mathbb{Q}$ such that $a < tr < b$.

The next result generalizes the well-known fact that $\sqrt{2}$ is irrational.

Proposition 1.3.16. *Let p be any prime. Then there exists no rational number r such that $r^2 = p$.*

Proof. Let $r = m/n \in \mathbb{Q}$ be such that $r^2 = p$. We assume that $m, n \in \mathbb{N}$ and they do not have common factors (other than 1). We have $m^2 = pn^2$. The prime number p divides the RHS[1] and hence p divides m^2. Since p is a prime, this means that p divides m. If we write $m = pk$, then we obtain $p^2k^2 = pn^2$ or $pk^2 = n^2$. We conclude that p divides RHS and hence p divides n. Thus, p is a common factor of m and n, a contradiction. We therefore conclude that no such r exists. $\square$

Remark 1.3.17. We can use the fundamental theorem of arithmetic to prove $\sqrt{2}$ is irrational and extend the argument to show that $\sqrt{n}$ is irrational where n is not a square (that is, $n \neq m^2$, for any integer m).

Thus there exists no solution **in** $\mathbb{Q}$ to the equations $X^2 = n$ where n is not a square. We contrast this with the next result which says that for any positive real number and a positive integer n, n-th roots exists in $\mathbb{R}$.

Theorem 1.3.18 (Existence of n-th roots of positive real numbers). *Let $\alpha \in \mathbb{R}$ be nonnegative and $n \in \mathbb{N}$. Then there exists a unique non-negative $x \in \mathbb{R}$ such that $x^n = \alpha$.*

> **Strategy:** Look at Figure 1.9. What we are looking for is the intersection of the graphs of the functions $y = \alpha$ and $y = x^n$. The common point will have coordinates (x, x^n) and (x, α). Hence it will follow that $x^n = \alpha$.
>
> Now how do we plan to get such an x? We work backward. Let $b \geq 0$ be such that $b^n = \alpha$. Now $b = $ lub $[0, b)$ and any $t \in [0, b)$ satisfies $t^n < \alpha$. Hence b may be thought of as the LUB of the set $S := \{t \geq 0 : t^n < \alpha\}$. Now if S is non-empty and bounded above, let $c := $ lub S. We hope to show that $c^n = \alpha$.

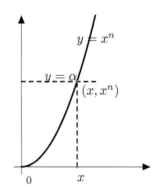

Figure 1.9: Graph of $y = x^n$.

---

[1]RHS stands for the *right-hand side*. What does LHS stand for?

We show that $c^n < \alpha$ or $c^n > \alpha$ cannot happen. See Figure 1.10. If $c^n < \alpha$, the picture shows that we can find $c_1 > c$, c_1 very near to c, such that we still have $c_1^n < \alpha$, $c_1 \in S$. This shows that $c_1 \in S$. Hence $c_1 \leq c$, a contradiction.

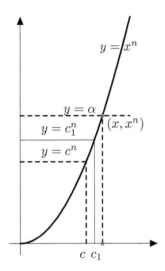

Figure 1.10: When $c^n < \alpha$.

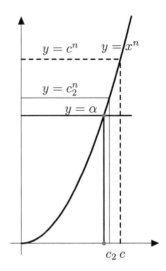

Figure 1.11: When $c^n > \alpha$.

If $c^n > \alpha$, Figure 1.11 shows that we can find a $c_2 < c$, c_2 very near to c but we still have $c_2^n > \alpha$. Since $c_2 < c$, there exists $t \in S$ such that $t > c_2$ and hence $t^n > c^n > \alpha$, a contradiction. Hence $c^n = \alpha$.

Now $c_1 > c$ and is very near to c and still retains $c_1^n < \alpha$. How do we look for such c_1? Naturally if $k \in \mathbb{N}$ is very large, we expect $c_1 = c + \frac{1}{k}$ is very near to c and $(c + \frac{1}{n})^n < \alpha$.

Similarly, for c_2 we look for a $k \in \mathbb{N}$ such that $(c - \frac{1}{k})^n > \alpha$.

In the first case, it behooves us to use the binomial expansion. We try to find an estimate of the form $(c + 1/k)^n \leq c^n + \frac{C}{k}$ for some constant C. So, it suffices to make sure $C/k < \alpha - c^n$, possible by Archimedean property.

In the second case, we try to find an estimate of the form $(c - 1/k)^n \geq c^n - \frac{C}{k}$. So, it suffices to make sure $c^n - \frac{C}{k} > \alpha$.

Proof. If $\alpha = 0$, the result is obvious, so we assume that $\alpha > 0$ in the following. We define
$$S := \{t \in \mathbb{R} : t \geq 0 \text{ and } t^n \leq \alpha\}.$$

Since $0 \in S$, we see that S is not empty. It is bounded above. For, by Archimedean property of $\mathbb{R}$, we can find $N \in \mathbb{N}$ such that $N > \alpha$. We claim that α is an upper bound for S. If this is false, then there exists $t \in S$ such that $t > N$. But, then we have

$$t^n > N^n \geq N > \alpha,$$

a contradiction, since for any $t \in S$, we have $t^n \leq \alpha$. Hence we conclude that N is an upper bound for S. Thus, S is a nonempty subset of $\mathbb{R}$, which is bounded above. By the LUB property of $\mathbb{R}$, there exists $x \in \mathbb{R}$ such that x is the LUB of S. We claim that $x^n = \alpha$.

Exactly one of the following is true: (i) $x^n < \alpha$, (ii) $x^n > \alpha$, or (iii) $x^n = \alpha$. We shall show that the first two possibilities do not arise.

Case (i): Assume that $x^n < \alpha$. For any $k \in \mathbb{N}$, we have

$$(x + 1/k)^n = x^n + \sum_{j=1}^{n} \binom{n}{j} x^{n-j}(1/k^j)$$

$$\leq x^n + \sum_{j=1}^{n} \binom{n}{j} x^{n-j}(1/k)$$

$$= x^n + C/k, \quad \text{where } C := \sum_{j=1}^{n} \binom{n}{j} x^{n-j}.$$

If we choose k such that $x^n + C/k < \alpha$, that is, for $k > C/(\alpha - x^n)$, it follows that $(x + 1/k)^n < \alpha$.

Case (ii): Assume that $x^n > \alpha$. We have $(-1)^j(1/k^j) > -1/k$ for $k \in \mathbb{N}$, $j \geq 1$. We use this below.

$$(x - 1/k)^n = x^n + \sum_{j=1}^{n} \binom{n}{j} (-1)^j x^{n-j}(1/k^j)$$

$$\geq x^n - \sum_{j=1}^{n} \binom{n}{j} x^{n-j}(1/k)$$

$$= x^n - C/k, \quad \text{where } C := \sum_{j=1}^{n} \binom{n}{j} x^{n-j}.$$

If we choose k such that $x^n - C/k > \alpha$, that is, if we take $k > C/(x^n - \alpha)$, it follows that $(x - 1/k)^n > \alpha$.

We now show that if x and y are non-negative real numbers such that $x^n = y^n = \alpha$, then $x = y$. Look at the following algebraic identity:

$$(x^n - y^n) \equiv (x - y) \cdot [x^{n-1} + x^{n-2}y + \cdots + xy^{n-2} + y^{n-1}].$$

If x and y are nonnegative with $x^n = y^n$ and if $x \neq y$, say, $x > y$, then the left-hand side is zero while both the factors in brackets on the right are strictly positive, a contradiction.

This completes the proof of the theorem. $\qquad \square$

How to get the algebraic identity

$$(x^n - y^n) \equiv (x - y) \cdot [x^{n-1} + x^{n-2}y + \cdots + xy^{n-2} + y^{n-1}]?$$

Recall how to sum a geometric series: $s_n := 1 + t + t^2 + \cdots + t^{n-1}$. Multiply both sides by t to get $ts_n = t + \cdots + t^n$. Subtract one from the other and get the formula for s_n. In the formula for s_n substitute y/x for t and simplify.

Remark 1.3.19. The proof of the last theorem brings out some standard tricks in analysis. Analysis uses inequalities to prove equalities and also to compare "unknown" objects with "known" objects. If we want to establish $A < B$, we may try to "simplify" the expression for A so that we get $A \leq C$ and $C \leq D$ and so on. Finally we may end up with $A \leq C \leq D \cdots \leq G$. G may be simple enough to put the appropriate conditions so as to make it less than B. Consequently, we obtain $A \leq B$. Go through the way the estimates for $(x + 1/k)^n$ and $(x - 1/k)^n$ were obtained. Do not be discouraged if you do not understand these vague remarks. As you go along, you will begin to appreciate this.

Remark 1.3.20. Observe that the argument of the proof in Theorem 1.3.18 can be applied to the set $A := \{x \in \mathbb{Q} : x^2 < 2\} \subset \mathbb{Q}$ (which is bounded above in $\mathbb{Q}$ by 2) to conclude that if $x \in \mathbb{Q}$ is the LUB of A, then $x^2 = 2$. (In the quoted proof, the numbers $x \pm \frac{1}{k} \in \mathbb{Q}$!) This contradicts Proposition 1.3.16. Hence we conclude that the ordered field $\mathbb{Q}$ does not enjoy the LUB property.

The interested reader should write down a complete proof of this.

Definition 1.3.21. A subset $J \subset \mathbb{R}$ is said to be an *interval* if $a, b \in J$ and if $a < x < b$, we then have $x \in J$:

$$\forall a, b \in J \text{ and } a < x < b \implies x \in J.$$

A real number x such that $a < x < b$ is said to be between a and b.

Example 1.3.22. Let $a \leq b$ be real numbers. We define

$$[a, b] := \{x \in \mathbb{R} : a \leq x \leq b\}$$
$$(a, b) := \{x \in \mathbb{R} : a < x < b\}$$
$$[a, b) := \{x \in \mathbb{R} : a \leq x < b\}$$
$$(a, b] := \{x \in \mathbb{R} : a < x \leq b\}$$
$$[a, \infty) := \{x \in \mathbb{R} : x \geq a\}$$
$$(a, \infty) := \{x \in \mathbb{R} : x > a\}$$
$$(-\infty, b] := \{x \in \mathbb{R} : x \leq b\}$$
$$(-\infty, b) := \{x \in \mathbb{R} : x < b\}$$
$$(-\infty, \infty) := \mathbb{R}.$$

Each subset of the list above is an interval. We shall prove this for $J = (a, b]$ as an example. Let $x, y \in J$ and $x < z < y$. We are required to prove that $z \in (a, b]$.

We observe that $a < x < z < y \le b$. Hence it follows that $a < z < b$ and hence $a < z \le b$.

Note that we have used ∞ and $-\infty$ as a place holder in the definitions of the last five subsets. In the definition of (a, ∞), there is one and only condition for a real number x to be in (a, ∞), namely, $x > a$. Compare and contrast this for $x \in \mathbb{R}$ to be in (a, b). We require that x satisfies two conditions $x > a$ and $x < b$. In the last case, there is no condition to be imposed on $x \in \mathbb{R}$ for it to be a member of $(-\infty, \infty)$!

Any interval which lies in the first four types is called a finite or bounded interval. The last five are called infinite or unbounded intervals.

Note that unlike many textbooks we have not insisted on $a < b$ while defining bounded intervals. In particular, if $a = b$, then $[a, b] = \{a\}$ is an interval and $(a, a) = \emptyset$ is an interval.

The intervals $[a, b]$, $[a, \infty)$, $(-\infty, b]$ are called closed intervals. The intervals of the form $[a, b)$, $(a, b]$ are called semi-open (or semi-closed) intervals.

The intervals (a, b), (a, ∞) and $(-\infty, b)$ are called open intervals.

It can be shown that any interval must be one of the types listed above.

Let $[a, b]$ and $[c, d]$ be intervals such that $[c, d] \subset [a, b]$. Then we have $a \le c \le d \le b$. Look at Figure 1.12. (How to prove this? Since $c \in [c, d] \subset [a, b]$, $c \in [a, b]$. That is $a \le c \le b$. Similarly, $a \le d \le b$.)

$$a \qquad c \qquad\qquad\qquad\qquad d \qquad b$$

Figure 1.12: $[c, d] \subset [a, b]$.

The next result deals with a sequence (J_n) of nested intervals, that is, $J_{n+1} \subset J_n$ for $n \in \mathbb{N}$. We assume that each one of them is bounded and closed. Let us look at some examples.

Let $J_n := [a - \frac{1}{n}, b + \frac{1}{n}]$. What is $\cap_n J_n$? Using the Archimedean property, we see that $\cap_n = [a, b]$.

Let $J_n := [a, b + \frac{1}{n}]$. Here too we obtain $\cap_n J_n = [a, b]$.

Let $J_n := [-\frac{1}{n}, \frac{1}{n}]$. We find that $\cap_n = \{0\}$. If you experiment a bit more, you will be convinced that under the stated conditions, $\cap_n J_n \ne \emptyset$. This is the content of the next theorem.

Theorem 1.3.23 (Nested Interval Theorem). *Let $J_n := [a_n, b_n]$ be intervals in $\mathbb{R}$ such that $J_{n+1} \subseteq J_n$ for all $n \in \mathbb{N}$. Then $\cap J_n \ne \emptyset$.*

Strategy: Assume that the result is true. Then there exists $c \in \mathbb{R}$ such that $a_n \le c \le b_n$. This says that c is an upper bound of $A := \{a_n : n \in \mathbb{N}\}$ and each b_k is an upper bound of A. Hence c is less than or equal to many upper bounds. An obvious choice for c is therefore lub A.

Proof. Let A be the set of left endpoints of J_n. Thus, $A := \{a \in \mathbb{R} : a = a_n \text{ for some } n\}$. A is nonempty.

We claim that b_k is an upper bound for A for each $k \in \mathbb{N}$, that is, $a_n \le b_k$ for all n and k.

If $k \le n$, then $[a_n, b_n] \subseteq [a_k, b_k]$ and hence $a_n \le b_n \le b_k$. See Figure 1.13.

Figure 1.13: Nested interval: $k \le n$.

If $k > n$, then $a_n \le a_k \le b_k$. See Figure 1.14. Thus the claim is proved.

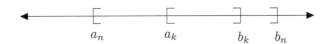

Figure 1.14: Nested interval: $k > n$.

By the LUB axiom there exists $c \in \mathbb{R}$ such that $c = \sup A$. We claim that $c \in J_n$ for all n. Since c is an upper bound for A, we have $a_n \le c$ for all n. Since each b_n is an upper bound for A and c is the least upper bound for A, we have $c \le b_n$. Thus we conclude that $a_n \le c \le b_n$ or $c \in J_n$ for all n. Hence $c \in \cap J_n$. $\square$

Definition 1.3.24. Let $x_1, x_2, \ldots, x_n \in \mathbb{R}$ be given. A *convex linear combination* of $\{x_j : 1 \le j \le n\}$ is any real number of the form $t_1 x_1 + \ldots + t_n x_n$ where (i) $t_j \ge 0$ for $1 \le j \le n$ and (ii) $t_1 + \ldots + t_n = 1$.

Note that if $n = 2$, we can write any convex linear combination of $x, y \in \mathbb{R}$ as $(1 - t)x + ty$ for $t \in [0, 1]$. Observe that $(1 - t)x + ty = x + t(y - x)$.

Definition 1.3.25. A subset $J \subset \mathbb{R}$ is said to be *convex* if for any $x, y \in J$ and $t \in [0, 1]$ we have $(1 - t)x + ty \in J$.

Exercise Set 1.3.26.

(1) Show that $J \subset \mathbb{R}$ is convex iff for any $n \in \mathbb{N}$, $x_j \in J$, $1 \le j \le n$ and $t_j \in [0, 1]$ with $t_1 + \ldots + t_n = 1$, we have $t_1 x_1 + \ldots + t_n x_n \in J$.

That is, J is convex iff any convex linear combination of any finite set of elements of J lies in J.

(2) Show that a subset $J \subset \mathbb{R}$ is an interval iff J is convex.

(3) Let $a_j \in \mathbb{R}$, $1 \le j \le n$. Let $m := \min\{a_j : 1 \le j \le n\}$ and $M := \max\{a_j : 1 \le j \le n\}$. Let $t_1, \ldots, t_n$ be non-negative real numbers such that $t_1 + \cdots + t_n = 1$. Show that $t_1 a_1 + \cdots + t_n a_n \in [m, M]$.

Exercise Set 1.3.27 (Exercises on LUB and GLB properties of $\mathbb{R}$).

(1) What can you say about A if lub $A =$ glb A?

(2) Prove that $\alpha \in \mathbb{R}$ is the lub of A iff (i) α is an upper bound of A and (ii) for any $\varepsilon > 0$, there exists $x \in A$ such that $x > \alpha - \varepsilon$.

Formulate an analogue for glb.

(3) Let A, B be nonempty subsets of $\mathbb{R}$ with $A \subset B$. Prove

$$\text{glb } B \le \text{glb } A \le \text{lub } A \le \text{lub } B.$$

(4) Let $A = \{a_i : i \in I\}$ and $B = \{b_i : i \in I\}$ be nonempty subsets of $\mathbb{R}$ indexed by I. Assume that for each $i \in I$, we have $a_i \le b_i$. Prove that glb $A \le$ glb B if A and B are bounded below. What is the analogous result if A and B are bounded above?

(5) Let A, B be nonempty subsets of $\mathbb{R}$. Assume that $a \le b$ for all $a \in A$ and $b \in B$. Show that lub $A \le$ glb B.

(6) Let $A, B \subset \mathbb{R}$ be bounded above. Find a relation between lub $(A \cup B)$, lub A and lub B.

Formulate an analogous question when the sets are bounded below and answer it.

(7) For $A \subset \mathbb{R}$, we define

$$-A := \{y \in \mathbb{R} : \exists x \in A \text{ such that } y = -x\}.$$

Thus $-A$ is the set of all negatives of elements of A.

  (a) Let $A = \mathbb{Z}$. What is $-A$?

  (b) If $A = [-1, 2]$, what is $-A$?

Assume that A is bounded above and $\alpha :=$ lub A. Show that $-A$ is bounded below and that glb $(-A) = -\alpha$.

Can you formulate the analogous result (for $-B$) if $\beta =$ glb B?

(8) Formulate the GLB property of $\mathbb{R}$ (in a way analogous to the LUB property of $\mathbb{R}$).

(9) Show that LUB property holds iff the GLB property holds true in $\mathbb{R}$.

(10) Let $A, B \subset \mathbb{R}$ be nonempty. Define

$$A + B := \{x \in \mathbb{R} : \exists\, (a \in A, b \in B) \text{ such that } x = a + b\}$$
$$= \{a + b : a \in A, b \in B\}.$$

(a) Let $A = [-1, 2] = B$. What is $A + B$?

(b) Let $A = B = \mathbb{N}$. What is $A + B$?

(c) Let $A = B = \mathbb{Z}$. What is $A + B$?

(d) Let $A = B = \mathbb{Q}$. What is $A + B$?

(e) Let $A = B$ be the set of all irrational numbers. What is $A + B$?

(11) Let $\alpha := \operatorname{lub} A$. Let $b \in \mathbb{R}$. Let $b + A := \{b + a : a \in A\}$. Find $\operatorname{lub}(b + A)$.

(12) Let $\alpha = \operatorname{lub} A$ and $\beta = \operatorname{lub} B$. Show that $A + B$ is bounded above and that $\operatorname{lub}(A + B) = \alpha + \beta$.

(13) Let $\alpha := \operatorname{lub} A$. Let $b \in \mathbb{R}$ be positive. Let $bA := \{ba : a \in A\}$. Find $\operatorname{lub}(bA)$. Investigate what result is possible when $b < 0$.

(14) Let A, B be nonempty subsets of positive real numbers. Let $\alpha := \operatorname{lub} A$ and $\beta := \operatorname{lub} B$. Define $A \cdot B := \{ab : a \in b \in B\}$. Show that $\operatorname{lub}(A \cdot B) = \alpha \cdot \beta$.

(15) Let $A \subset \mathbb{R}$ with $\operatorname{glb} A > 0$. Let $B := \{x^{-1} : x \in A\}$. Show that B is bounded above and relate its lub with the glb of A.

(16) Let $\emptyset \neq A \subset \mathbb{R}$ be bounded above in $\mathbb{R}$. Let B be the set of upper bounds of A. Show that B is bounded below and that $\operatorname{lub} A = \operatorname{glb} B$.

(17) Show that $\operatorname{glb} \{1/n : n \in \mathbb{N}\} = 0$.

(18) Show that $\operatorname{lub} \{1 - \frac{1}{n^2} : n \in \mathbb{N}\} = 1$.

(19) Let $A := \{x \in \mathbb{R} : x^2 - 5x + 6 < 0\}$. Find the lub and glb of A.

(20) Find the glb of $\{x + x^{-1} : x > 0\}$. Is the set bounded above?

(21) Let $A := \{\frac{1}{3} \pm \frac{n}{3n+1} | n \in \mathbb{N}\}$. Show that $\operatorname{lub} A = 2/3$ and $\operatorname{glb} A = 0$.

(22) Find the lub and glb of $\{\frac{m+n}{mn} : m, n \in \mathbb{N}\}$.

(23) Find the glb and lub of the set of real numbers in $(0, 1)$ whose decimal expansion contains only 0's and 1's.

(24) Let $x, y \in \mathbb{R}$ be such that $x \leq y + \frac{1}{n}$ for all $n \in \mathbb{N}$. Show that $x \leq y$.

(25) What is the relevance of (1) and (2) of Exercise 1.3.8 to the nested interval theorem?

1.4   Absolute Value and Triangle Inequality

Definition 1.4.1 (Absolute value of a real number). For $x \in \mathbb{R}$, we define

$$|x| = \begin{cases} x, & \text{if } x > 0 \\ -x, & \text{if } x \leq 0. \end{cases}$$

Note that $|x| = \max\{x, -x\}$. If we draw the graphs of the functions $f : x \mapsto x$ and $g : x \mapsto -x$, the graph of $x \mapsto |x|$ is $\max\{f(x), g(x)\}$. Look at Figures 1.15–1.16.

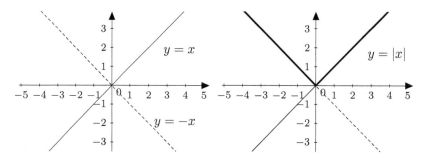

Figure 1.15: Graph of $y = \pm x$.   Figure 1.16: Graph of $y = |x|$.

The next result deals with all the important properties of the absolute value function on $\mathbb{R}$.

Theorem 1.4.2. *The following are true:*

(1) $|ab| = |a|\,|b|$ *for all* $a, b \in \mathbb{R}$.

(2) $|a|^2 = a^2$ *for any* $a \in \mathbb{R}$. *In particular,* $|x| = \sqrt{x^2}$, *the unique nonnegative square root of* x^2.

(3) $\pm a \leq |a|$ *for all* $a \in \mathbb{R}$.

(4) $-|a| \leq a \leq |a|$ *for all* $a \in \mathbb{R}$.

(5) $|x| < \varepsilon$ *iff* $x \in (-\varepsilon, \varepsilon)$.

(6) $|x - a| < \varepsilon$ *iff* $x \in (a - \varepsilon, a + \varepsilon)$.

(7) **Triangle Inequality.** $|a + b| \leq |a| + |b|$ *for all* $a, b \in \mathbb{R}$. *Equality holds iff both a and b are on the same side of* 0.

(8) $||a| - |b|| \leq |a - b|$ *for all* $a, b \in \mathbb{R}$.

(9) $\max\{a, b\} = \frac{1}{2}(a + b + |a - b|)$ *and* $\min\{a, b\} = \frac{1}{2}(a + b - |a - b|)$ *for any* $a, b \in \mathbb{R}$.

Proof. To prove (1), observe that $|ab|$ is either ab or $-ab$. If one of them is zero, the result is obvious. So, we assume that both are nonzero and hence $|ab| > 0$.

Case (1): $|ab| = ab$. Since $|ab|$ is positive, both a and b are either negative or positive. In any case, $ab = (-a)(-b)$. If both are positive, then $a = |a|$ and $b = |b|$ so that $ab = |a|\,|b|$. If both are negative, then $(-a)(-b) = |a|\,|b|$.

Case (2): $|ab| = -ab$. In this case, one of them is negative and the other is positive. Assume that $a < 0$. Then $|a| = -a$. Thus $|a|\,|b| = (-a)b = -ab = |ab|$.

We have proved (1).

We now prove (4). Since $-a \leq |a|$, we have $-|a| \leq a$. Combined with the inequality $a \leq |a|$, we get (4).

To prove (5), assume $|x| < \varepsilon$. Then $\max\{-x, x\} < \varepsilon$. Hence $x < \varepsilon$ and $-x < \varepsilon$, or $x > -\varepsilon$. Thus, we obtain $-\varepsilon < x < \varepsilon$.

(6) follows easily from (5):

$$|x - a| < \varepsilon \iff -\varepsilon < x - a < \varepsilon \iff a - \varepsilon < x < a + \varepsilon.$$

We prove (7). Consider $|x + y|$. Observe the following:

$$x + y \leq |x| + |y| \text{ by (iii)}$$
$$-x - y \leq |x| + |y| \text{ by (iii)}.$$

Thus, $\max\{x + y, -(x + y)\} \leq |x| + |y|$. Since $|x + y| = \max\{x + y, -(x + y)\}$, the triangle inequality follows.

Let us analyze when equality arises. Assume that $|x + y| = |x| + |y|$.

Case 1): $|x + y| = x + y$ so that $x + y = |x| + |y|$. It follows that $0 = (|x| - x) + (|y| - y)$. By (3), we see that $|x| - x \geq 0$ and $|y| - y \geq 0$. Since the sum of two nonnegative numbers is zero, we deduce that each of the terms is zero. Hence $x = |x|$ and $y = |y|$. In particular, both are nonnegative.

Case 2): $|x + y| = -(x + y)$. Then $-x - y = |x| + |y|$. Hence $0 = (|x| + x) + (|y| + y)$. Since the LHS, the sum of two nonnegative numbers (by (iii)), is zero, we deduce that each of the terms is zero. Hence $-x = |x|$ and $-y = |y|$. In particular, both x and y are non positive.

To summarize, if the equality holds in the triangle inequality, then both x and y are of the same sign.

The converse is easy to see. If $x < 0$ and $y < 0$, then $|x| = -x$, $|y| = -y$. Also, $x + y < 0$ so that $|x + y| = \max\{x + y, -x - y\} = -x - y = |x| + |y|$. The other case is similar.

Thus we have proved that $|x + y| = |x| + |y|$ if and only if both x and y are of the same sign.

We prove (8). $|a| = |(a - b) + b| \leq |a - b| + |b|$. From this, we obtain, $|a| - |b| \leq |a - b|$. Interchanging a and b in this we get $|b| - |a| \leq |b - a|$. Hence, $||a| - |b|| := \max\{|a| - |b|, |b| - |a|\} \leq |a - b|$.

(9). How does one get such an expression for $\max\{a, b\}$ or $\min\{a, b\}$? Note that $(a + b)/2$ is the midpoint of a and b. The distance between them is $|a - b|$. Hence, if we move to the left half the distance from the midpoint, then we must get the minimum. That is, $\min\{a, b\} = (a + b)/2 - |a - b|/2$. Similarly, if we move to the right half the distance from the midpoint, then we must get the maximum. That is, $\max\{a, b\} = (a + b)/2 + |a - b|/2$.

Having guessed this, it is easy to verify it. Let $b > a$. Then $\max\{a, b\} = b$. and hence $|a - b|/2 = (b-a)/2$. We have $(a+b)/2+|a - b|/2 = (a+b)/2+(b-a)/2 = b$. $\qquad\square$

Remark 1.4.3. It is very useful to think of $|a - b|$ as the distance between a and b. In particular, $|x|$ is the distance of x from 0.

Definition 1.4.4. A subset $A \subset \mathbb{R}$ is said to be *bounded* in $\mathbb{R}$ iff it is both bounded above and bounded below. That is, there exist $\alpha \in \mathbb{R}$ and $\beta \in \mathbb{R}$ such that for all $x \in A$, we have $\alpha \le x \le \beta$.

Geometrically this means that $A \subset (\alpha, \beta)$.

Example 1.4.5. The set of rationals in $[-5, 1]$ is bounded in $\mathbb{R}$.

For $a, b \in \mathbb{R}$, each interval of the form $[a, b]$, $(a, b]$, (a, b), and $[a, b)$ is bounded.

Look at $A_1 := [-3, -2]$, $A_2 := [-3, 5]$, $A_3 := [3, 5]$. Can you find an $M_i > 0$, $1 \le i \le 3$ such that the following holds?

$$\forall x \in A_i, \text{ we have } |x| \le M_i.$$

Do you observe any pattern?

Proposition 1.4.6. *A subset $A \subset \mathbb{R}$ is bounded in $\mathbb{R}$ iff there exists $M > 0$ such that $-M \le x \le M$ for all $x \in A$, that is, A is bounded iff there exists $M > 0$ such that $|x| \le M$ for all $x \in A$.*

Proof. Let $A \subset \mathbb{R}$ be bounded. The examples investigated above should have led you to conclude the following: If α and β are the lower and upper bounds, respectively, then we may take M as $\max\{|\alpha|, |\beta|\}$.

We need to estimate $|x|$, for $x \in A$. Since $|x| = \max\{x, -x\}$, we need to find estimates for x and $-x$. Now it is given that $\alpha \le x \le \beta$. From this it follows that $-\alpha \ge -x \ge -\beta$.

Now, if $|x| = x$, then

$$|x| = x \le \beta \le |\beta| \le M.$$

On the other hand, if $|x| = -x$, then we have

$$|x| = -x \le -\alpha \le |\alpha| \le M.$$

Hence we have proved that for each $x \in A$, we have $|x| \le M$.

The converse is true, since for each $x \in A$, we have $-M \le x \le M$, that is, A is both bounded above and bounded below. $\qquad\square$

Example 1.4.7. Subsets of $\mathbb{R}$ are defined by equalities and inequalities. We do the following as samples.

(1) $A := \{x \in \mathbb{R} : |x - a| = |x - b|\}$ (where $a \neq b$) $= \{\frac{a+b}{2}\}$.

$|x - a| = |x - b|$ says that the distances of x from a and b are the same. Thinking geometrically, we see that there is only one such real number, namely the midpoint $c := (a + b)/2$ of a and b. See Figure 1.17. How do we prove this?

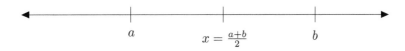

Figure 1.17: Example 1.4.7: Figure 1.

Assume $a < b$. If $x < a$, then $0 < |x - a| = a - x < b - x = |x - b|$ and hence x is not in the set A. See Figure 1.18.

Figure 1.18: Example 1.4.7: Figure 2.

If $x > b$, then $0 < |x - b| = x - b < x - a = |x - a|$. Hence $x \notin A$. See Figure 1.19.

Figure 1.19: Example 1.4.7: Figure 3.

Let $a \leq x \leq b$. Then by law of trichotomy, exactly one of the following is true. (i) $x < c$, (ii) $x > c$, or (iii) $x = c$.

In the first case, $|x - a| = x - a < c - a < b - a = |b - a|$. Hence such an $x \notin A$.

In the second, $|x - b| = b - x > b - c = (a + b)/2$ whereas $|x - a| = x - a > c - a = (a + b)/2$. Hence $x \notin A$.

Clearly, $c \in A$. Hence we conclude that $A = \{(a + b)/2\}$.

What happens when $a = b$? In this case, $A = \mathbb{R}$.

(2) $A := \{x \in \mathbb{R} : \frac{x+2}{x-1} < 4\} = (-\infty, 1) \cup (2, \infty)$. The temptation would be to clear off the fraction by multiplying both sides of the inequality by $x - 1$. We need to take care of the sign of $x - 1$.

Case 1. $x - 1 > 0$. Then we obtain

$$\frac{x+2}{x-1} < 4 \iff x + 2 < 4(x-1) \iff 6 < 3x \iff x > 2.$$

Case 2. $x - 1 < 0$. We obtain

$$\frac{x+2}{x-1} < 4 \iff x + 2 > 4(x-1) \iff 6 > 3x \iff x < 2.$$

Note that this condition is superfluous, since $x < 1$!

Thus, we get $A \subset (-\infty, 1) \cup (2, \infty)$. Equality is easily checked.

(3) $\{x \in \mathbb{R} : \left|\frac{2x-3}{3x-2}\right| = 2\} = \{1/4, 7/8\}$.

Case 1. $\left|\frac{2x-3}{3x-2}\right| = \frac{2x-3}{3x-2}$. We have

$$\frac{2x-3}{3x-2} = 2 \text{ iff } 2x - 3 = 6x - 4 \text{ iff } 1 = 4x \text{ iff } x = 1/4.$$

Case 2. $\left|\frac{2x-3}{3x-2}\right| = -\frac{2x-3}{3x-2}$.

$$\frac{2x-3}{3x-2} = -2 \text{ iff } 2x - 3 = 4 - 6x \text{ iff } -7 = -8x \text{ iff } x = 7/8.$$

(4) Identify $A := \{x \in \mathbb{R} : \left|\frac{3-2x}{2+x}\right| < 2\} = (-1/4, \infty)$.

A real $x \in \mathbb{R}$ lies in A iff

$$-2 < \frac{3-2x}{2+x} < 2.$$

Before we clear the fraction, we need to be aware of two cases (a) $2 + x > 0$ and (b) $2 + x < 0$. Accordingly, $A = A_+ \cup A_-$ where

$$A_+ := \{x \in \mathbb{R} : -2 < \frac{3-2x}{2+x} < 2 \text{ and } x + 2 > 0\}$$

$$A_- := \{x \in \mathbb{R} : -2 < \frac{3-2x}{2+x} < 2 \text{ and } x + 2 < 0\}.$$

We identify A_+ first. $x \in A_+$ iff $x > -2$ and $-2 < \frac{3-2x}{2+x} < 2$, that is, iff $x > -2$ and $-2(2+x) < 3 - 2x < 2(2+x)$.

$$x \in A_+ \iff -4 - 2x < 3 - 2x \text{ and } 3 - 2x < 4 + 2x \text{ and } x > -2$$
$$\iff -4 < 3 \text{ and } -1 < 4x \text{ and } x > -2$$
$$\iff -4 < 3 \text{ and } x > -1/4 \text{ and } x > -2.$$

Hence $x \in A_+$ iff x satisfies all the three conditions. Note that the first condition does not involve x and is always true. Hence, any $x \in \mathbb{R}$ satisfies the first condition. If x satisfies the second condition $x > -1/4$, it satisfies the third. Hence we conclude that $A_+ = \{x \in \mathbb{R} : x > -1/4\}$.

We identify A_-.

$$
\begin{aligned}
x \in A_- &\iff -4 - 2x > 3 - 2x \text{ and } 3 - 2x > 4 + 2x \text{ and } x < -2 \\
&\iff -4 > 3 \text{ and } -1 > 4x \text{ and } x > -2 \\
&\iff -4 < 3 \text{ and } x < -1/4 \text{ and } x > -2.
\end{aligned}
$$

If $x \in \mathbb{R}$ lies in A_-, all three conditions are met. But the first condition can never be satisfied. That is, no $x \in \mathbb{R}$ can be in A_-. Hence we conclude that $A_- = \emptyset$. Therefore, $A = \{x \in \mathbb{R} : x > -1/4\}$.

(5) $\{x \in \mathbb{R} : x^4 - 5x^2 + 4 < 0\} = (-2, -1) \cup (1, 2)$.

We rewrite $x^4 - 5x^2 + 4 = (x^2 - 4)(x^2 - 1)$. Hence the product is negative iff the factors of opposite signs.

Case 1: $x^2 - 4 > 0$ and $x^2 - 1 < 0$. Hence $x^2 > 4$ and $x^2 < 1$. This is impossible. Hence this case does not arise.

Case 2: $x^2 - 4 < 0$ and $x^2 - 1 > 0$. Hence $x^2 < 4$ and $x^2 > 1$. The first condition says $x \in (-2, 2)$ and the second says $x > 1$ or $x < -1$. We therefore conclude the set is $(-2, -1) \cup (1, 2)$.

The main purpose of the following set of problems is to make you acquire facility in dealing with inequalities.

Exercise Set 1.4.8. Identify the following subsets of $\mathbb{R}$:

(1) $\{x \in \mathbb{R} : |3x + 2| > 4|x - 1|\}$.

(2) $\{x \in \mathbb{R} : \left|\frac{x}{x+1}\right| > \frac{x}{x+1} \text{ where } x \neq -1\}$.

(3) $\{x \in \mathbb{R} : \left|\frac{x+1}{x+5}\right| < 1 \text{ where } x \neq -5\}$.

(4) $\{x \in \mathbb{R} : x^2 > 3x + 4\}$.

(5) $\{x \in \mathbb{R} : 1 < x^2 < 4\}$.

(6) $\{x \in \mathbb{R} : 1/x < x\}$.

(7) $\{x \in \mathbb{R} : 1/x < x^2\}$.

(8) $\{x \in \mathbb{R} : |4x - 5| < 13\}$.

(9) $\{x \in \mathbb{R} : |x^2 - 1| < 3\}$.

(10) $\{x \in \mathbb{R} : |x + 1| + |x - 2| = 7\}$.

(11) $\{x \in \mathbb{R} : |x| + |x + 1| < 2\}$.

(12) $\{x \in \mathbb{R} : 2x^2 + 5x + 3 > 0\}$.

(13) $\{x \in \mathbb{R} : \frac{2x}{3} - \frac{x^2 - 3}{2x} + \frac{1}{2} < \frac{x}{6}\}$.

Chapter 2

Sequences and Their Convergence

Contents

Sequences arise naturally when we want to approximate quantities. For instance, when wish to use decimal expansion for the rational number $1/3$ we get a sequence $0.3, 0.33, 0.333, \ldots$. We also understand that each term is approximately equal to $1/3$ up to certain level of accuracy. What do we mean by this? If we want the difference between $1/3$ and the approximation to be less than, say, 10^{-3}, we may take any one of the decimal numbers $0.\overbrace{3\ldots3}^{n-\text{times}}$ where $n > 3$. Or, if we want to use decimal expansion for $\sqrt{2}$, we look at

$$1.4, 1.41, 1.414, 1.4142, 1.41421, \ldots, 1.41421356237309504880016887242\ldots.$$

The idea of a sequence (x_n) and its convergence to $x \in \mathbb{R}$ is another way of saying that we give a sequence of approximations x_n to x in such a way that if one prescribes a level of accuracy, we may ask him to take any x_n after some N-th term onward. Each such will be near to x to a desired level of accuracy.

It is our considered opinion that students who master this chapter will begin to appreciate analysis and the way the proofs are considered.

2.1   Sequences and Their Convergence

Definition 2.1.1. Let X be a nonempty set. A *sequence* in X is a function $f\colon \mathbb{N} \to X$. We let $x_n := f(n)$ and call x_n the n-th term of the sequence. One usually denotes f by (x_n) or as an infinite tuple $(x_1, x_2, \ldots, x_n, \ldots)$.

Remark 2.1.2. Suppose (x_n) is a sequence in $\mathbb{R}$. If we plot the points of this sequence, as the graph of the function $f\colon \mathbb{N} \to \mathbb{R}$, it is a set of points $\{(k, x_k) \equiv (k, f(k)) : k \in \mathbb{N}\}$ in the plane. Look at Figure 2.1.

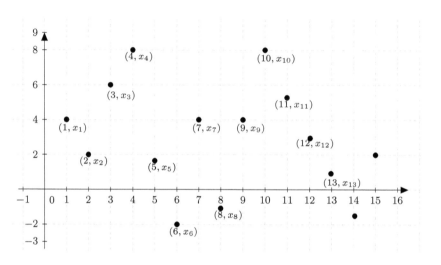

Figure 2.1: Graph of a sequence.

Example 2.1.3. Let us look at some examples of sequences. (Readers are encouraged to plot the points of these sequences.)

(1) Fix a real number $c \in \mathbb{R}$ and define $x_n = c$ for all n. The sequence is $(c, c, c, \ldots)$. This is called a constant sequence.

(2) Let $x_n = \frac{1}{n}$, the sequence is $\left(1, \frac{1}{2}, \frac{1}{3}, \frac{1}{4}, \ldots, \frac{1}{n}, \ldots\right)$.

(3) Let $x_n = \frac{(-1)^{n+1}}{n}$, the sequence is $\left(1, -\frac{1}{2}, \frac{1}{3}, -\frac{1}{4}, \ldots, \frac{1}{2k-1}, -\frac{1}{2k}, \ldots\right)$.

(4) Let $x_n = (-1)^n$, the sequence is $(-1, 1, -1, 1 \ldots)$.

(5) Let $x_n = n$, the sequence is $(1, 2, 3, 4, \ldots)$.

(6) Let $x_n = 2^n$, the sequence is $(2, 4, 8, 16, \ldots)$.

(7) Let $x_n = \frac{1}{2^n}$, the sequence is $\left(\frac{1}{2}, \frac{1}{4}, \frac{1}{8}, \frac{1}{16}, \ldots\right)$.

(8) Let $x_1 = x_2 = 1$ and define $x_n = x_{n-1} + x_{n-2}$ for all $n \geq 3$. The sequence is $(1, 1, 2, 3, 5, 8, 13, 21, 34, 53, \ldots)$.

(9) Let $x_n = \frac{2n+1}{n^2-3}$. Then the first few terms of this sequence are $(-\frac{3}{2}, 5, \frac{7}{6}, \frac{9}{13}, \frac{11}{22}, \frac{13}{33}, \ldots)$.

(10) Let $x_1 = 0.3, x_2 = 0.33, x_3 = 0.333, \ldots, x_n = 0.\underbrace{33\ldots3}_{n-\text{times}}$.

Item 10 of Example 2.1.3 explains the decimal expansion of $\frac{1}{3}$. Note that $\frac{1}{3} - x_n < \frac{1}{10^n}$ for all n. Also x_n approximates $\frac{1}{3}$, in the sense that if someone wants to approximate $\frac{1}{3}$ with an error less than $\frac{1}{10^{10}}$, then we can take x_n for $n \geq 10$. This example captures the essence of the next definition.

Let $y_n := 0.\underbrace{99\ldots9}_{n-\text{times}}$. What can you say about (y_n)?

Definition 2.1.4. Let (x_n) be a real sequence. We say that (x_n) *converges* to $x \in \mathbb{R}$ if for any given $\varepsilon > 0$, there exists $N \in \mathbb{N}$ such that for all $k \geq N$, we have $x_k \in (x - \varepsilon, x + \varepsilon)$, that is, for $k \geq N$, we have $|x - x_k| < \varepsilon$. The number x is called a *limit* of the sequence (x_n). We then write $x_n \to x$. We also say that (x_n) is *convergent* to x. We write this as $\lim_n x_n = x$.

We say that a sequence (x_n) is *divergent* if it is not convergent.

If $f \colon \mathbb{N} \to \mathbb{R}$ is a sequence, denoted by (x_n), its restriction to the subset $\{k \in \mathbb{N} : k \geq N\}$ is denoted by $(x_n)_{n \geq N}$. It is called a *tail* of the sequence (x_n). Thus if (x_n) converges to x, then we want the entire tail after N to lie in the interval $(x - \varepsilon, x + \varepsilon)$. Look at Figure 2.2.

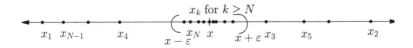

$$x_k \text{ for } k \geq N$$

$x_1 \quad x_{N-1} \quad\quad x_4 \quad\quad\quad x_N \quad x \quad\quad x_3 \quad\quad x_5 \quad\quad\quad\quad x_2$

$x - \varepsilon \quad\quad x + \varepsilon$

Figure 2.2: Tail of a convergent sequence.

Also note that if we take $N_1 > N$, then for for all $k \geq N_1$, $x_k \in (x - \varepsilon, x + \varepsilon)$. Therefore, N is not unique. Thus if N "works" for $\varepsilon > 0$, then any $N_1 > N$ also will do.

Remark 2.1.5. Suppose (x_n) is a sequence of real numbers converging to ℓ. Geometrically, it means that if we plot the graph of this sequence and draw an ε-band around $y = \ell$, say, $y = \ell - \varepsilon$ to $y = \ell + \varepsilon$, then there exists an integer N such that for all $n \geq N$ the points (n, x_n) lie inside this band. See Figure 2.3.

The definition of convergence of (x_n) can be written in terms of quantifiers as follows:

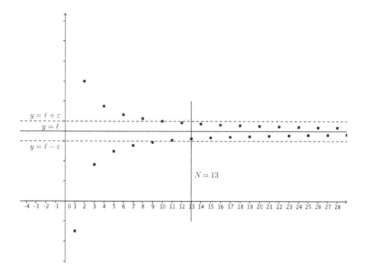

Figure 2.3: Graph of a convergent sequence.

We say that (x_n) converges to x if

$$\forall \varepsilon > 0 \left(\exists n_0 \in \mathbb{N} \left(\forall n \geq n_0 \left(x_n \in (x - \varepsilon, x + \varepsilon) \right) \right) \right).$$

Equivalently, (x_n) converges to x if

$$\forall \varepsilon > 0 \left(\exists n_0 \in \mathbb{N} \left(\forall n \geq n_0 \left(|x - x_n| < \varepsilon \right) \right) \right).$$

Exercise 2.1.6. What does it mean to say that a sequence (x_n) does not converge to x? Write it in words and then in terms of quantifiers.

Definition 2.1.7. We say that a sequence (x_n) of real numbers is convergent if there exists $x \in \mathbb{R}$ such that $x_n \to x$. Note that if we want to say that (x_n) is convergent, we need to find $x \in \mathbb{R}$ and then show $x_n \to x$. In terms of quantifiers, this definition may be written as

(x_n) is convergent if $\exists x \in \mathbb{R} \left(\forall \varepsilon > 0 \left(\exists N \in \mathbb{N} \left(\forall k \geq N (x_n \in (x - \varepsilon, x + \varepsilon)) \right) \right) \right).$

If a sequence is not convergent, we also say that it is *divergent*.

Let us look at convergence and nonconvergence of the sequences discussed earlier. A crucial point in convergence of a sequence is to first make a guess for a limit x and then estimate $|x - x_n|$.

Example 2.1.8. Let us consider the constant sequence $x_n = c$ for all n. It is easy to see that sequence converges and the limit should be c. Note that $|x_n - c| = 0$ for all n. This means, for any $\varepsilon > 0$, $|x_n - c| < \varepsilon$ for all n. So, if $\varepsilon > 0$ is given, we may take $N = 1$. Then $|x_n - c| = 0 < \varepsilon$ for all $n \geq 1 = N$.

Example 2.1.9. Let us consider the sequence (x_n) where $x_n = \frac{1}{n}$. If you plot the points of this sequence on the real line it is easy to see that this sequence should have limit $x = 0$. We claim that $\frac{1}{n} \to 0$.

Let $\varepsilon > 0$ be given. Let us estimate $|x_n - x| = \left|\frac{1}{n} - 0\right| = \frac{1}{n}$. We want to choose N such that for all $n \geq N$, $|x_n - x| < \varepsilon$, that is, $\frac{1}{n} < \varepsilon$ or $n > \frac{1}{\varepsilon}$. Such an n exists, by the Archimedean property.

Thus, choose an integer $N > \frac{1}{\varepsilon}$ by the Archimedean property. Then for all $n \geq N$, we have

$$|x_n - x| = \left|\frac{1}{n} - 0\right| = \frac{1}{n} \leq \frac{1}{N} < \varepsilon.$$

Hence $\frac{1}{n} \to 0$.

What can you say about a sequence defined as $x_n = \frac{1}{an+b}$ where $a, b \in \mathbb{R}$ and $a \neq 0$?

Example 2.1.10. Consider $x_n = \frac{1}{2^n}$. Once again, if we look at the points of this sequence on the real line, it is easy to see that this sequence should converge to 0.

Let $\varepsilon > 0$ be given. We have to find $N \in \mathbb{N}$ such that for all $n \geq N$

$$|x_n - x| = \left|\frac{1}{2^n} - 0\right| = \frac{1}{2^n} < \varepsilon.$$

Note that for all $n \in \mathbb{N}$, $2^n > n$ (Item 2 of Exercise 1.3.9). Hence $\frac{1}{2^n} < \frac{1}{n}$ for all $n \in \mathbb{N}$. Thus for all $n \in \mathbb{N}$, we have

$$|x_n - x| = \left|\frac{1}{2^n} - 0\right| = \frac{1}{2^n} < \frac{1}{n}.$$

Again, as in the last example, we choose an integer N such that $N > \frac{1}{\varepsilon}$ by the Archimedean property, then for all $n \geq N$, we have

$$|x_n - x| = \left|\frac{1}{2^n} - 0\right| = \frac{1}{2^n} < \frac{1}{n} < \varepsilon.$$

Remark 2.1.11. The reader must have observed that the natural number N, while not unique, depends on the given $\varepsilon > 0$ while checking for convergence. In view of this, when we want to be precise or when there are more than one ε, we may denote N by $N(\varepsilon)$ to show its dependence on ε.

Example 2.1.12. Consider the sequence $x_n = n$. It is not difficult to see intuitively that this sequence does not converge. Assume the contrary. Let x_n converge to a real number x. For $\varepsilon = 1$, there exists a natural number N such that the tail of the sequence $(x_n)_{n \geq N}$ lies in the interval $(x - 1, x + 1)$. In particular, $n \in (x - 1, x + 1)$ for all $n \geq N$. If we let $M := \max\{x + 1, N - 1\}$, then M is an upper bound for $\mathbb{N}$. This contradicts the Archimedean property. Hence we conclude that (x_n) is not convergent.

Example 2.1.13. Consider the sequence $x_n = (-1)^n$. Suppose this sequence converges to a real number x. Let $\varepsilon > 0$. Then there exists a natural number N such that $x_n \in (x - \varepsilon, x + \varepsilon)$ for all $n \geq N$. In particular, 1 and -1, both must be in the interval $(x - \varepsilon, x + \varepsilon)$. (Why? For example, if $k = 2N, 2N + 1$, then $x_k = -1$, $x_k = 1$.) Look at Figure 2.4. What does it mean for ε? The picture suggests that $2\varepsilon > 2$. The interval $[-1, 1] \subset (x - \varepsilon, x + \varepsilon)$ and hence the length 2ε of $(x - \varepsilon, x + \varepsilon)$ must be at least that of the subinterval $[-1, 1]$. But what we have shown is that this must happen for any $\varepsilon > 0$. This is absurd, if $\varepsilon \leq 1$.

How does a textbook proof go now?

Let $x_n \to x$. Choose $\varepsilon > 0$ such that $\varepsilon < 1$. Let $N \in \mathbb{N}$ be such that for all $k \geq N$, we have $x_k \in (x - \varepsilon, x + \varepsilon)$. In particular, $-1 = x_{2N}$, $1 = x_{2N+1} \in (x - \varepsilon, x + \varepsilon)$. Since $1 < x + \varepsilon$ and $-1 > x - \varepsilon$ and hence $-(-1) < -(x - \varepsilon)$, we obtain

$$2 = 1 - (-1) < x + \varepsilon - (x - \varepsilon) = 2\varepsilon.$$

That is, $1 < \varepsilon$. This is a contradiction. Hence we conclude that (x_n) does not converge to x. Since $x \in \mathbb{R}$ is arbitrary, it follows that (x_n) is divergent.

Figure 2.4: Divergence of $(-1)^n$.

The reader should observe that we translated our geometric idea about the lengths of $[-1, 1]$ and $(x - \varepsilon, x + \varepsilon)$ into the inequality above. Though a diligent reader may follow the logic and find the proof complete, he may be puzzled how we know that we have to arrive such an inequality and get a contradiction. It is one of the aims of the book to show you how many such thought processes hide behind a carefully executed proof.

Example 2.1.14. Let $x_n = \frac{2n+1}{n^2-3}$.

In this sequence, both numerator and denominator go to ∞. (We will explain this in detail in later section.) However, the denominator being quadratic in n goes to ∞ much faster than the numerator. Therefore, we can expect this sequence to converge to zero. Thus, we need to estimate $\left| \frac{2n+1}{n^2-3} - 0 \right| = \frac{2n+1}{n^2-3}$ for $n \geq 2$.

The main idea is to get an estimate of the form $\left| \frac{2n+1}{n^2-3} \right| \leq C\frac{1}{n}$ for some constant $C > 0$. Note that $\frac{C}{n}$ can be made as small as possible. There is no unique way of achieving this. One can obtain different bounds, depending upon the estimates used. We shall look at least two ways of estimating $\left| \frac{2n+1}{n^2-3} \right|$.

Note that $2n + 1 \leq 2n + n$ for all n and also $n^2 - 3 \geq n^2 - n$ for all $n > 3$.

Hence for $n > 3$, we have

$$\left|\frac{2n+1}{n^2-3}\right| = \frac{2n+1}{n^2-3} \leq \frac{2n+n}{n^2-n} = \frac{3}{n-1}.$$

Note that the right-hand side can be made as small as possible for sufficiently large n.

Let $\varepsilon > 0$ be given. Choose $N > 3 \in \mathbb{N}$ such that $\frac{3}{N-1} < \varepsilon$. That is, $N > \frac{3}{\varepsilon}+1$. Such an N exists by the Archimedean property. For any $n > N$, we have

$$\left|\frac{2n+1}{n^2-3}\right| \leq \frac{3}{n-1} \leq \frac{3}{N-1} < \varepsilon.$$

We can also estimate $\left|\frac{2n+1}{n^2-3}\right|$ as follows:

$$\left|\frac{2n+1}{n^2-3}\right| \leq \left|\frac{2n+1}{n^2-4}\right| \leq \left|\frac{2n+4}{n^2-4}\right| = \frac{2}{n-2}.$$

In this case we choose $N > \frac{2}{\varepsilon} + 2$.

An observant reader must have noticed that in the cases when (x_n) is convergent, the moment we guessed a possible limit, we stopped looking for other real numbers y such that $x_n \to y$. Why did we do so? Is it possible for a sequence x_n to converge to two distinct real numbers x and y? Look at Figure 2.5. It should convince you that it is not possible.

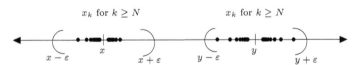

Figure 2.5: Uniqueness of limit.

Assume $x < y$. Motivated by the picture, we are looking for an ε such that $x + \varepsilon < y - \varepsilon$. That is, $2\varepsilon < y - x$. Assume $\varepsilon = \frac{y-x}{2}$. We claim that $(x - \varepsilon, x + \varepsilon) \cap (y - \varepsilon, y + \varepsilon) = \emptyset$. For, if $z \in (x - \varepsilon, x + \varepsilon) \cap (y - \varepsilon, y + \varepsilon)$, then

$$|y - x| \leq |y - z| + |z - x| < \frac{y-x}{2} + \frac{y-x}{2} = |y - x|,$$

which is a contradiction.

Since $x_n \to x$ and $x_n \to y$, for given $\varepsilon = \frac{y-x}{2}$, there exist natural numbers n_1 and n_2 such that $|x_n - x| < \varepsilon/2$ for all $n \geq n_1$ and $|x_n - y| < \varepsilon/2$ for all $n \geq n_2$.

For $n = n_1 + n_2$, x_n must be in each of the intervals $(x - \varepsilon, x + \varepsilon)$ and $(y - \varepsilon, y + \varepsilon)$. This contradicts the fact that the intervals $(x - \varepsilon, x + \varepsilon)$ and $(y - \varepsilon, y + \varepsilon)$ are disjoint.

We now give a second proof to show that if $x_n \to x$ and $x_n \to y$, then $x = y$.

Proposition 2.1.15 (Uniqueness of the limit). *If $x_n \to x$ and $x_n \to y$, then $x = y$.*

Strategy: By Theorem 1.3.5, it is enough to show that for given any $\varepsilon > 0$, $|x - y| < \varepsilon$. This means we need to estimate $|x - y|$, and we know how to estimate $|x_n - x|$ and $|x_n - y|$. Now the triangle inequality comes to our help.

Proof. Let $\varepsilon > 0$ be given. Since $x_n \to x$ and $x_n \to y$, there exist integers n_1, n_2 such that for $k \geq n_1$, we have $|x_k - x| < \varepsilon/2$ and for $k \geq n_2$, we have $|x_k - y| < \varepsilon/2$. Consider $N = \max\{n_1, n_2\}$. Then for all $k \geq N$, we have

$$|x - y| = |x - x_k + x_k - y| \leq |x_k - x| + |x_k - y| < \frac{\varepsilon}{2} + \frac{\varepsilon}{2} = \varepsilon.$$

$\square$

Exercise 2.1.16. Prove that each of the following sequences (a_n) converges to a limit a. Given $\varepsilon > 0$, find an $n_0 \in \mathbb{N}$ such that $|a_n - a| < \varepsilon$ for $n \geq n_0$.

(1) $a_n = 1/(n+1)$

(2) $a_n = (n+1)/(2n+3)$

(3) $a_n = n/(n^2 - n + 1)$

(4) $a_n = 1/2^n$

(5) $a_n = 2/\sqrt{n}$

(6) $a_n = \sqrt{n+1} - \sqrt{n}$

Lemma 2.1.17. *Let $x_n \to x$, $x_n, x \in \mathbb{R}$. Fix $N \in \mathbb{N}$. Define a sequence (y_n) such that $y_n := x_n$ if $n \geq N$ while $y_k \in \mathbb{R}$ could be any real number for $1 \leq k < N$. Then $y_n \to x$.*

Thus, if we alter a finite number of terms of a convergent sequence, the new sequence still converges to the limit of the original sequence.

Strategy: To prove $y_k \to x$, we need to estimate $|y_k - x|$.
If $k \geq N$, then $y_k = x_k$; therefore we need to estimate $|x_k - x|$. Since $x_k \to x$, we know how to estimate $|x_k - x|$.

Proof. Let $\varepsilon > 0$ be given. There exists an integer n_1 such that for $k \geq n_1$, $|x_k - x| < \varepsilon$. Thus if $n_0 = \max\{n_1, N\}$, then $y_k = x_k$ for $k \geq n_0$. Hence we have

$$k \geq n_0 \implies |y_k - x| = |x_k - x| < \varepsilon.$$

Since $\varepsilon > 0$ is arbitrary, it follows that $y_n \to x$. $\square$

Proposition 2.1.18. *Let (x_n) be a sequence of real numbers.*
  (i) *If $x_n \to x$, then $|x_n| \to |x|$. However the converse is not true.*
  (ii) *The sequence $x_n \to 0$ iff $|x_n| \to 0$.*
  (iii) *The sequence $x_n \to x$ iff $x_n - x \to 0$ iff $|x_n - x| \to 0$.*

Proof. To prove (i), we need to estimate $||x_n| - |x||$. Since we know how to estimate $|x_n - x|$, we use the inequality (8) on page 20. Therefore, we obtain

$$||x_n| - |x|| \leq |x_n - x|$$

and we know how to estimate $|x_n - x|$.

Let $\varepsilon > 0$ be given. Then there exists $N \in \mathbb{N}$ such that for all $n \geq N$, we have $|x_n - x| < \varepsilon$. Hence for all $n \geq N$, we have

$$||x_n| - |x|| \leq |x_n - x| < \varepsilon.$$

To see that the converse is not true, consider the sequence $x_n = (-1)^n$. Look at Example 2.1.13.

(ii) We only need to prove that if $|x_n| \to 0$, then $x_n \to 0$. Let $\varepsilon > 0$ be given. Since $|x_n| \to 0$, there exists $N \in \mathbb{N}$ such that for all $k \geq N$, $||x_k| - 0| = |x_k| < \varepsilon$. So for $k \geq N$, we have $|x_k - 0| = |x_k| < \varepsilon$.

(iii) follows from (ii).   □

Exercise 2.1.19. Let $b_n \geq 0$ and $b_n \to 0$. Assume that there exists an integer N such that $|a_n - a| \leq b_n$ for all $n \geq N$. Prove that $a_n \to a$.

Proposition 2.1.20. *Let (x_n) be a sequence of real numbers. Let $x_n \to x$ and $x > 0$. Then there exists $N \in \mathbb{N}$ such that $x_k > \frac{x}{2}$ for $k \geq N$.*

Proof. Look at Figure 2.6. As suggested by the picture, we take $\varepsilon = \frac{x}{2}$. Since $x_n \to x$, for this ε, there exists $N \in \mathbb{N}$ such that for $k \geq N$, $x_k \in (x - \varepsilon, x + \varepsilon) = \left(\frac{x}{2}, \frac{3x}{2}\right)$. In particular, for $k \geq N$, we have $x_k > \frac{x}{2}$.   □

Figure 2.6: Figure for Proposition 2.1.20.

Remark 2.1.21. The last result says that if $x_n \to x$ and if $x > 0$, then except for a finite number of terms, all $x_n > 0$. Note that this is a corollary of the proposition but weaker than the proposition.

What is the analogue of the above proposition if $x < 0$?

Let (x_n) be a sequence of real numbers. Let $x_n \to x < 0$. Then there exists $N \in \mathbb{N}$ such that $x_k < \frac{x}{2}$ for $k \geq N$. In particular, $x_n < 0$ for all large n.

Can you think of a (single) formulation which encompasses both these results?

Proposition 2.1.22. *Let (x_n) be a real sequence such that $x_n \to x$. Assume that $x \neq 0$. Then there exists N such that for all $k \geq N$, we have $|x_k| \geq |x|/2$.*

Proof. This is done above in Proposition 2.1.20 (and the remarks following it) in a geometric way for the case of real sequences.

To do the general case, let N correspond to $\varepsilon := |x|/2$. Then for $k \geq N$, we have

$$|x| \leq |x - x_k| + |x_k| < \varepsilon + |x_k| \text{ so that } |x_k| > |x| - \varepsilon = \frac{|x|}{2}.$$

This proves the proposition. $\qquad\square$

Definition 2.1.23. A sequence (x_n) of real numbers is said to be *bounded* if there exists $C > 0$ such that $|x_n| \leq C$ for all $n \in \mathbb{N}$.

If the sequence (x_n) is the function $f \colon \mathbb{N} \to \mathbb{R}$, then the set $\{x_n : n \in \mathbb{N}\}$ is the image of f. Note that (x_n) is bounded iff the image set $\{x_n : n \in \mathbb{N}\}$ of the sequence is a bounded subset of $\mathbb{R}$.

Example 2.1.24. Let us look at Example 2.1.3. Except items (5), (6), and (8), all other sequences are bounded.

Proposition 2.1.25. *Every convergent sequence of real numbers is bounded.*

Strategy: Let (x_n) be a sequence of real numbers converging to x. To show the boundedness of (x_n), we need to estimate $|x_k|$. But we know how to estimate $|x_k - x|$ for large k. We write

$$|x_k| = |x_k - x + x| \leq |x - x_k| + |x|.$$

Proof. Let $\varepsilon = 1$. Since $x_n \to x$, there exists $N \in \mathbb{N}$ such that, for $k \geq N$, we have $|x_k - x| < 1$. Then

$$|x_k| < |x - x_k| + |x| = 1 + |x| \text{ for } k \geq N.$$

To get an estimate for all x_n, we let $C := \max\{|x_1|, \ldots, |x_{N-1}|, 1 + |x|\}$. Then it is easy to see that C is an upper bound of (x_n). $\qquad\square$

Is the converse of the above proposition true? Consider the sequence $((-1)^n)$. It is bounded but not convergent. See Example 2.1.13.

Given two sequences (x_n) and (y_n) in $\mathbb{R}$ and $\lambda \in \mathbb{R}$, we can construct three new sequences as follows. The sum (z_n) is a new sequence such that $z_n = x_n + y_n$, that is, the n-th term of the new sequence is the sum $x_n + y_n$ of the n-th terms. We denote (z_n) by $(x_n) + (y_n)$.

The product (t_n) is a new sequence such that $t_n = x_n y_n$. We denote the product sequence by $(x_n y_n)$.

If $\lambda \in \mathbb{R}$, we can define a new sequence (u_n) such that $u_n = \lambda x_n$. We denote this new sequence by $\lambda(x_n)$.

We may now ask: If (x_n) and (y_n) are convergent, can we conclude the newly constructed sequences are also convergent? The next result answers this in the affirmative.

Theorem 2.1.26 (Algebra of Convergent Sequences). *Let* $x_n \to x$, $y_n \to y$ *and* $\alpha \in \mathbb{R}$. *Then:*

(1) $x_n + y_n \to x + y$.

(2) $\alpha x_n \to \alpha x$.

(3) $x_n \cdot y_n \to xy$.

(4) $\frac{1}{x_n} \to \frac{1}{x}$ *provided that* $x \neq 0$. *(Note that by Proposition 2.1.22, there exists a natural number* N *such that for all* $n \geq N$ *the terms* $x_n \neq 0$ *and hence* $1/x_n$ *makes sense.)*

Proof.

Strategy for (1): To prove (1), we need to estimate, $|(x_n + y_n) - (x + y)|$. Using the triangle inequality, we get,

$$|(x_n + y_n) - (x - y)| = |(x_n - x) + (y_n - y)| \leq |x_n - x| + |y_n - y|.$$

Since $x_n \to x$, $y_n \to y$ we know how to estimate each of this term on the right-hand side.

Let $\varepsilon > 0$ be given. Since $x_n \to x$, there exists a natural number n_1, such that for all $k \geq n_1$, we have $|x_k - x| < \frac{\varepsilon}{2}$. Similarly, there exists a natural number n_2, such that for $k \geq n_2$, we have $|y_k - y| < \frac{\varepsilon}{2}$. Now choose, $N := \max\{n_1, n_2\}$. Then for all $k \geq N$, we have $|x_k - x| < \frac{\varepsilon}{2}$ and $|y_k - y| < \frac{\varepsilon}{2}$. Therefore, for all $k \geq N$,

$$|(x_k - y_k) - (x - y)| \leq |x_k - x| + |y_k - y| < \frac{\varepsilon}{2} + \frac{\varepsilon}{2} = \varepsilon.$$

Strategy for (2): In order to prove (2), we need to estimate $|\alpha x_n - \alpha x| = |\alpha| |x_n - x|$. Since we know how to estimate $|x_n - x|$, the natural temptation is to choose $N \in \mathbb{N}$ such that $|x_n - x| < \frac{\varepsilon}{|\alpha|}$. However, if $\alpha = 0$, then this does not make sense. Note that if $\alpha = 0$, then (αx_n) is a constant sequence and hence $\alpha x_n \to 0 = \alpha x$. So we may assume $\alpha \neq 0$.

Let $\varepsilon > 0$ be given and $\alpha \neq 0$. Since $x_n \to x$, there exists $N \in \mathbb{N}$ such that for all $k \geq N$, we have $|x_k - x| < \frac{\varepsilon}{|\alpha|}$. Hence for all $k \geq N$, we have

$$|\alpha x_k - \alpha x| = |\alpha| |x_k - x| < |\alpha| \times \frac{\varepsilon}{|\alpha|} = \varepsilon.$$

Strategy for (3): To prove (3), we need to estimate $|x_n y_n - xy|$. Since $x_n \to x$ and $y_n \to y$, we know how to estimate $|x_n - x|$ and $|y_n - y|$. Somehow we need to bring in these two terms in the estimate of $|x_n y_n - xy|$. This is achieved by adding and subtracting the cross term $x_n y$ (or you can also add xy_n). Thus we have

$$|x_n y_n - xy| = |x_n y_n - x_n y + x_n y - xy| \leq |x_n| |y_n - y| + |y| |x_n - x|.$$

Since, (x_n) is convergent, by Proposition 2.1.25, it is bounded. Therefore, there exists $C > 0$ such that $|x_n| \leq C$ for all $n \in N$. Now the above inequality can be written as

$$|x_n y_n - xy| \leq |x_n|\,|y_n - y| + |y|\,|x_n - x| \leq C\,|y_n - y| + (|y| + 1)\,|x_n - x|\,.$$

Now it is easy to estimate $|x_n y_n - xy|$.

Now we go for a textbook proof of (3).

Let $\varepsilon > 0$ be given. Since (x_n) is convergent, there exists $C > 0$ such that

$$|x_n| \leq C \text{ for all } n \in \mathbb{N}. \tag{2.1}$$

Since $x_n \to x$, there exists a natural number n_1 such that

$$k \geq n_1 \implies |x_k - x| < \frac{\varepsilon}{2(|y| + 1)}. \tag{2.2}$$

Similarly, there exists a natural number n_2 such that

$$k \geq n_2 \implies |y_k - y| < \frac{\varepsilon}{2C}. \tag{2.3}$$

Choose, $N = \max\{n_1, n_2\}$. Then for all $k \geq N$, we have

$$
\begin{aligned}
|x_k y_k - xy| &= |x_k y_k - x_k y + x_k y - xy| \\
&= |x_k(y_k - y) + y(x_k - x)| \\
&\leq |x_k|\,|y_k - y| + |y|\,|x_k - x| \\
&\leq C\,|y_k - y| + (|y| + 1)\,|x_k - x|\,, && \text{by (2.1)} \\
&\leq C\frac{\varepsilon}{2C} + (|y| + 1)\frac{\varepsilon}{2(|y| + 1)}, && \text{by (2.2) and (2.3)} \\
&= \varepsilon.
\end{aligned}
$$

In the above set of equations we have used $|y| + 1$ as an upper bound for $|y|$. Do you understand why?

Strategy for (4): To prove (4), we need to estimate $\left|\frac{1}{x_n} - \frac{1}{x}\right|$. We have

$$
\begin{aligned}
\left|\frac{1}{x_n} - \frac{1}{x}\right| &= \left|\frac{x - x_n}{x x_n}\right| \\
&= \frac{1}{|x_n|}\frac{1}{|x|}|x - x_n| \\
&\leq \frac{2}{|x|}\frac{1}{|x|}|x - x_n|\,, \text{ say, for } n \geq n_1 \text{ using Proposition 2.1.22.}
\end{aligned}
$$

This inequality gives us an idea how to estimate $\left|\frac{1}{x_n} - \frac{1}{x}\right|$ to complete the proof.

Let $\varepsilon > 0$ be given. Since $x_n \to x$ and $x \neq 0$, by Proposition 2.1.22, there exists $n_1 \in \mathbb{N}$ such that

$$k \geq n_1 \implies |x_n| > \frac{|x|}{2}. \tag{2.4}$$

Also there exists $n_2 \in \mathbb{N}$, such that

$$k \geq n_2 \implies |x_k - x| < \frac{\varepsilon |x|^2}{2}. \tag{2.5}$$

Now choose $N = \max\{n_1, n_2\}$, Then, for all $k \geq N$, we have

$$
\begin{aligned}
\left| \frac{1}{x_k} - \frac{1}{x} \right| &= \left| \frac{x - x_k}{x x_k} \right| \\
&= \frac{1}{|x_k|} \frac{1}{|x|} |x - x_k| \\
&< \frac{2}{|x|} \frac{1}{|x|} |x - x_k|, \qquad \text{by (2.4)} \\
&< \frac{2}{|x|} \frac{1}{|x|} \times \frac{\varepsilon |x|^2}{2}, \qquad \text{by (2.5)} \\
&= \varepsilon.
\end{aligned}
$$

This completes the proof. $\qquad\qquad\qquad\qquad\qquad\qquad\qquad\qquad\square$

The above theorem gives rise to a natural vector space structure on the set of all real convergent sequences and a linear transformation from it to $\mathbb{R}$. This is the content of the next proposition.

Proposition 2.1.27. *The set $\mathcal{C}$ of convergent sequences of real numbers form a real vector space under the operations: $(x_n) + (y_n) := (x_n + y_n)$ and $\alpha \cdot (x_n) := (\alpha x_n)$.*

Moreover, the map $(x_n) \mapsto \lim x_n$ from $\mathcal{C}$ to $\mathbb{R}$ is a linear transformation.

Proof. We shall only sketch the argument. By the algebra of convergent sequences, the sequence $(x_n) + (y_n) := (x_n + y_n)$ is convergent. Hence $\mathcal{C}$ is closed under the addition. Similarly, if $\lambda \in \mathbb{R}$, the sequence $\lambda(x_n) := (\lambda x_n)$ is convergent and hence $\mathcal{C}$ is closed under scalar multiplication. That $\mathcal{C}$ is a vector space under these operations is easy to check. Also we have

$$\lim[(x_n) + (y_n)] = \lim(x_n + y_n) = x + y = \lim x_n + \lim y_n.$$

Similarly

$$\lim \lambda(x_n) = \lim(\lambda x_n) = \lambda x = \lambda \lim x_n.$$

The displayed equations show that the map $(x_n) \mapsto \lim x_n$ is a linear transformation. $\qquad\square$

Exercise Set 2.1.28.

(1) Given that $x_n \to 1$, identify the limits of the sequences whose n-th terms are
(a) $1 - x_n$, (b) $2x_n + 5$, (c) $(4 + x_n^2)/x_n$.

(2) Let $x_n \to x$. Assume that $x_n \geq 0$ for all n. Then show that $x \geq 0$.

(3) Let $a_n \leq b_n$ for $n \in \mathbb{N}$. Assume that $a_n \to a$ and $b_n \to b$. Show that $a \leq b$.

(4) Let $a \leq x_n \leq b$ for $n \in \mathbb{N}$. If $x_n \to x$, show that $a \leq x \leq b$.

(5) Let (x_n) and (y_n) be convergent. Let $s_n := \min\{x_n, y_n\}$ and $t_n := \max\{x_n, y_n\}$. Are the sequences (s_n) and (t_n) convergent?

(6) Show that the set of bounded (real) sequences form a real vector space.

(7) True or False: If (x_n) and $(x_n y_n)$ are bounded, then (y_n) is bounded.

(8) Let $x_n \geq 0$, $x_n \to x$. Prove that $\sqrt{x_n} \to \sqrt{x}$.

(9) True or False: If (x_n) and (y_n) are sequences such that $x_n y_n \to 0$, then one of the sequences converges to 0.

(10) Let (x_n) be a sequence. Prove that $x_n \to 0$ iff $x_n^2 \to 0$.

(11) Let $a_n \to 0$. What can you say about the sequence (a_n^n)?

(12) Let (x_n) and (y_n) be two real sequences. Let (z_n) be a new sequence defined $(x_1, y_1, x_2, y_2, \ldots)$. (Can you write down explicit expression for z_n?) Show that (z_n) is convergent iff both the sequences converge to the same limit.

(13) Let (x_n) be given. Let $s_n := x_{2n-1}$ and $t_n := x_{2n}$, $n \geq 1$. We thus get two new sequences. Note that $(s_n) = (x_1, x_3, x_5, \ldots)$. (What is (t_n)?) Show that (x_n) is convergent iff both (s_n) and (t_n) converge to the same limit.

   (This is same as the last exercise, packaged differently!)

(14) Let (x_n) be a sequence. Assume that $x_n \to 0$. Let $\sigma \colon \mathbb{N} \to \mathbb{N}$ be a bijection. Define a new sequence $y_n := x_{\sigma(n)}$. Show that $y_n \to 0$.

(15) Let $x_n := \left(1 - \frac{1}{2}\right)\left(1 - \frac{1}{3}\right) \cdots \left(1 - \frac{1}{n+1}\right)$. Show that (x_n) is convergent.

(16) Given any real number a, show that there exists a sequence, say (x_n) of rationals such that $x_n \to a$. Similarly, there exists a sequence of irrationals, say (y_n) such that $y_n \to a$.

2.2   Cauchy Sequences

Let (x_n) be a convergent sequence converging to x. Let $\varepsilon > 0$, be given. Let $N \in \mathbb{N}$ such that for all $k \geq N$, we have $|x_k - x| < \varepsilon$. Look at Figure 2.7. When $m, n \geq N$, what can you say about $|x_m - x_n|$?
   This motivates the following definition.

Definition 2.2.1 (Cauchy Sequence). A sequence (x_n) in $\mathbb{R}$ is said to be *Cauchy* if for each $\varepsilon > 0$ there exists $N \in \mathbb{N}$ such that for all $m, n \geq N$ we have $|x_n - x_m| < \varepsilon$.

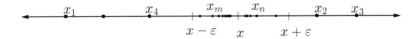

Figure 2.7: Cauchy sequence.

Example 2.2.2. Any real convergent sequence is Cauchy.
Assume that $x_n \to x$. We want to prove that (x_n) is Cauchy. That is, we need to estimate $|x_n - x_m|$. Since $x_n \to x$, we know how to estimate $|x_n - x|$ and $|x - x_m|$. We use the triangle inequality to estimate $|x_n - x_m|$.

Let $\varepsilon > 0$ be given. Since $x_n \to x$, there exists $N \in \mathbb{N}$ such that for all $k \geq N$ we have $|x_k - x| < \frac{\varepsilon}{2}$. Let $m, n \geq N$, then

$$|x_m - x_n| = |(x_m - x) + (x_n - x)| \leq |x_m - x| + |x_n - x| < \frac{\varepsilon}{2} + \frac{\varepsilon}{2} = \varepsilon.$$

One may ask now whether the converse of the previous example is true. The answer is yes. In fact, the only examples of Cauchy sequences in $\mathbb{R}$ are convergent sequences. This is the context of the next theorem.

Theorem 2.2.3 (Cauchy Completeness of $\mathbb{R}$). *A real sequence (x_n) is Cauchy iff it is convergent.*

Proof. We have already proved that every convergent sequence is a Cauchy sequence. We now prove its converse. That is, if (x_n) is Cauchy, then it is convergent.

Strategy: This is more like a motivation than a strategy. Assume that $x_n \to x$. Then $x = \text{lub}\,(-\infty, x)$. Thus for any $y < x$, there exists $N \in \mathbb{N}$ such that $x_k > y$ for $k \geq N$. We collect $y \in \mathbb{R}$ with the property that $x_k > y$ for all large values of k into set E. We show this set is nonempty, and bounded above. Given $\varepsilon > 0$, let N correspond to the Cauchy condition of (x_n). We show $x_N - \varepsilon \in E$ and $x_N + \varepsilon$ is an upper bound of E. If $x = \text{lub}\,E$, we use the facts that $x_N - \varepsilon \in E$ and $x_N + \varepsilon$ is an upper bound of E to show $|x - x_N| < \varepsilon$. Since we know how to estimate $|x_n - x_N|$, the result will follow by triangle inequality.

Let $E := \{x \in \mathbb{R} : \exists\, N \text{ such that } n \geq N \implies x < x_n\}$.
Let $\delta > 0$. Since (x_n) is Cauchy, there exists $n_0 = n_0(\delta)$ such that for all $m, n \geq n_0$, $|x_m - x_n| < \delta$. In particular, for all $n \geq n_0$, $|x_{n_0} - x_n| < \delta$. That is, we have

$$n \geq n_0 \implies x_n \in (x_{n_0} - \delta, x_{n_0} + \delta), \text{ in particular}, x_n < x_{n_0} + \delta. \qquad (2.6)$$

Claim 1. $x_{n_0} - \delta \in E$. For, if we take $N = n_0(\delta)$, then $n \geq n_0 \implies x_n > x_{n_0} - \delta$.

Claim 2. $x_{n_0} + \delta$ is an upper bound of E. If not, let $x \in E$ be such that $x > x_{n_0} + \delta$. This means that there exists some N such that for all $n \geq N$ $x_n \geq x > x_{n_0} + \delta$. In particular, for all $n \geq \max\{n_0, N\}$, we have $x_n > x_{n_0} + \delta$. This contradicts (2.6).

Claims (1) and (2) show that E is a nonempty set and is bounded above.

Let $\ell := \text{lub } E$.

Claim 3. $x_n \to \ell$.

Let $\varepsilon > 0$ be given. We have to estimate $|x_n - \ell|$ using the fact that (x_n) is Cauchy and $\ell = \text{lub } E$. Since (x_n) is Cauchy, there exists, $n_0 = n_0(\varepsilon)$ such that for all $n \geq n_0$, we have $|x_n - x_{n_0}| < \frac{\varepsilon}{2}$.

By claim (1), $x_{n_0} - \varepsilon/2 \in E$. This implies $x_{n_0} - \varepsilon/2 \leq \ell$. On the other hand $\ell \leq x_{n_0} + \varepsilon/2$ by Claim (2). Therefore $|x_{n_0} - \ell| \leq \varepsilon/2$.

Now for all $n \geq n_0$, we have

$$|x_n - \ell| \leq |x_n - x_{n_0}| + |x_{n_0} - \ell| < \varepsilon/2 + \varepsilon/2 = \varepsilon.$$

This completes the proof. $\qquad\square$

Lemma 2.2.4. *Any Cauchy sequence is bounded.*

Proof. This is obvious since any Cauchy sequence is convergent and convergent sequences are bounded.

Let us give a direct proof adapting the proof of Proposition 2.1.25.

If (x_n) is Cauchy, for $\varepsilon = 1$, there exists N such that $k, m \geq N$, we have $|x_k - x_m| < \varepsilon = 1$. In particular, if we take $m = N$, we obtain for $k \geq N$, $|x_k - x_N| < 1$. Hence it follows that $|x_k| \leq |x_k - x_N| + |x_N| < 1 + |x_N|$. Let $C := \max\{|x_1|, \ldots, |x_{N-1}|, 1 + |x_N|\}$. Then it is easy to show that $|x_n| \leq C$ for all n. $\qquad\square$

Exercise Set 2.2.5.

(1) Prove that the sum of two Cauchy sequences and the product of two Cauchy sequences are Cauchy.

(2) Let (x_n) be a sequence such that $|x_n| \leq \frac{1+n}{1+n+2n^2}$ for all $n \in \mathbb{N}$. Prove that (x_n) is Cauchy.

(3) If (x_n) is a Cauchy sequence of integers, what can you say about the sequence?

(4) Let (x_n) be a sequence and let $a > 1$. Assume that $|x_{k+1} - x_k| < a^{-k}$ for all $k \in \mathbb{N}$. Show that (x_n) is Cauchy.

(5) Let (x_n) be a sequence such that

$$|x_{n+1} - x_n| \leq c\,|x_n - x_{n-1}|,$$

for some constant c with $0 < c < 1$. Show that (x_n) is convergent.

2.3 Monotone Sequences

Definition 2.3.1. We say a sequence (x_n) of real numbers is *increasing* if for each n, we have $x_n \le x_{n+1}$. Clearly, any increasing sequence is bounded below by x_1. Hence such a sequence is bounded iff it is bounded above.

We say that (x_n) is strictly increasing if $x_n < x_{n+1}$ for all $n \in \mathbb{N}$.

Define decreasing sequences. When is it bounded? A sequence (x_n) is said to be *monotone* if it is either increasing or decreasing.

Proposition 2.3.2. *Let (x_n) be an increasing sequence. Then it is convergent iff it is bounded above.*

Proof. Let (x_n) be increasing and bounded above. We need to show that (x_n) is convergent.

Let $x(\mathbb{N}) := \{x_n : n \in \mathbb{N}\}$ be the *image* of the sequence x. Note that $x(\mathbb{N})$ is nonempty and bounded above. Let ℓ be the lub of this set. We claim that $x_n \to \ell$. Look at Figure 2.8.

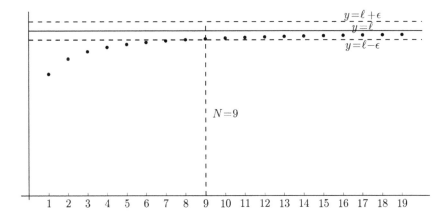

Figure 2.8: Increasing and bounded above sequence.

Let $\varepsilon > 0$. Note that $\ell - \varepsilon$ is not an upper bound of $x(\mathbb{N})$. Hence there exists $N \in \mathbb{N}$ such that $x_N > \ell - \varepsilon$. Since the sequence is increasing, for all $n \ge N$, we have $x_N \le x_n$ and hence $\ell - \varepsilon < x_N \le x_n \le \ell < \ell + \varepsilon$, that is, $x_n \to \ell$.

The converse of this result is very easy, since any convergent sequence is bounded. $\qquad\square$

What is the analogous result in the case of decreasing sequences?

Let (x_n) be a decreasing sequence. Then it is convergent iff it is bounded below.

Notation: The symbol $x_n \searrow x$ stands for the statement that the sequence (x_n) is decreasing and convergent to x. What should the symbol $x_n \nearrow x$ mean?

Example 2.3.3. We shall give two important recurring examples.

(1) Let $0 \le r < 1$ and $x_n := r^n$. If $r = 0$, the sequence is the zero sequence and hence is convergent. So we shall assume that $0 < r < 1$. Since $r^n > r^{n+1}$ for $n \in \mathbb{N}$, the sequence (x_n) is decreasing. It is bounded below by zero. So we conclude that it is convergent. Let $x_n \to \ell$. Now by the algebra of convergent sequences $rx_n \to r\ell$. But $rx_n = x_{n+1}$ and hence $rx_n \to \ell$. Hence by the uniqueness of the limit, we conclude $r\ell = \ell$. Since $0 < r < 1$, we deduce that $\ell = 0$. Hence for $0 \le r < 1$, the sequence (r^n) converges to 0.

(2) Let the notation be as in the last example. Consider now $s_n := 1 + r + \cdots + r^n$. If $r = 0$, the sequence (s_n) is the constant sequence 1 and is convergent. Assume $0 < r < 1$. Since $s_{n+1} = s_n + r^{n+1} > s_n$, the sequence (s_n) is increasing. It is bounded above:

$$s_n = \frac{1 - r^{n+1}}{1 - r} \le \frac{1}{1 - r}.$$

We therefore conclude that (s_n) is convergent. Since

$$s_n = \frac{1 - r^{n+1}}{1 - r} = \frac{1}{1 - r} + \frac{r^{n+1}}{1 - r},$$

it follows from the algebra of convergent sequences that $s_n \to \frac{1}{1-r}$.

Exercise Set 2.3.4.

(1) Let $x_n := \frac{1}{n+1} + \frac{1}{n+2} + \cdots + \frac{1}{2n}$. Show that (x_n) is convergent to a limit at most 1.

(2) Let (x_n) be a sequence of positive real numbers. Assume that $\frac{x_{n+1}}{x_n} \to \ell$ with $\ell < 1$. Show that $x_n \to 0$.

(3) Let $a_n := \frac{n!}{n^n}$. Show that $a_n \to 0$.

(4) Let (a_n) be bounded. Assume that $a_{n+1} \ge a_n - 2^{-n}$. Show that (a_n) is convergent.

The Number e

We were lucky in Examples 2.3.3 to find the limits explicitly. In general it may not be possible. In fact, some real numbers are defined as the limit of such sequences. For instance, consider $x_n := \left(1 + \frac{1}{n}\right)^n$. We shall show that (x_n) is increasing and bounded above. Therefore, it is convergent. The real number which is the limit of this sequence is denoted by e and called the Euler number.

We shall outline the existence of limit of the sequence $x_n = \left(1 + \frac{1}{n}\right)^n$ using the following steps:

(1) By binomial theorem

$$x_n = 1 + \sum_{k=1}^{n} \frac{n!}{k! \, (n-k)!} n^{-k}$$

$$= 1 + \sum_{k=1}^{n} \frac{1}{k!} \frac{n(n-1) \cdots (n-k+1)}{n^k}$$

$$= 1 + \sum_{k=1}^{n} \frac{1}{k!} \left(1 - \frac{1}{n}\right) \left(1 - \frac{2}{n}\right) \cdots \left(1 - \frac{k-1}{n}\right). \qquad (2.7)$$

(2) We claim that $x_n < x_{n+1}$. That is, x_n is an increasing sequence.

$$x_n = 1 + \sum_{k=1}^{n} \frac{1}{k!} \left(1 - \frac{1}{n}\right) \left(1 - \frac{2}{n}\right) \cdots \left(1 - \frac{k-1}{n}\right)$$

$$< 1 + \sum_{k=1}^{n} \frac{1}{k!} \left(1 - \frac{1}{n+1}\right) \left(1 - \frac{2}{n+1}\right) \cdots \left(1 - \frac{k-1}{n+1}\right)$$

$$< x_{n+1}.$$

(3) From (2.7), we see that $x_n \leq 1 + \sum_{k=1}^{n} \frac{1}{k!}$.

(4) $1 + \sum_{k=1}^{n} \frac{1}{k!} < 1 + 1 + \sum_{k=1}^{n-1} \frac{1}{2^k} = 1 + \frac{1-2^{-n}}{1-1/2} < 1 + \frac{1}{1/2} = 3$. From (3), it follows that $x_n < 3$ for $n \in \mathbb{N}$.

(5) Thus (x_n) is increasing and bounded above and hence by the Proposition 2.3.2, $\lim x_n$ exists. Let $e := \lim x_n$.

Proposition 2.3.5. *Let* $y_n := \sum_{k=0}^{n} \frac{1}{k!}$. *Then* $\lim y_n = e$.

Proof. We take $x_n = \left(1 + \frac{1}{n}\right)^n$. The proof follows from the following steps.

(1) From Step 3 of the last example, we know that $x_n \leq y_n$ and hence $e = \lim x_n \leq \lim y_n$ (by item 3 of Exercise 2.1.28.)

(2) For $n > m$ omitting terms for $k \geq m + 1$ from (2.7) we get:

$$x_n \geq 1 + 1 + \frac{1}{2!} \left(1 - \frac{1}{n}\right) + \cdots + \frac{1}{m!} \left(1 - \frac{1}{n}\right) \cdots \left(1 - \frac{m-1}{n}\right).$$

Fix m. Then for any $n > m$, we have $x_n > y_m$ and hence $e := \text{lub } \{x_n\} > y_m$. Since $y_m < e$ for all m, we deduce that $\text{lub } \{y_m\} \leq e$.

(3) Hence $e := \lim_{n \to \infty} (1 + \frac{1}{n})^n = \lim_n \left(\sum_{k=0}^{n} \frac{1}{k!}\right)$.

$\square$

Proposition 2.3.6. *e is irrational.*

Proof. Suppose e is a rational number. Let $e = \frac{p}{q}$ with $q \in \mathbb{N}$. Note that $q > 1$. (Why?)

Let $s_q := \sum_{k=0}^{q} \frac{1}{k!}$. Since e is the lub of $\{\sum_{k=0}^{m} \frac{1}{k!} : m \in \mathbb{N}\}$, it follows that $e \geq \sum_{k=0}^{N} \frac{1}{k!}$ for any $N \in \mathbb{N}$. In particular, $e - s_q \geq \sum_{k=0}^{N} \frac{1}{k!} - \sum_{k=0}^{q} \frac{1}{k!}$ for any $N \geq q$. We conclude that

$$e = s_q + \text{lub} \left\{ \sum_{k=q+1}^{N} \frac{1}{k!} : N > q \right\} = s_q + \text{lub } R_q^N,$$

where $R_q^N := \sum_{k=q+1}^{N} \frac{1}{k!}$. We observe that for $N > q$,

$$\sum_{k=q+1}^{N} \frac{1}{k!} = \frac{1}{(q+1)!} + \frac{1}{(q+1)!}\frac{1}{q+2} + \frac{1}{(q+1)!}\frac{1}{(q+2)(q+3)} + \cdots$$

$$+ \frac{1}{(q+1)!}\frac{1}{(q+2)\cdots N}$$

$$\leq \frac{1}{(q+1)!}\left(1 + \frac{1}{(q+1)} + \frac{1}{(q+1)^2} + \cdots + \frac{1}{(q+1)^{N-q-1}}\right)$$

$$= \frac{1}{(q+1)!} \sum_{r=0}^{N-q-1} \frac{1}{(q+1)^r}$$

$$\leq \frac{1}{(q+1)!}\frac{1}{q}.$$

(Can you justify the last inequality above?) Hence we conclude that $e \leq s_q + \frac{1}{(q+1)!}\frac{1}{q}$.

We multiply both sides of the inequality $e - s_q < \frac{1}{(q+1)!}\frac{1}{q}$ by $(q+1)!$. Since $e = p/q$, we deduce $(q+1)!e \in \mathbb{N}$. It is clear that $(q+1)!(s_q) \in \mathbb{N}$. Hence it follows that $(q+1)!(e - s_q) < \frac{1}{q}$. This is absurd. So we are forced to conclude that e is irrational. $\qquad\square$

2.4   Sandwich Lemma

An easy and very useful result is the following. If we guess that a sequence (z_n) converges to α and we want to prove it rigorously, the lemma suggests an approach. Find lower and upper bounds x_n and y_n for z_n such that $x_n \to \alpha$ and $y_n \to \alpha$. Notice that this is typical of analysis.

Lemma 2.4.1 (Sandwich Lemma). *Let (x_n), (y_n) and (z_n) be sequences such that (i) $x_n \to \alpha$ and $y_n \to \alpha$ and (ii) $x_n \leq z_n \leq y_n$ for all n. Then $z_n \to \alpha$.*

Strategy: Let $\varepsilon > 0$ be given. Note that for sufficiently large k, both x_k and y_k lie inside the interval $(\alpha - \varepsilon, \alpha + \varepsilon)$. Since $x_k \leq z_k \leq y_k$ for all k, z_k must also lie in the interval $(\alpha - \varepsilon, \alpha + \varepsilon)$ for sufficiently large k. In particular, z_n must also converge to α. This is the basic idea of the proof.

Proof. Look at Figure 2.9.

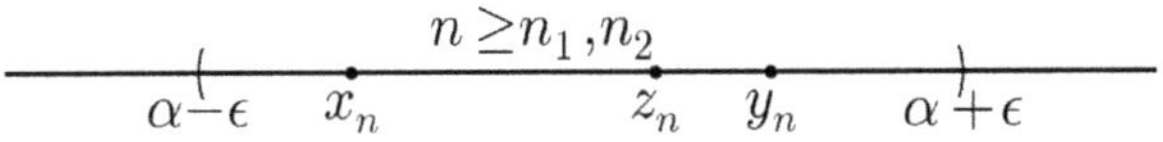

Figure 2.9: Sandwich lemma.

For given $\varepsilon > 0$, choose n_1, n_2 such that for all $k \geq n_1$, we have $x_k \in (\alpha - \varepsilon, \alpha + \varepsilon)$ and for all $k \geq n_2$, we have $y_k \in (\alpha - \varepsilon, \alpha + \varepsilon)$. Let $N = \max\{n_1, n_2\}$. Then for $k \geq N$, we observe

$$\alpha - \varepsilon < x_k \leq z_k \text{ and } z_k \leq y_k < \alpha + \varepsilon.$$

That is, $z_k \in (\alpha - \varepsilon, \alpha + \varepsilon)$ for all $k \geq N$. Hence $z_k \to \alpha$.   □

Example 2.4.2 (Typical uses of the sandwich lemma).

(1) Let $a \in \mathbb{R}$. For each $n \in \mathbb{N}$, select any element $x_n \in (a - \frac{1}{n}, a + \frac{1}{n})$. Then $x_n \to a$. (Note that this simple observation will be repeatedly used in the book!)

For, $a - \frac{1}{n} < x_n < a + \frac{1}{n}$. The result follows from (the algebra of convergent sequences and) the sandwich lemma.

(2) We have $\frac{\sin n}{n} \to 0$, as $-1/n \leq (\sin n)/n \leq 1/n$.

(3) Given any real number x, there exist sequences (x_n) of real numbers $x_n \to x$

Hint: For each $n \in \mathbb{N}$, take x_n with $x - 1/n < x_n < x$.

(4) Given any real number x, there exist sequences (s_n) of rational numbers and (t_n) of irrational numbers such that $s_n \to x$ and $t_n \to x$.

Hint: By density of rationals there exists r such that $x - 1/n < r < x$. Call this r as r_n.

(5) Let $\alpha := \text{lub } A \subset \mathbb{R}$. Then there exists a sequence (a_n) in A such that $a_n \to \alpha$.

Hint: $\alpha - 1/n$ is not an upper bound of A. Let $a_n \in A$ be such that $\alpha - 1/n < a_n \leq \alpha$.

Formulate the analogous result for glb.

Exercise Set 2.4.3. Use the sandwich lemma to solve the following.

(1) Let (a_n) be a bounded (real) sequence and (x_n) converge to 0. Then show that $a_n x_n \to 0$.

(2) The sequence $\sqrt{n+1} - \sqrt{n} \to 0$.

(3) $x_n := \frac{1}{\sqrt{n^2+1}} + \frac{1}{\sqrt{n^2+2}} + \cdots + \frac{1}{\sqrt{n^2+n}} \to 1$.

(4) Let $0 < a < b$. The sequence $((a^n + b^n)^{1/n}) \to b$.

Theorem 2.4.4 (Nested Interval Theorem – Standard Version). *Let $J_n := [a_n, b_n]$ be intervals in $\mathbb{R}$ such that $J_{n+1} \subseteq J_n$ for all $n \in \mathbb{N}$. Assume further that $b_n - a_n \to 0$, that is, the sequence of lengths of J_n's goes to zero. Then there exists a unique c such that $\cap_n J_n = \{c\}$.*

Proof. We have already seen (Theorem 1.3.23) that there exists at least one $c \in \cap_n J_n$. Let $c, d \in \cap_n J_n$. Assume that $c \leq d$. Then, since $a_n \leq c \leq d \leq b_n$, we see that $0 \leq d - c \leq b_n - a_n$. Since $b_n - a_n \to 0$, it follows from the sandwich lemma that $d - c = 0$. ☐

Exercise 2.4.5. If $x_n \to x$ and $x_n \geq 0$, then $x \geq 0$. (If $x < 0$, use Proposition 2.1.20 to arrive at $x_n < 0$ for $n > n_0$.) However, if each $x_n > 0$ and if $x_n \to x$, then x need not be positive. Can you give an example?

2.5   Some Important Limits

Analysis deals with unknown sequences or functions by trying to compare their behavior with known things. If we want to be adept in this technique, it is of paramount importance that we have a quite good command over commonly occurring sequences and functions. In this section, we deal with some of the most often used sequences and their convergence questions.

Theorem 2.5.1. *We have the following important limits:*

(1) *Let $0 \leq r < 1$ and $x_n := r^n$. Then $x_n \to 0$.*

(2) *Let $-1 < t < 1$. Then $t^n \to 0$.*

(3) *Let $|r| < 1$. Then $nr^n \to 0$.*

(4) *Let $a > 0$. Then $a^{1/n} \to 1$.*

(5) $n^{1/n} \to 1$.

(6) *Fix $a \in \mathbb{R}$. Then $\frac{a^n}{n!} \to 0$.*

Proof. We start with the proof of 1. We have already seen this in Example 2.3.3.

Let us look at another proof of this result. If $0 < r < 1$, then we can write $r = 1/(1+h)$ for some $h > 0$. Using binomial theorem, we have

$$(1+h)^n = 1 + nh + \frac{n(n-1)}{2} + \cdots + h^n > nh, \qquad (2.8)$$

since all terms are positive.

In particular, $r^n = \frac{1}{(1+h)^n} \leq \frac{1}{nh}$ for all n. Thus we have $0 \leq r^n \leq \frac{1}{nh}$ for all n. Hence by the sandwich lemma, we have $r^n \to 0$.

Proof of 2: Let $-1 < t < 1$. Then $t^n \to 0$.

$t^n \to 0$ iff $|t^n| = |t|^n \to 0$ in view of Proposition 2.1.18. Now the result follows from the last item.

Proof of 3: Let $|r| < 1$. Then $nr^n \to 0$.

Notice that the sequence n is unbounded, whereas $r^n \to 0$. In fact, n diverges to ∞ as we shall see later. Thus if we say $nr^n \to 0$, it means that r^n goes to 0 much faster that n diverging to ∞. For example, if $x_n = n$ and $y_n = \frac{1}{n^2}$, then $x_n y_n \to 0$. On the other hand, if $z_n = \frac{1}{\sqrt{n}}$ then $x_n z_n \to \infty$.

It is enough to prove the result for $0 < r < 1$.

If you try to use the estimate, as in Equation (2.8), we end up $0 \leq nr^n \leq \frac{1}{h}$ from which we cannot conclude $nr^n \to 0$. Therefore, to take care of the presence of n in nr^n, we need an estimate of the form $r^n \leq C\frac{1}{n^2}$ for some constant C. This is achieved by retaining the quadratic term in the binomial expansion of $(1+h)^n$. More precisely

$$(1+h)^n = 1 + nh + \frac{n(n-1)}{2} + \cdots + h^n > \frac{n(n-1)}{2}h^2, \qquad (2.9)$$

since all terms are positive. Thus we have $0 \leq nr^n \leq \frac{2}{h^2(n-1)}$ for all $n \geq 2$. Hence using the Sandwich lemma, we have $nr^n \to 0$.

What can you say about the sequence $(n^2 r^n)$? Is there any way to generalize this?

Proof of 4: Let $a > 0$. Then $a^{1/n} \to 1$.

If $a > 1$, then we claim that $a^{1/n} > 1$. Note that we have seen if $0 \leq a \leq b$, then $0 \leq a^n \leq b^n$ for any natural number n. Also when $a > 0$, $a^{1/n}$ is uniquely defined. (See Item 3 of Exercise 1.1.3.) Suppose, $a^{1/n} \leq 1$, then $a \leq 1$ which is a contradiction.

If $a > 1$, then we can write $a^{1/n} = 1 + h_n$, with $h_n > 0$. (Do you understand why we are using $a^{1/n} = 1 + h_n$ and not $a^{1/n} = 1 + h$ unlike the previous proof?) This implies $a = (1 + h_n)^n \geq nh_n$, and hence $0 \leq h_n \leq \frac{a}{n}$. This means, $h_n \to 0$. Therefore, $a^{1/n} \to 1$ as desired.

When $0 < a < 1$, we apply the result to $b^{1/n}$ where $b = 1/a > 1$. Observe that $a^{1/n} = 1/(b^{1/n})$. By the first case, we have $b^{1/n} \to 1$, and hence using the algebra of limits, $1/(b^{1/n}) \to 1$. This proves that $\lim a^{1/n} = \lim 1/(b^{1/n}) = 1$.

Proof of 5: $n^{1/n} \to 1$.

For $n > 1$, we have $n^{1/n} > 1$ as seen in the last item. So we can write $n^{1/n} = 1 + h_n$ with $h_n > 0$. It is enough to show that $h_n \to 0$. From (2.9), we have

$$n = (1 + h_n)^n \geq \frac{n(n-1)}{2} h_n^2.$$

This implies $0 \leq h_n^2 \leq \frac{2}{n-1}$. Hence $h_n^2 \to 0$ by the sandwich lemma. Hence $h_n \to 0$ using Item 10 of Exercise 2.1.28.

Now we can give another proof of the item (4).

Note that $1 \leq a^{1/n} \leq n^{1/n}$ for $n \geq a$. Hence by item (5) and the sandwich lemma, we have $a^{1/n} \to 1$.

Proof of 6: Fix $a \in \mathbb{R}$. Then $\frac{a^n}{n!} \to 0$. Note that it is enough to show that $\frac{|a|^n}{n!} \to 0$. In particular, it is enough to show that $\frac{a^n}{n!} \to 0$ for $a > 0$.

Assume $a > 0$. By the Archimedean property, there exists $N \in \mathbb{N}$ such that $N > a$. Then, for $n \geq N$, we have

$$\frac{a^n}{n!} = \left(\frac{a}{1} \frac{a}{2} \cdots \frac{a}{N} \right) \frac{a}{N+1} \cdots \frac{a}{n}$$

$$\leq C r^{-N} r^n, \text{ where } C := \left(\frac{a}{1} \frac{a}{2} \cdots \frac{a}{N} \right) \text{ and } r := \frac{a}{N}.$$

Thus we get $0 \leq \frac{a^n}{n!} \leq C r^{-N} r^n$. Since $0 < r < 1, r^n \to 0$ and hence $C r^{-N} r^n \to 0$. Hence by the sandwich lemma we have $\frac{a^n}{n!} \to 0$. $\qquad \square$

Exercise 2.5.2. Let $a > 1$. Show that $\frac{n}{a^n} \to 0$.

The next result employs an often used trick in analysis, which we call the Divide and Conquer trick. Suppose that we want to estimate a sum of infinite terms or a definite integral of the form $\int_a^b f(x)\, dx$. We split the sum into two parts, first consisting of terms which are well behaved and second consisting of the remaining terms. In the case of the integral, we split the domain into two parts, say, the set A of points at which the function is well-behaved and B the rest. We then try to get control over the set B and use a crude estimate of f over B. In the case of sums, we try to get some crude estimate for the terms in the second part and try to control the size of the second part. For example, suppose, $x_n \to 0$, then for any $\varepsilon > 0$, there exists $N \in \mathbb{N}$ such that for $k \geq N$, $|x_k| < \varepsilon$. However, we do not have any control over $x_1, \ldots, x_{N-1}$. But the number of such terms is at the most $N - 1$.

Theorem 2.5.3. *Let $x_n \to 0$. Let (s_n) be the sequence of arithmetic means (or averages) defined by $s_n := \frac{x_1 + \cdots + x_n}{n}$. Then $s_n \to 0$.*

Proof. We need to estimate $|s_n|$. As suggested prior to the proof, we employ the divide and conquer trick. We break s_n into two parts.

$$s_n = \left[\frac{x_1 + \cdots + x_{N-1}}{n} \right] + \left[\frac{x_N + \cdots + x_n}{n} \right].$$

The terms in the second bracket are "well behaved" and terms in the first bracket are only finitely many in number.

Since (x_n) is convergent, there exists $M > 0$ such that $|x_n| \leq M$ for all n. Hence we have an estimate for $|s_n|$:

$$|s_n| \leq \frac{(N-1)M}{n} + \frac{(n-N+1)}{n}\varepsilon \leq \frac{(N-1)M}{n} + \varepsilon.$$

By taking n sufficiently large, the first term can also be made less than ε.

Now we shall write a detailed proof.

Given $\varepsilon > 0$, choose N such that for $k \geq N$, $|x_k| < \varepsilon/2$. Since (x_n) is convergent, it is bounded. Let M be such that $|x_k| \leq M$ for all k. Choose n_1 such that $n \geq n_1$ implies $(MN)/n < \varepsilon/2$. Observe that for $n \geq \max\{n_1, N\}$

$$|s_n| = \frac{|(x_1 + \cdots + x_N) + (x_{N+1} + \cdots + x_n)|}{n}$$
$$\leq \frac{MN}{n} + \frac{(n-N)}{n}\frac{\varepsilon}{2} = \varepsilon.$$

$\square$

Corollary 2.5.4. *Let $x_n \to x$. Then applying the last result to the sequence $y_n := x_n - x$, we conclude that the sequence (s_n) of arithmetic means converges to x.* $\square$

Corollary 2.5.5. *Let $x_n \to x$ and $y_n \to y$. Then*

$$\frac{x_1 y_n + x_2 y_{n-1} + \cdots + x_n y_1}{n} \to xy.$$

Proof. Let $u_n := x_1 y_n + x_2 y_{n-1} + \cdots + x_n y_1$. Let $s_n := \sum_{k=1}^{n} x_k$. Adding and subtracting the term $(x_1 + \cdots + x_n)y$ in u_n, we get

$$u_n = (x_1 + \cdots + x_n)y + [x_1(y_n - y) + \cdots + x_n(y_1 - y)]$$
$$= s_n y + v_n, \text{ say.}$$

If M is a bound for (x_n), then we have the estimate

$$|v_n| \leq M(|y_n - y| + \cdots + |y_1 - y|).$$

Since $y_n \to y$, $(y_n - y) \to 0$, hence by Theorem 2.5.3, we conclude that $\frac{v_n}{n} \to 0$. We now observe

$$\frac{u_n}{n} = \frac{s_n y}{n} + \frac{v_n}{n} \to yx + 0, \text{ by Corollary (2.5.4).} \qquad \square$$

2.6   Sequences Diverging to $\pm\infty$

Definition 2.6.1. Let (x_n) be a real sequence. We say that (x_n) *diverges* to $+\infty$ (or simply diverges to ∞) if for any $R \in \mathbb{R}$ there exists $N \in \mathbb{N}$ such that $n \geq N$ implies $x_n > R$.

Note that ∞ is just a symbol and it is not a real number. The symbol "$x_n \to \infty$" means exactly what is defined above.

If a sequence (x_n) diverges to ∞, geometrically, this means if we look at the points of the sequence and take line $y = R$, then there exists a natural number N such that all the points (n, x_n) for $n \geq N$ lie above the line $y = R$. Look at Figure 2.10.

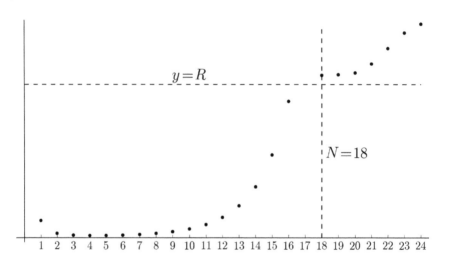

Figure 2.10: Sequence diverging to ∞.

Formulate an analogous notion of a sequence diverging to $-\infty$.

Note that any sequence diverging to ∞ or ∞ is unbounded.

It is easy to see if a sequence of real numbers diverges to ∞ (or to $-\infty$), then it is not convergent. (Can you see why?)

Example 2.6.2. Let $x_n := n$, and $y_n := 2^n$. Then the sequences (x_n) and (y_n) diverge to infinity.

Hint: First one follows directly from the Archimedean property. For the second one, use induction on n to show that $2^n > n$ for all n.

Example 2.6.3. Let $x_{2k-1} := 1$ and $x_{2k} = 2k$. This sequence is unbounded; however, it does not diverge to ∞.

For if $R = 1$, then given any $N \in \mathbb{N}$, $2N - 1 \geq N$ but $x_{2N-1} = 1$ is not greater than R.

Example 2.6.4.

(1) Let $a > 1$. Then $a^n \to \infty$.

Since $a > 0$, we can write $a = a + h$ for some $h > 0$. For any n, we have $a^n = (1 + h)^n > nh$. Let $R \in \mathbb{R}$. By the Archimedean property there exists, $n \in \mathbb{N}$ such that $nh > R$.

(2) $(n!)^{1/n}$ diverges to ∞.

Let $R > 0$ be given. Since $\frac{R^n}{n!} \to 0$ by Item 6 of Exercise 2.5.1, there exists N, such that for all $n \geq N$, $R^n/n! < 1$, That is, $R^n < n!$. Hence $(n!)^{1/n} > R$ for $n \geq N$. This proves the result.

(3) Consider the sequence (x_n) where $x_n = (-1)^n n$. This sequence is divergent, (that is, not convergent!) but divergent neither to ∞ nor to $-\infty$.

Exercise Set 2.6.5.

(1) Let $x_n > 0$. Then $x_n \to 0$ iff $1/x_n \to +\infty$. What happens if $x_n < 0$ and $\lim x_n = 0$?

The sequence $x_n := \frac{(-1)^n}{n} \to 0$ but the sequence of reciprocals is $((-1)^n n)$. Refer to Item 3 of Example 2.6.4.

(2) Let $x_n := \sum_{k=1}^n \frac{1}{k}$. Show that the sequence (x_n) diverges to ∞.

(3) Let (x_n) be a sequence in $(0, \infty)$. Let $y_n := \sum_{k=1}^n (x_k + \frac{1}{x_k})$. Show that (y_n) diverges to ∞.

(4) Let (x_n) and (y_n) be sequences of positive reals. Assume that $\lim x_n/y_n = A > 0$. Show $\lim x_n = +\infty$ iff $\lim y_n = +\infty$.

(5) Show that $\lim \frac{an^2 + b}{cn + d} = \infty$ if $ac > 0$.

(6) Let (a_n) be a sequence of positive reals. Assume that $\lim \frac{a_{n+1}}{a_n} = \alpha$. Then show that $\lim (a_n)^{\frac{1}{n}} = \alpha$.

(7) Use the last item to find the "limit" of $\frac{n}{(n!)^{\frac{1}{n}}}$.

(8) Find the limit of $\left(\frac{a^n - b^n}{a^n + b^n}\right)$ where $a, b \in (0, \infty)$.

2.7 Subsequences

Definition 2.7.1. Let $x : \mathbb{N} \to \mathbb{R}$ be a sequence. Then a *subsequence* is the restriction of x to an infinite subset S of $\mathbb{N}$.

For example, if S is a set of even integers, then the subsequence is

$$(x_2, x_4, \ldots, x_{2n}, \ldots).$$

Similarly, if S is a set of prime numbers, then the subsequence is

$$(x_2, x_3, x_5, x_7, x_{11}, x_{13}, \ldots).$$

If we restrict a sequence to the set of prime numbers, is it a subsequence? Can we exhibit its terms explicitly?

We suggest that the reader understand the statement of the next result and return to its proof later. We need to use the well-ordering principle of $\mathbb{N}$ thrice in the proof. Let us state it.

Well-Ordering Principle. If $S \subset \mathbb{N}$ is nonempty, then it has a least element, that is, there exists $\ell \in S$ such that $\ell \leq x$ for all $x \in S$.

Proposition 2.7.2. *An infinite subset $S \subset \mathbb{N}$ can be listed as $\{n_1 < n_2 < \cdots < n_k < n_{k+1} \cdots \}$.*

Proof. Let n_1 be the least element of S. Now choose n_2 as the least element of $S \setminus \{n_1\}$. Note that $n_1 < n_2$. Assume that we have chosen $n_1, \ldots, n_k \in S$ such that $n_1 < n_2 < \cdots < n_k$. Define $S_k := S \setminus \{n_1, \ldots, n_k\}$. Then it is easy to see that $S_k \neq \emptyset$. (Why?) Let n_{k+1} be the least element of S_k. Thus we have a recursively defined sequence of integers. Observe that by our choice $n_k \geq k$ for each $k \in \mathbb{N}$.

We claim that this process exhausts S. Suppose $T := S \setminus \{n_k : k \in \mathbb{N}\} \neq \emptyset$ and m be the least element of T. (Note that m is an element of S!) Consider $A := \{k \in \mathbb{N} : n_k \geq m\}$. Since $n_m \geq m$, we deduce that $A \neq \emptyset$. Let k be the least element of A. Then we must have $n_{k-1} < m$. Since $m \notin S_{k-1}$, since $m \leq n_k$ and since n_k is the least element of S_{k-1}, we conclude that $m = n_k$. But this is a contradiction to the fact that $m \in T$. $\square$

In view of the last proposition, the standard practice is to denote the subsequence as (x_{n_k}) where $n_1 < n_2 < \cdots < n_k < n_{k+1} < \cdots$.

> Most useful/handy observation: $n_k \geq k$ for all k.

For, $n_1 \geq 1$. Now, n_2 is the least element of $S \setminus \{n_1\}$ and hence $n_2 > n_1 \geq 1$. Hence $n_2 \geq 2$. We conclude that $n_k \geq k$ by induction.

Let (x_n) be a sequence and (x_{n_k}) be a subsequence. What does it mean to say that the subsequence converges to x?

Let us define a new sequence (y_k) where $y_k := x_{n_k}$. Then we say $x_{n_k} \to x$ iff $y_k \to x$. That is, for a given $\varepsilon > 0$ there exists $k_0 \in \mathbb{N}$ such that for $k \geq k_0$, we must have $|y_k - x| < \varepsilon$, which is the same as saying that

$$\text{for } k \geq k_0, \text{ we have } |x_{n_k} - x| < \varepsilon.$$

Lemma 2.7.3. *If $x_n \to x$, and if (x_{n_k}) is a subsequence, then $x_{n_k} \to x$ as $k \to \infty$.*

Proof. Let $\varepsilon > 0$ be given. Since $x_n \to x$, there exists $N \in \mathbb{N}$ such that for $k \geq N$, $|x_k - x| < \varepsilon$. Note that if $k \geq N$, then $n_k \geq k \geq N$. Hence, for $k \geq N$, we have $|x_{n_k} - x| < \varepsilon$. This implies that (x_{n_k}) converges to x.   □

Theorem 2.7.4 (Existence of a monotone subsequence of a real sequence). *Given any real sequence (x_n) there exists a monotone subsequence.*

> **Motivation:** Imagine the following scenario. Scientists have predicted the eruption of a dormant volcano in an island in the Indian ocean. There are lots of tourists pouring into the island to watch the spectacular show of Nature. The local business has constructed an infinite number of towers lined in front of the volcano, numbered serially. The n-th tower is of height x_n. See Figure 2.11. As tourists, we would like to observe the event from the tallest tower and at the same time the one farthest from the volcano. If we observe from the n-th tower, we want the heights x_m of the towers that are in front of the n-th tower to satisfy $x_m < x_n$ for $m > n$. This suggests us to consider the set S of the serial numbers of the towers that are suitable for observation.

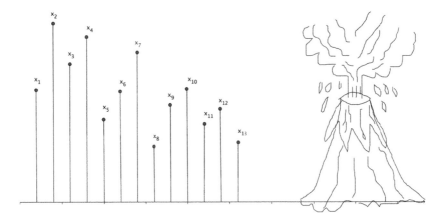

Figure 2.11: Observation towers.

Proof. Consider the set S defined by

$$S := \{n \in \mathbb{N} : x_m < x_n \text{ for } m > n\}.$$

There are two cases: S is finite or infinite.

Case 1. S is finite. Let N be any natural number such that $k \leq N$ for all $k \in S$. Let $n_1 > N$. Then $n_1 \notin S$. Hence there exists $n_2 > n_1$ such that $x_{n_2} \geq x_{n_1}$. Since $n_2 > n_1 > N$, $n_2 \notin S$. Hence we can find an $n_3 > n_2$ such that $x_{n_3} \geq x_{n_2}$. This way, we can find a monotone nondecreasing (increasing) subsequence, (x_{n_k}).

Case 2. S is infinite. Let n_1 be the least element of S. Let n_2 be the least element of $S \setminus \{n_1\}$ and so on. We thus have a listing of S:

$$n_1 < n_2 < n_3 < \cdots$$

Since n_{k-1} is an element of S and since $n_{k-1} < n_k$, we see that $x_{n_k} < x_{n_{k-1}}$, for all k. We now have a monotone decreasing sequence. □

Theorem 2.7.5 (Bolzano-Weierstrass Theorem). *If (x_n) is a bounded real sequence, it has a convergent subsequence.*

Proof. By Theorem 2.7.4, (x_n) has a monotone subsequence, say (x_{n_k}). Since (x_n) is bounded, (x_{n_k}) is also bounded. Thus (x_{n_k}) is monotone and bounded. Hence it is convergent. □

Theorem 2.7.6. *Let (x_n) be Cauchy. Let a subsequence (x_{n_k}) converge to x. Then $x_n \to x$.*

Strategy: We need to estimate $|x_n - x|$. What we know is how to estimate $|x_n - x_m|$ and $|x_{n_k} - x|$. So we use triangle inequality.

$$|x_n - x| \le |x - x_m| + |x_m - x_n|.$$

If $m = n_k$ for some large k, we can estimate $|x - x_m|$, since $x_{n_k} \to x$. If m and n are very large, since the sequence (x_n) is Cauchy, we can estimate $|x_m - x_n|$. So we need to choose the "intermediary/curry leaf" m so that we can estimate both the terms as we wish.

Proof. Let $\varepsilon > 0$ be given. Since $x_{n_k} \to x$, there exists k_1 such that

$$k \ge k_1 \implies |x_{n_k} - x| < \varepsilon.$$

For the same ε, since (x_n) is Cauchy, there exists k_2 such that

$$m, n \ge k_2 \implies |x_m - x_n| < \varepsilon.$$

Let $N := \max\{k_1, k_2\}$. Fix an $k \ge N$. Let $m = n_k$. Then $m = n_k \ge k \ge k_1$ and hence $|x - x_m| < \varepsilon$. If $n \ge N$, then $n \ge k_2$. It follows that $|x_n - x_m| < \varepsilon$, since $m = n_k \ge k \ge k_2$. We are now ready to complete the proof. For $n \ge N$, we have

$$|x - x_n| \le |x - x_m| + |x_m - x_n| < 2\varepsilon.$$

It follows that $x_n \to x$. □

The proof above uses a trick which we shall refer to as the "curry leaf trick." In Indian cooking, curry leaves are used to enhance the aroma and to garnish the dishes, but they are mostly thrown out while one eats the dishes. The integer m in the proof is one such. Go through the proof of Proposition 2.1.15. Did we employ the curry leaf trick there?

We can now give a second proof of the Cauchy completeness of $\mathbb{R}$.

Theorem 2.7.7. *If (x_n) is a Cauchy sequence in $\mathbb{R}$, then there exists $x \in \mathbb{R}$ such that $x_n \to x$.*

Proof. Let (x_n) be a Cauchy sequence in $\mathbb{R}$. By Lemma 2.2.4 it is bounded. Now using the Bolzano-Weierstrass Theorem 2.7.5, it has a convergent subsequence, say, (x_{n_k}). Hence the original sequence (x_n) is convergent by Theorem 2.7.6. $\square$

Do subsequences arise naturally in mathematics?

Exercise 2.7.8. Let (a_n) be a sequence. Prove that (a_n) is divergent iff for each $a \in \mathbb{R}$, there exists an $\varepsilon > 0$ and a subsequence (x_{n_k}) such that $|a - a_{n_k}| \geq \varepsilon$ for all k.

Example 2.7.9. Some typical uses of subsequences.

(1) Consider the sequence $a^{1/n}$ where $a > 1$.

   We proved in Item (4) on page 48, that $a^{1/n}$ is bounded below by 1. It is easy to show that $a^{1/n}$ is decreasing. Hence it is convergent. Let us assume that $a^{1/n} \to \ell$.

   Then subsequence $(a^{1/2n})$ is also convergent to ℓ by Lemma 2.7.3. Hence $(a^{1/2n})^2 = a^{1/n} \to \ell^2$. Thus by the uniqueness of limit, we have $\ell^2 = \ell$. This implies $\ell = 1$. (Why?)

(2) Assume that the sequence $(n^{1/n})$ is convergent. We wish to find its limit.

   Let $(n^{1/n})$ converge to ℓ. Then $((2n)^{1/2n}) \to \ell$ by Lemma 2.7.3. Therefore, $((2n)^{1/2n})^2 \to \ell^2$. But $((2n)^{1/2n})^2 = 2^{1/n}n^{1/n}$. From the last item, $2^{1/n} \to 1$. Hence $2^{1/n}n^{1/n} \to \ell$ by algebra of limits. Thus by the uniqueness of limit, we have $\ell^2 = \ell$. That is, $\ell = 1$.

(3) Show that the sequence $((-1)^n)$ is divergent.

   Suppose $((-1)^n)$ is convergent. Then by Lemma 2.7.3, every subsequence of $((-1)^n)$ converges to the same limit. The subsequence of even terms converges to 1, whereas the subsequence of odd terms converges to -1. Hence we conclude that $((-1)^n)$ is not convergent.

Exercise Set 2.7.10.

(1) Prove that the sequence (x_n) where $x_n := \frac{(n^2+13n-41)\cos(2^n)}{n^2+2n+1}$ has a convergent subsequence.

(2) True or false: For any sequence (x_n), the sequence $y_n := \frac{x_n}{1+|x_n|}$ has a convergent subsequence.

(3) True or false: A sequence (x_n) is bounded iff every subsequence of (x_n) has a convergent subsequence.

(4) Prove that a sequence (x_n) is unbounded iff there exists a subsequence (x_{n_k}) such that $|x_{n_k}| \geq k$ for each $k \in \mathbb{N}$.

(5) Let (a_n) be a sequence. Prove that (a_n) is divergent iff for each $a \in \mathbb{R}$, there exists an $\varepsilon > 0$ and a subsequence (x_{n_k}) such that $|a - a_{n_k}| \geq \varepsilon$ for all k.

(6) Show that if a monotone sequence has a convergent subsequence, then it is convergent.

(7) Let $\{r_n\}$ be an enumeration of all rationals in $[0, 1]$. Show that $\{r_n\}$ is not convergent.

Exercise Set 2.7.11 (Typical uses of Bolzano-Weierstrass theorem).

(1) The sequence $(\sin(n))$ has a convergent subsequence.

(2) True or false: A sequence (x_n) is bounded iff every subsequence of (x_n) has a convergent subsequence.

2.8   Sequences Defined Recursively

If you recall, the sequence defined in Item 8 of Example 2.1.3, the n-th term of this sequence is not defined explicitly in terms of n. Rather it is defined in terms of previous two terms. Such sequences whose n-th term is defined in terms of previous terms are called recursive sequences or sequences defined recursively. This type of sequence occurs naturally. In this section we shall look at some examples of sequences defined recursively and find their limits if they are convergent.

Example 2.8.1. Let a and b be any two distinct real numbers. Let $x_1 = a$ and $x_2 = b$. Define $x_{n+2} := \frac{x_n + x_{n+1}}{2}$. Mark points of this sequence on the real line. What can you say about this sequence? See Figure 2.12..

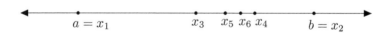

$$a = x_1 \qquad\qquad x_3 \qquad x_5\ x_6\ x_4 \qquad b = x_2$$

Figure 2.12: Figure for Example 2.8.1.

We claim that (x_n) is convergent. Note that it is enough to show that it is a Cauchy sequence. We need to estimate $|x_n - x_m|$. First of all, let us look at

$$|x_n - x_{n-1}| = \left| \frac{x_{n-1} + x_{n-2}}{2} - x_{n-1} \right| = \left| \frac{x_{n-2} - x_{n-1}}{2} \right| = \left| \frac{x_{n-2} - x_{n-3}}{2^2} \right|.$$

Continuing this way inductively, we get

$$|x_n - x_{n-1}| = \frac{x_2 - x_1}{2^{n-2}} = \frac{b - a}{2^{n-2}}.$$

Now we can estimate $|x_n - x_m|$. Let $n > m$. Then

$$|x_n - x_m| \leq |x_n - x_{n-1}| + |x_{n-1} - x_{n-2}| + \cdots + |x_{m+1} - x_m|$$
$$= \frac{|b - a|}{2^{n-2}} + \frac{b - a}{2^{n-3}} + \cdots + \frac{b - a}{2^{m-1}}$$
$$= \frac{|b - a|}{2^{m-1}} \left[1 + \frac{1}{2} + \cdots + \frac{1}{2^{n-m-1}} \right]$$
$$\leq \frac{|b - a|}{2^{m-1}} \times \frac{1}{2}$$
$$= \frac{|b - a|}{2^m}.$$

(In the above we made use of the fact that the sequence $(\sum_{k=0}^{n} \frac{1}{2^k})$ is increasing and converges to $1/2$, the lub of its terms. See Item 2 of Example 2.3.3 on page 44.) Now let us show that (x_n) is Cauchy. Let $\varepsilon > 0$ be given. Since the sequence $\frac{1}{2^k} \to 0$, there exists $N \in \mathbb{N}$ such that for all $k \geq N$, we have $\frac{1}{2^k} < \frac{\varepsilon}{|b-a|}$. Therefore, for $m, n \geq N$, we have

$$|x_n - x_m| \leq \frac{|b - a|}{2^m} < |b - a| \frac{\varepsilon}{|b - a|} = \varepsilon.$$

Thus (x_n) is a Cauchy sequence and hence it converges by the Cauchy completeness theorem. However, we do not have its limit explicitly.

It can be shown that the limit is $\frac{1}{3}(a + 2b)$.

Example 2.8.2. Square Roots. Let $x_1 = 2$, define

$$x_{n+1} = \frac{1}{2} \left(x_n + \frac{2}{x_n} \right).$$

We claim that this sequence is convergent and it converges to $\sqrt{2}$.

Note that $x_{n+1} = \frac{1}{2} \left(x_n + \frac{2}{x_n} \right)$ implies $x_n^2 - 2x_{n+1} + 2 = 0$. This is a quadratic in x_n. Since $x_n \in \mathbb{R}$. This quadratic has a real solution. This means that its discriminant is nonnegative. That is, $4x_{n+1}^2 - 8 \geq 0$. This implies $x_n^2 \geq 2$ for all n. In particular, x_n is bounded below.

Next we claim that x_n is a decreasing sequence. We have

$$x_{n+1} - x_n = \frac{1}{2} \left(x_n + \frac{2}{x_n} \right) - x_n$$
$$= \frac{2 - x_n^2}{2x_n} \leq 0 \text{ from the last claim.}$$

Thus we have proved that (x_n) is a decreasing sequence which is bounded below. Hence it is convergent.

Let $x_n \to \ell$. Then by the algebra of limits, the left-hand side converges to ℓ where as the right-hand side converges to $\frac{1}{2}(\ell + \frac{2}{\ell})$. Hence, ℓ is a positive solution of $\ell^2 - 2 = 0$. This mean $\ell = \sqrt{2}$.

Given $a > 0$, can you modify the sequence to produce one which converges to $\sqrt{a}$?

Exercise Set 2.8.3. Find the limits (if they exist) of the following recursively defined sequences.

(1) $x_1 = \sqrt{2}$, $x_n = \sqrt{2 + \sqrt{x_{n-1}}}$ for $n \geq 2$.

(2) $x_1 = 1$, $x_n = \sqrt{2x_{n-1}}$ for $n \geq 2$.

(3) For $a > 0$, let x_1 be any positive real number and $x_{n+1} = \frac{1}{2}\left(x_n + \frac{a}{x_n}\right)$.

(4) Let $0 < a \leq x_1 \leq x_2 \leq b$. Define $x_n = \sqrt{(x_{n-1}x_{n-2})}$ for $n \geq 3$. Show that $a \leq x_n \leq b$ and $|x_{n+1} - x_n| \leq \frac{b}{a+b}|x_n - x_{n-1}|$ for $n \geq 2$. Prove (x_n) is convergent.

(5) Let $0 < y_1 < x_1$. Define

$$x_{n+1} = \frac{x_n + y_n}{2} \text{ and } y_{n+1} = \sqrt{x_n y_n}, \text{ for } n \in \mathbb{N}.$$

(a) Prove that (y_n) is increasing and bounded above while (x_n) is decreasing and bounded below.

(b) Prove that $0 < x_{n+1} - y_{n+1} < 2^{-n}(x_1 - y_1)$ for $n \in \mathbb{N}$.

(c) Prove that x_n and y_n converge to the same limit.

(6) Let $x_1 \geq 0$, and define recursively $x_{n+1} = \sqrt{2 + x_n}$ for $n \in \mathbb{N}$. Show that if the sequence is convergent, then the limit is 2.

(7) Let $0 \leq x_1 \leq 1$. Define $x_{n+1} := 1 - \sqrt{1 - x_n}$ for $n \in \mathbb{N}$. Show that if the sequence is convergent, then the limit is either 0 or 1.

Exercise Set 2.8.4 (Euler's constant). Let

$$\gamma_n = \sum_{k=1}^{n} \frac{1}{k} - \log n = \sum_{k=1}^{n} \frac{1}{k} - \int_{1}^{n} t^{-1} dt.$$

(1) Show that γ_n is a decreasing sequence.

(2) Show that $0 < \gamma_n \leq 1$ for all n.

(3) $\lim \gamma_n$ exists and is denoted by γ.

The real number γ is called the *Euler's constant*. At the time of writing this book, it is not known whether γ is rational or not!

Exercise Set 2.8.5 (Fibonacci's sequence). Let $x_0 = 1$, $x_1 = 1$. Define (x_n) recursively by $x_n = x_{n-1}+x_{n-2}$, $n \geq 2$. This (x_n) is called the *Fibonacci sequence*. Let $\gamma_n := \frac{x_n}{x_{n-1}}$, $n \geq 1$. Then prove the following:

(1) (x_n) is divergent.

(2) (i) $1 \leq \gamma_n \leq 2$,   (ii) $\gamma_{n+1} = 1 + \frac{1}{\gamma_n}$.

(3) (γ_{2n}) is decreasing.

(4) (γ_{2n+1}) is increasing.

(5) (γ_{2n}) and (γ_{2n+1}) are convergent. The limits of both of these sequences satisfy the equation $\ell^2 - \ell - 1 = 0$.

(6) $\lim \gamma_n = \frac{1+\sqrt{5}}{2}$. This limit is called the *golden ratio*.

Exercise 2.8.6. Let (a_n) be a sequence such that $|a_n - a_m| < \varepsilon$ for all $m, n \geq N$. If $a_n \to a$, show that $|a_n - a| \leq \varepsilon$ for all $n \geq N$. (An easy but often used result.)

Exercise Set 2.8.7 (Miscellaneous Exercises).

(1) Decide for what values of x, the sequences whose n-th term is $x_n := \frac{x+x^n}{1+x^n}$ is convergent.

(2) Find the limit of the sequence whose n-th term is $\frac{1+a+a^2+\cdots+a^{n-1}}{n!}$.

(3) Let $a_n := \frac{n}{2^n}$. Show that $\lim a_n = 0$.

(4) Let $a \in \mathbb{R}$. Consider $x_1 = a$, $x_2 = \frac{1+a}{2}$, and by induction $x_n := \frac{1+x_{n-1}}{2}$. Can you guess what (x_n) converges to? Draw pictures and guess the limit and prove your guess.

(5) Consider the sequence

$$\sqrt{2}, \sqrt{2+\sqrt{2}}, \sqrt{2+\sqrt{2+\sqrt{2}}}, \sqrt{2+\sqrt{2+\sqrt{2+\sqrt{2}}}},\ldots,$$

Show that $x_n \to 2$.

(6) Prove that the sequence $(\sin n)$ is divergent.

Chapter 3

Continuity

Contents

In this chapter, we shall define continuity of functions and study their properties. Unlike books at this level, we shall start with a definition which makes use of our knowledge of convergent sequences. Later we shall give an equivalent definition. At the end we shall give the definition of a limit of a function at a point and use it to give a third definition of continuity of a function at a point. First two definitions allow us to define the continuity at a point of a function defined on an arbitrary subset of $\mathbb{R}$.

3.1 Continuous Functions

Definition 3.1.1. Let $J \subset \mathbb{R}$. (An important class of subsets J are intervals of any kind.)

Let $f : J \to \mathbb{R}$ be a function and $a \in J$. We say that f is *continuous* at a if for *every* sequence (x_n) **in** J with $x_n \to a$, we have $f(x_n) \to f(a)$.

We say that f is *continuous* on J if it is continuous at every point $a \in J$.

Remark 3.1.2. The crucial point of the definition is that (i) the sequences are in the domain of f converging to a point of the domain and (ii) we need to verify the condition of the definition for *each* such sequence in the domain converging to a. See Item 3 of Example 3.1.3.

The second crucial point is that even if $x_n \to a$, it may happen that $(f(x_n))$ may converge to a limit other than $f(a)$ or worse, $(f(x_n))$ may not converge at all! See Item 4 of Example 3.1.3.

Example 3.1.3. We now look at examples of continuous and non-continuous functions.

(1) Let f be a constant function on J, say, $f(x) = c$ for $x \in J$. Then f is continuous on J.

Let $a \in J$. Let (x_n) be *any* sequence in J such that $x_n \to a$. Then $f(x_n) = c$ for each n, that is, the sequence $(f(x_n))$ is a constant sequence and hence $f(x_n) \to c = f(a)$. Therefore we conclude that f is continuous at a. Since a is arbitrary, f is continuous on J.

(2) Let $f(x) := x$ for all $x \in J$. Then f is continuous on J.

Let $a \in J$ and (x_n) be a sequence in J converging to a. Then $f(x_n) = x_n$ and hence $f(x_n) = x_n \to a = f(a)$. We conclude that f is continuous at a. Since $a \in J$ is arbitrary, f is continuous on J.

More generally, $f(x) := x^n$ is continuous on $\mathbb{R}$, $n \in \mathbb{N}$.

(3) Let $f \colon \mathbb{R} \to \mathbb{R}$ be given by $f(x) = 1$ if $x \in \mathbb{Q}$ and 0 otherwise. Then f is not continuous at any point of $\mathbb{R}$. This is known as *Dirichlet's function*.

Let $a = 0$. Let $x_n = 1/n$. Then $x_n \to 0$. Since $f(x_n) = 1$ for $n \in \mathbb{N}$, the sequence $(f(x_n))$ is a constant sequence converging to $1 = f(0)$. But on the other hand, if we let $y_n = \frac{\sqrt{2}}{n}$, then (y_n) is a sequence in the domain of f converging to 0. We have $f(x_n) = 0$ so that $f(x_n) \to 0$. By the uniqueness of the limit, $f(x_n)$ does not converge to $1 = f(0)$. Hence we conclude that f is not continuous at 0.

Consider the sequence (x_n) defined by $x_n = \frac{1}{n}$, if n is odd and $x_n = \frac{\sqrt{2}}{n}$, if n is even. Then by Exercise 13 on page 40, $x_n \to 0$. But the sequence $(f(x_n))$ is $(1, 0, 1, 0, \ldots)$, a sequence in which 1 and 0 alternate and hence does not even converge, let alone converge to $f(0) = 1$. Hence f is not continuous at $a = 0$. Now go back to Remark 3.1.2 and understand what it says.

We claim that f is not continuous at any $a \in \mathbb{R}$. We know from Item 4 of Example 2.4.2 on page 47 that there exist sequences (x_n) of rationals and (y_n) a sequence of irrationals such that $x_n \to a$ and $y_n \to a$. We have $f(x_n) = 1$ and $f(y_n) = 0$. Hence $f(x_n) \to 1$ whereas $f(y_n) \to 0$. Now, $f(a) = 1$ if $a \in \mathbb{Q}$ and 0 otherwise. Let us assume the latter for definiteness sake. Then $f(x_n)$ does not converge to $f(a)$. We conclude that f is not continuous at a. Similarly, if $a \in \mathbb{Q}$, then $f(y_n)$ does not converge to $f(a) = 1$. Hence f is continuous nowhere on $\mathbb{R}$.

(4) Let $f: \mathbb{R} \to \mathbb{R}$ be given by $f(x) = \begin{cases} 0, & \text{if } x < 0 \\ 1, & \text{if } x \geq 0. \end{cases}$

Look at Figure 3.1 of the graph of this function.

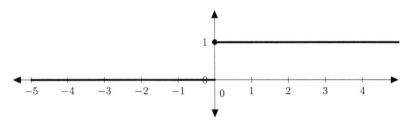

Figure 3.1: Figure for item 4 of Example 3.1.3.

Your earlier encounter with the notion of continuity at school level should lead you to infer that f is continuous at all nonzero a but not at 0. Your guess is correct and we shall prove it.

Let $a \neq 0$. Either $a > 0$ or $a < 0$. Let us assume $a < 0$. If (x_n) is real sequence converging to a, then we know (from Proposition 2.1.22) that there exists $N \in \mathbb{N}$ such that $x_k < 0$ for $k \geq N$. Hence the sequence $(f(x_n))$ is such that $f(x_k) = 0$ for $k \geq N$ (Remark 2.1.21). Hence $f(x_n) \to 0 = f(a)$. We conclude that f is continuous at a. Similarly, we can conclude that f is continuous at $a > 0$.

What happens at $a = 0$? Consider $x_n = 1/n$, $y_n = -1/n$, and $z_n = (-1)^n/n$. All three sequences converge to 0. But we have $f(x_n) \to 1$, $f(y_n) \to -1$ and the worst of all is $(f(z_n))$, which does not converge at all! Therefore, f is not continuous at 0.

Thus f is continuous at all nonzero elements of $\mathbb{R}$ and is not continuous at 0.

(5) Let $f: \mathbb{R} \to \mathbb{R}$ be given by $f(x) = \begin{cases} x^2, & \text{if } x \geq 0 \\ x, & \text{if } x < 0. \end{cases}$

Look at the graph of f. (See Figure 3.2.) You will be convinced of the continuity of f.

If $a \neq 0$, the continuity at a is established by the same argument as in the last example. At $a = 0$, we need to show that if $x_n \to 0$, then $f(x_n) \to 0$. Now x_n could be non-negative or negative so that $f(x_n)$ could be x_n^2 or x_n. If $x_n \to 0$, there exists N such that for $k \geq N$, we have $|x_k| \leq 1$ and hence $|x_k^2| \leq |x_k|$. For such k, we have $|f(x_k)| \leq |x_k|$. Since $x_n \to 0$ iff $|x_n| \to 0$, we see that $0 \leq |f(x_k)| \leq |x_k|$. By the sandwich lemma, it follows that $|f(x_k)| \to 0$ and hence $f(x_k) \to 0$. That is, f is continuous at 0.

Let $m, n \in \mathbb{N}$. Consider $f(x) = \begin{cases} x^n, & \text{if } x \geq 0 \\ x^m, & \text{if } x < 0. \end{cases}$

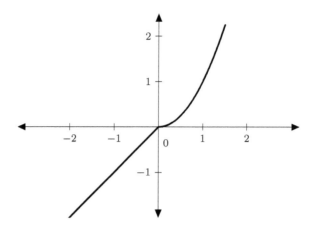

Figure 3.2: Figure for item 5 of Example 3.1.3.

What can you say about the continuity of f?

(6) Let $\mathbb{R}^* = \mathbb{R} \setminus \{0\}$. Let $f \colon \mathbb{R}^* \to \mathbb{R}^*$ be given by $f(x) = 1/x$. Then f is continuous on $\mathbb{R}^*$.

This is easy. If $a \in \mathbb{R}^*$ and (x_n) is a sequence in $\mathbb{R}^*$, then $x_n \to a$ implies $1/x_n \to 1/a$ by the algebra of convergent sequences, (Item 4 of Theorem 2.1.26). Hence f is continuous at a and hence on all of $\mathbb{R}^*$.

(7) Let $f \colon \mathbb{R} \to \mathbb{R}$ be given by

$$f(x) = \begin{cases} \alpha, & \text{if } x < 0 \\ ax^2 - bx + c, & \text{if } x \geq 0. \end{cases}$$

What value of α ensures the continuity of f at 0?

If (s_n) is a sequence such that $s_n < 0$ and $s_n \to 0$, then $f(s_n) \to \alpha$. If f has to be continuous at 0, then $f(s_n) \to f(0) = c$. Hence a necessary condition for the continuity of f at 0 is that $\alpha = c$.

To complete the proof, we need to show that if $\alpha = c$, then f is continuous at 0. Let $x_n \to 0$. We need to estimate $|f(x_n) - c|$. We have

$$|f(x_n) - c| = \begin{cases} 0, & \text{if } x_n < 0 \\ ax_n^2 - bx_n, & \text{otherwise.} \end{cases} \tag{3.1}$$

Since $x_n \to 0$, by the algebra of convergent sequences, $ax_n^2 - bx_n \to 0$. Given $\varepsilon > 0$, there exists $N \in \mathbb{N}$ such that for $k \geq N$, we have $\left|ax_k^2 - bx_k\right| < \varepsilon$. Hence (3.1) becomes $|f(x_k) - c| < \varepsilon$ for $k \geq N$. Hence, f is continuous at 0.

Exercise Set 3.1.4.

(1) Let $f \colon J \to \mathbb{R}$ be continuous. Let $J_1 \subset J$. Let g be the restriction of f to J_1. Show that g is continuous on J_1.

(2) Let $f(x) = 3x$ for $x \in \mathbb{Q}$ and $f(x) = x + 8$ for $x \in \mathbb{R} \setminus \mathbb{Q}$. Find the points at which f is continuous.

(3) Let $f(x) = x$ if $x \in \mathbb{Q}$ and $f(x) = 0$ if $x \notin \mathbb{Q}$. Then show that f is continuous only at $x = 0$.

(4) Let $f \colon \mathbb{R} \to \mathbb{R}$ be continuous. Assume that $f(r) = 0$ for $r \in \mathbb{Q}$. Show that $f = 0$.

(5) Let $f, g \colon \mathbb{R} \to \mathbb{R}$ be continuous. If $f(x) = g(x)$ for $x \in \mathbb{Q}$, then show that $f = g$.

(6) Let $f \colon \mathbb{R} \to \mathbb{R}$ be continuous which is also an additive homomorphism, that is, $f(x + y) = f(x) + f(y)$ for all $x, y \in \mathbb{R}$. Then $f(x) = \lambda x$ where $\lambda = f(1)$.

(7) Let

$$f(x) = \begin{cases} x \sin(1/x), & \text{if } x \neq 0 \\ 0, & \text{if } x = 0. \end{cases}$$

Show that f is continuous at 0.

(8) Let $f \colon \mathbb{R} \to \mathbb{R}$ be defined by $f(x) = x - [x]$, where $[x]$ stands for the greatest integer less than or equal to x. At what points is f continuous?

(9) Let $f \colon \mathbb{R} \to \mathbb{R}$ be defined by $f(x) = \min\{x - [x], 1 + [x] - x\}$, that is, the minimum of the distances of x from $[x]$ and $[x] + 1$. At what points is f continuous?

(10) Let $f \colon J \to \mathbb{R}$ be continuous. Let $\alpha \in \mathrm{Im}\,(f)$. Let $S := f^{-1}(\alpha)$. Show that if (x_n) is a sequence in S converging to an element $a \in J$, then $a \in S$.

Let f, g be real-valued functions defined on a subset $J \subset \mathbb{R}$ and $\alpha \in \mathbb{R}$. We define new functions $f + g$, fg, αf, $|f|$ and $1/h$ if $h(x) \neq 0$ for $x \in J$ as follows.

$$(f + g)(x) := f(x) + g(x), x \in J,$$
$$(fg)(x) := f(x)g(x), x \in J,$$
$$(\alpha f)(x) := \alpha f(x), x \in J,$$
$$(1/h)(x) := 1/h(x), x \in J,$$
$$|f|\,(x) := |f(x)|\,, x \in J.$$

Note that the expressions on the RHS of the definitions involve standard arithmetic operations on $\mathbb{R}$. For example, $f(x)g(x)$ is the product of the two real numbers $f(x)$ and $g(x)$.

Theorem 3.1.5 (Algebra of Continuous Functions). *Let $f, g\colon J \to \mathbb{R}$ be continuous at $a \in J$. Let $\alpha \in \mathbb{R}$. Then:*

1. *$f + g$ is continuous at a.*
2. *αf is continuous at a.*
3. *The set of functions from $J \to \mathbb{R}$ continuous at a is a real vector space.*
4. *The product fg is continuous at a.*
5. *Assume further that $f(a) \neq 0$. Then there exists $\delta > 0$ such that for each $x \in (a-\delta, a+\delta) \cap J \to \mathbb{R}$, we have $f(x) \neq 0$. The function $\frac{1}{f}\colon (a-\delta, a+\delta) \cap J \to \mathbb{R}$ is continuous at a.*
6. *$|f|$ is continuous at a.*
7. *Let $h(x) := \max\{f(x), g(x)\}$. Then h is continuous at a. Similarly, the function $k(x) := \min\{f(x), g(x)\}$ is continuous at a.*
8. *Let $f_i\colon J_i \to \mathbb{R}$ be continuous at $a_i \in J_i$, $i = 1, 2$. Assume that $f_1(J_1) \subset J_2$ and $a_2 = f_1(a_1)$. Then the composition $f_2 \circ f_1$ is continuous at a_1.*

Proof. You will see the advantage of our definition of continuity. All proofs are immediate applications of the analogous results in the theory of convergent sequences.

To prove (1), let (x_n) be a sequence in J such that $x_n \to a$. We need to prove that $(f+g)(x_n) \to (f+g)(a)$. By definition, $(f+g)(x_n) = f(x_n) + g(x_n)$. Since f and g are continuous at a, we have $f(x_n) \to f(a)$ and $g(x_n) \to g(a)$. By the algebra of convergent sequences, we have $f(x_n) + g(x_n) \to f(a) + g(a)$. That is, $(f+g)(x_n) \to f(a) + g(a)$. Since $(f+g)(a) = f(a) + g(a)$, we have proved that for any sequence $x_n \to a$, we have $(f+g)(x_n) \to (f+g)(a)$. Therefore, $f + g$ is continuous at a.

We were very elaborate here so that you could see how the definitions are used.

Proof of (2) is similar and left to the reader.

(3) follows from (1) and (2) and involves a detailed verification of the axioms of a vector space.

Let us prove (4). Let (x_n) be a sequence in J such that $x_n \to a$. We need to prove that $(fg)(x_n) \to (fg)(a)$. Now, $(fg)(x_n) = f(x_n)g(x_n)$. Since f and g are continuous at a, we have $f(x_n) \to f(a)$ and $g(x_n) \to g(a)$. By the algebra of convergent sequences, we have $f(x_n)g(x_n) \to f(a)g(a)$. That is, $(fg)(x_n) \to f(a)g(a)$. Since $(fg)(a) = f(a)g(a)$, we have proved that for any sequence $x_n \to a$, we have $(fg)(x_n) \to (fg)(a)$. Therefore, fg is continuous at a. (What we have done here is to cut and paste the proof of (1) and replaced addition by product!)

We now prove (5) by contradiction. To exploit the continuity at a, we need to generate a sequence (x_n) in J such that $x_n \to a$. As we have pointed out earlier in Item 1 of Example 2.4.2, we have to look at intervals of the form $(a - \frac{1}{n}, a + \frac{1}{n})$. Now if no $\delta > 0$ is as required in (5), then $\delta := 1/n$ will not be as required. In particular, for $\delta = 1/n$, there exists $x_n \in (a - \frac{1}{n}, a + \frac{1}{n}) \cap J$ such that $f(x_n) = 0$. By the sandwich lemma, $x_n \to a$. Since f is continuous at a, we must have $f(x_n) \to f(a)$. Since $f(x_n) = 0$ for all $n \in \mathbb{N}$, $f(x_n) \to 0$. By the uniqueness of the limit, we conclude $f(a) = 0$, a contradiction. We therefore conclude that there exists $\delta > 0$ such that $f(x) \neq 0$ for $x \in (a - \delta, a + \delta) \cap J$. Hence $1/f$ is defined on $(a - \delta, a + \delta) \cap J$. (Note that a lies in the intersection.)

To complete the proof of (5), we need to establish $1/f$ is continuous at $a \in J_a := (a - \delta, a + \delta) \cap J$. Let (x_n) be a sequence in J_a such that $x_n \to a$. Since $f(x_n) \neq 0$, by algebra of convergent sequences, we conclude that $1/f(x_n) \to 1/f(a)$. Since $1/f(x_n) = (1/f)(x_n)$ etc., we have shown that $1/f$ is continuous at a.

We now prove (6). With the notation established above in (1)–(4), we need to prove that $|f|(x_n) \to |f|(a)$. Since f is continuous at a, $f(x_n) \to f(a)$, and hence $|f(x_n)| \to |f(a)|$. That is, $|f|(x_n) \to |f|(a)$. Thus, $|f|$ is continuous at a.

To prove (7), we recall Item 9 of Theorem 1.4.2. We have

$$\max\{f, g\}(x) = \frac{(f(x) + g(x)) + |f(x) - g(x)|}{2} = \frac{1}{2}\left((f + g) + |f - g|\right)(x)$$

$$\min\{f, g\}(x) = \frac{(f(x) + g(x)) - |f(x) - g(x)|}{2} = \frac{1}{2}\left((f + g) - |f - g|\right)(x).$$

The functions on the rightmost side of these equations are continuous by (1), (2), and (6). Hence the result follows. $\qquad\square$

You might have noticed, while talking continuity of f, that we allowed the domain J to be any nonempty subset of $\mathbb{R}$. Let us look at some interesting examples of J which are not intervals.

Let $J = \{a\}$. (Technically this is an interval!) Let $f \colon J \to \mathbb{R}$ be any function. If (x_n) is a sequence in J, then necessarily $x_n = a$ for all $n \in \mathbb{N}$. Hence $x_n \to a$ and $(f(x_n))$ is the constant sequence $(f(a))$. It follows that f is continuous at a and hence on J. Thus we conclude that $f \colon J \to \mathbb{R}$ is continuous.

Let $J := \{a_1, \ldots, a_n\}$. Let $f \colon J \to \mathbb{R}$. Let $a := a_j$ for a fixed $1 \leq j \leq n$. If (x_n) is a sequence in J such that $x_n \to a$, then there exists $N \in \mathbb{N}$ such that $x_n = x_j$ for $n \geq N$. (Why? Take $\varepsilon := \min\{|a_k - a_j| : k \neq j, 1 \leq k \leq n\} > 0$. Then there exists $N \in \mathbb{N}$ such that $|x_n - a| < \varepsilon$. This can happen only when $x_n = a$ by our choice of ε.) Hence it turns out that the sequence $(f(x_n))$ is eventually constant, the constant being $f(a)$. Consequently, $f(x_n) \to f(a)$. Hence f is continuous at a. Since $a \in J$ was arbitrary, f is continuous on J. We conclude that $f \colon J \to \mathbb{R}$ is continuous on J.

You can argue in a similar way to prove that any function $f \colon \mathbb{Z} \to \mathbb{R}$ is continuous.

Remark 3.1.6. This remark may be skipped on first reading. Go through the proof in the last paragraph. What made the proof work?

A careful study will yield the following conclusion. If the set J is such that $\varepsilon := \inf\{|x - y| : x, y \in J : x \neq y\}$ is positive, then we can repeat the same argument to conclude that any $f \colon J \to \mathbb{R}$ is continuous.

Note, however, that this condition is a sufficient condition but not necessary. See Item 10 of Example 3.2.3 on page 77.

We could have also attacked the case of finite A as follows. If $A :=$ $\{a_1, \ldots, a_n\}$, let $A_k := \{a_k\}$, $1 \le k \le n$. Then $d_A = \min\{d_{A_k} : 1 \le k \le n\}$ is continuous by item 5 of Theorem 3.1.5.

The general case of A is a little difficult. We shall return to this later; see Example 3.2.5 on page 77.

Example 3.1.7. If $A \subset \mathbb{R}$ is a nonempty subset, define $f(x) := \mathrm{glb}\ \{|x - a| : a \in A\}$. Then f is continuous on $\mathbb{R}$ and is usually denoted by d_A. Look at Figure 3.3 for $A = \{a\}$, $A = \{a, b\}$ and $A = [a, b], (a, b)$.

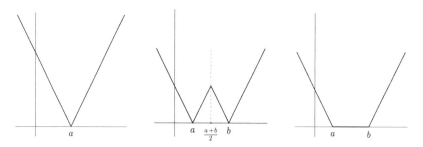

Figure 3.3: Graph of distance function.

Exercise 3.1.8. Let $A = \mathbb{Z}$. Write down d_A explicitly and draw its graph. This function and functions in Example 3.1.7 are typical "saw-tooth" functions. Do you understand why they are called so?

Exercise Set 3.1.9.

(1) Let $f(x) := x$ and $g(x) := x^2$ for $x \in \mathbb{R}$. Find $\max\{f, g\}$ and draw its graph.

(2) Let $f, g \colon [-\pi, \pi] \to \mathbb{R}$ be given by $f(x) := \cos x$ and $g(x) := \sin x$. Draw the graph of $\min\{f, g\}$. See Figure 3.4.

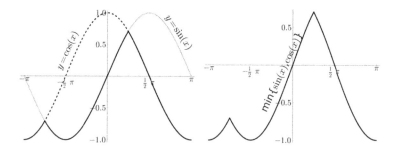

Figure 3.4: $\min\{\sin x, \cos x\}$.

(3) Any polynomial function $f\colon J \to \mathbb{R}$ of the form $f(x) := a_0 + a_1 x + \cdots + a_n x^n$ is continuous on J.

(4) A rational function is a function of the form $f(x) = \frac{p(x)}{q(x)}$ where p, q are polynomial functions. The domain of a rational function is the complement (in $\mathbb{R}$) of the set of points at which q takes the value 0. The rational functions are continuous on their domains of definition.

3.2 ε-δ Definition of Continuity

We now give the standard definition of continuity using ε - δ. To understand this definition, consider the following situation. Think of f as a process which will produce a new material y if x is its ingredient. If the customer wants an output y_0, we know that if we can input x_0, our process is so reliable that we shall get y_0. In real life, we cannot be sure of the purity or genuineness of the input x_0. The customer is aware of this and he sets his limit of error tolerance $\varepsilon > 0$. He says that he is ready to accept y provided it is "ε-close" to y_0. Since our process f is reliable, we know if we can ensure an input x which is δ-close to x_0, we shall end up with an output $y = f(x)$, which will be ε-close to the desired output. Do not worry too much if you find all this too vague. Even thinking along these lines will make you appreciate the next definition.

Definition 3.2.1. Let $f\colon J \to \mathbb{R}$ be given and $a \in J$. We say that f is continuous at a if for any given $\varepsilon > 0$, there exists $\delta > 0$ such that

$$x \in J \text{ and } |x - a| < \delta \implies |f(x) - f(a)| < \varepsilon. \tag{3.2}$$

We offer two pictures (Figures 3.5 and 3.6) to visualize this definition.

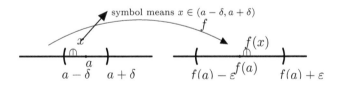

Figure 3.5: Continuity of f at a.

Theorem 3.2.2. *The definitions of continuity in Definition 3.1.1 and Definition 3.2.1 are equivalent.*

Proof. Let f be continuous at a according to Definition 3.1.1. We now show that f is continuous according to ε-δ definition. We prove this by contradiction. Assume the contrary. Then there exists $\varepsilon > 0$ for which no δ as required exists. This means that if a $\delta > 0$ is given, there exists $x \in J \cap (a - \delta, a + \delta)$ such that $|f(x) - f(a)| \geq \varepsilon$.

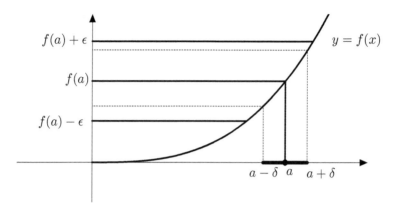

Figure 3.6: ε-δ definition of continuity.

As said earlier, we want to find a sequence (x_n) in J converging to a. For each $\delta = 1/k$, we have $x_k \in J \cap (a - 1/k, a + 1/k)$ with $|f(x_k) - f(a)| \geq \varepsilon$. Since $x_k \to a$, we must have $f(x_k) \to f(a)$, that is, $|f(x_k) - f(a)| \to 0$. This cannot happen, since $|f(x_k) - f(a)| \geq \varepsilon$ for all $k \in \mathbb{N}$. This contradiction shows that our assumption that no such δ as required in Definition 3.2.1 is wrong.

Assume that f is continuous according to the ε-δ definition. Let $x_n \in J$ be such that $x_n \to a$. To prove $f(x_n) \to f(a)$, let $\varepsilon > 0$ be given. Then by the ε-δ definition applied to f, there exists $\delta > 0$ such that (3.2) holds. Since $x_n \to a$, for the $\delta > 0$, there exists N such that

$$n \geq N \implies |x_n - a| < \delta.$$

It follows that if $n \geq N$, $x_n \in (a - \delta, a + \delta)$ and hence $f(x_n) \in (f(a) - \varepsilon, f(a) + \varepsilon)$. That is, $f(x_n) \to f(a)$. $\qquad\square$

Example 3.2.3. We look at some examples and prove their continuity or otherwise using the ε - δ definition. This will give us enough practice to work with the ε-δ definition. The basic idea to show the continuity of f at a is to obtain an estimate of the form

$$|f(x) - f(a)| \leq C_a |x - a|,$$

where $C_a > 0$ may depend on a. There are situations when this may not work. See Item 9 of Example 3.2.3 below. In Items 3 and 6, one can choose C_a independent of a.

(1) Let $f \colon \mathbb{R} \to \mathbb{R}$ be defined by $f(x) = x$.

To check continuity at a, we need to estimate $|f(x) - f(a)| = |x - a|$. We wish to have $|f(x) - f(a)| = |x - a| < \varepsilon$, whenever $|x - a| < \delta$. This suggests that if $\varepsilon > 0$ is given, we may take $\delta = \varepsilon$. Now for a textbook proof.

Let $a \in \mathbb{R}$. Let $\varepsilon > 0$ be given. Let $\delta = \varepsilon$. Then for any x with $|x - a| < \delta$, we have

$$|f(x) - f(a)| = |x - a| < \delta = \varepsilon.$$

(2) Let $f \colon \mathbb{R} \to \mathbb{R}$ be defined by $f(x) = x^2$.

We need to estimate $|f(x) - f(a)|$ to check of continuity of f at a. Now

$$|f(x) - f(a)| = |x^2 - a^2| = |(x + a)(x - a)| = |x + a|\,|x - a|\,.$$

Since we want an estimate of the form $|f(x) - f(a)| \le C_a\,|x - a|$, an obvious thing to do is to form the estimate

$$|f(x) - f(a)| \le (|x| + |a|)\,|x - a|\,.$$

We need to estimate $|x|$ in terms of a. Assume that we have found a δ. Then $|x| = |x - a + a| \le |x - a| + |a| < \delta + |a|$. If a δ works, any $\delta' \le \delta$ also works, and we may as well assume that $\delta < 1$. Hence we get an estimate of the form $|x| < 1 + |a|$. That is, we are restricting x in the interval $(a - 1, a + 1)$. The final estimate therefore is of the form

$$|f(x) - f(a)| \le (1 + 2\,|a|)\,|x - a|\,.$$

So, if we want $|f(x) - f(a)| < \varepsilon$, it is enough to make sure $(1 + 2\,|a|)\,|x - a| < \varepsilon$. That is, we need to ensure $|x - a| < \varepsilon/(1 + 2\,|a|)$. This suggests that we take $\delta < \varepsilon/(1 + 2\,|a|)$. Since we also wanted the first estimate, namely, $|x| < |a| + 1$, so we need to take $\delta < \min\{1, \frac{\varepsilon}{1 + 2|a|}\}$. Now we are ready for a textbook proof.

Fix $a \in \mathbb{R}$. Let $\varepsilon > 0$ be given. Choose $\delta < \min\{1, \frac{\varepsilon}{1 + 2|a|}\}$. Let $x \in \mathbb{R}$ be such that $|x - a| < \delta$. Then $|x - a| < 1$ so that $|x| = |x - a + a| \le |x - a| + |a| < 1 + |a|$. Consider

$$\begin{aligned}
|f(x) - f(a)| = |x^2 - a^2| &= |x + a|\,|x - a| \\
&\le (|x| + |a|)\,|x - a| \\
&\le (1 + 2\,|a|)\,|x - a| \\
&< (1 + 2\,|a|)\delta < \varepsilon.
\end{aligned}$$

Thus, f is continuous at $a \in \mathbb{R}$ and hence on $\mathbb{R}$, since a was an arbitrary element of $\mathbb{R}$.

(3) Let $R > 0$ and $f \colon [-R, R] \to \mathbb{R}$ be defined as $f(x) = x^2$.

Here it is a very straightforward estimate. Proceeding as in the last example, we arrive at $|f(x) - f(a)| \le (|x| + |a|)\,|x - a|$. Since both $x, a \in [-R, R]$, we have obvious estimates $|x| \le R$ and $|a| \le R$. Hence we have $|f(x) - f(a)| \le 2R\,|x - a|$. So we need to ensure $2R\,|x - a| < \varepsilon$. This leads us to take any positive $\delta < \frac{\varepsilon}{2R}$.

We now write down a textbook proof. Let $a \in [-R, R]$ and let $\varepsilon > 0$ be given. Choose $0 < \delta < \frac{\varepsilon}{2R}$. Let $|x - a| < \delta$ and $x \in [-R, R]$. We then have

$$
\begin{aligned}
|f(x) - f(a)| &\leq |x + a|\,|x - a| \\
&\leq (|x| + |a|)\,|x - a| \\
&\leq (R + R)\,|x - a| \\
&< 2R\frac{\varepsilon}{2R} = \varepsilon.
\end{aligned}
$$

(4) Consider $f\colon \mathbb{R} \to \mathbb{R}$ defined by $f(x) = x^n$, very similar to item 2 of Example 3.2.3 above. We need to estimate

$$
\begin{aligned}
&|f(x) - f(a)| \\
&= |x^n - a^n| \\
&= |(x - a)(x^{n-1} + x^{n-2}a + \cdots + xa^{n-2} + a^{n-1})| \\
&\leq |x - a|\,(|x^{n-1}| + |x^{n-2}a| + \cdots + |a^{n-1}|) \\
&\leq |x - a|\,[(1 + |a|)^{n-1} + |a|\,(1 + |a|)^{n-2} + \cdots + a^{n-1}] \\
&\leq |x - a|\,[(1 + |a|)^{n-1} + (1 + |a|)^{n-1} + \cdots + (1 + |a|)^{n-1}] \\
&= n(1 + |a|^{n-1})\,|x - a|\,.
\end{aligned}
$$

Let $C_a := n(1 + |a|^{n-1})$ so that $|f(x) - f(a)| \leq C_a\,|x - a|$. We therefore are led to take $\delta < \min\left\{1, \frac{\varepsilon}{C_a(1+|a|^{n-1})}\right\}$.

Can you write a textbook proof on your own now?

(5) Let $f\colon (0, \infty) \to \mathbb{R}$ be given by $f(x) = x^{-1}$. Go through the proof of $1/x_n \to 1/x$ on page 39. Let $a > 0$ be given. We need to estimate

$$
\begin{aligned}
\left|\frac{1}{x} - \frac{1}{a}\right| &= \left|\frac{x - a}{xa}\right| \\
&= \frac{1}{x}\frac{1}{a}\,|x - a| \\
&\leq \frac{2}{a}\frac{1}{a}\,|x - a| \quad \text{for } x > \frac{a}{2}.
\end{aligned}
$$

So, for a given $\varepsilon > 0$, , we take $\delta < \min\{\frac{a}{2}, \frac{a^2\varepsilon}{2}\}$.

Write down a textbook proof now.

(6) Fix $\alpha > 0$. Let $f\colon (\alpha, \infty) \to \mathbb{R}$ be given by $f(x) = x^{-1}$.

The twist here is similar to item 3 of Example 3.2.3.

Proceeding as in the last example, we arrive at

$$
|f(x) - f(a)| \leq \frac{1}{xa}\,|x - a| \leq \frac{1}{\alpha^2}\,|x - a|\,,
$$

since $x, a \geq \alpha$. Hence we may choose $\delta < \frac{\alpha^2\varepsilon}{2}$.

(7) Let $f\colon (0, \infty) \to \mathbb{R}$ be given by $f(x) := x^{1/n}$ for a fixed $n \in \mathbb{N}$.

If our aim is to find an estimate of the form

$$|f(x) - f(a)| = \left| x^{1/n} - a^{1/n} \right| \le C_a \, |x - a|,$$

the way out seems to be to use the identity

$$t^n - s^n = (t - s)(t^{n-1} + \cdots + s^{n-1}).$$

Let $t := x^{1/n}$ and $s := a^{1/n}$. Then

$$|x - a| = |t^n - s^n| = |t - s| \, \left| t^{n-1} + \cdots + s^{n-1} \right|.$$

What we want is a lower bound for the RHS. If we restrict x so that $x > a/2$, then the RHS is greater than or equal to $|t - s| \, n(a/2)^{n-1}$. Let $C_a := n(a/2)^{n-1}$. Then we obtain an estimate of the form

$$|x - a| \ge C \, |t - s| \quad \text{or} \quad |t - s| \le \frac{1}{C_a} \, |x - a|.$$

Since we need to ensure that $\frac{1}{C_a} |x - a| < \varepsilon$, we may take a positive $\delta < C_a \varepsilon$. Since we also wanted $x > a/2$, the choice of δ is any positive number less than $\min\{\frac{a}{2}, C_a \varepsilon\}$. Now you may write down a textbook proof.

(8) Let $f\colon \mathbb{R} \to \mathbb{R}$, $f(x) = \begin{cases} x^2, & \text{if } x < 0 \\ x, & \text{if } x \ge 0. \end{cases}$

If $a \ne 0$, we may restrict x so that the sign of x is the same as that of a. For example, we may confine ourselves to $(-\infty, a/2)$ if $a < 0$ or to $(a/2, \infty)$ if $a > 0$. This means that we need to take $\delta < |a|/2$.

Let $a > 0$ and $x > a/2$. Then

$$|f(x) - f(a)| = |x - a|.$$

This suggests that we take $\delta < \min\{\frac{a}{2}, \varepsilon\}$.

Let $a < 0$ and $x < 0$. Then

$$|f(x) - f(a)| = \left| x^2 - a^2 \right| = |x + a| \, |x - a| \le (|x| + |a|) \, |x - a|.$$

If we restrict x to an $-a$ interval around a, that is, if $x \in (-2a, 0)$, then the estimate above leads us to $|f(x) - f(a)| \le 3 \, |a| \, |x - a|$. Thus we may choose δ such that $0 < \delta < \frac{\varepsilon}{3|a|}$.

Let $a = 0$. Then $f(x)$ could be x^2 or x. Hence $|f(x) - f(a)|$ could therefore be either $\left| x^2 - a^2 \right| = \left| x^2 \right|$ or $|x - a| = |x|$. If we restrict x so that $|x| < 1$, we have $\left| x^2 \right| \le |x|$. Hence for $|x| < 1$, we have

$$|f(x) - f(a)| \le |x|.$$

This suggests that we may take $\delta < \min\{1, \varepsilon\}$.

Needless to say, you are expected to write a textbook proof of all the cases.

(9) **Thomae's function:** Let $f : (0, 1) \to \mathbb{R}$ given by

$$f(x) = \begin{cases} 0, & x \text{ is irrational} \\ 1/q, & x = p/q \text{ with } p, q \in \mathbb{N}, \, p, \, q \text{ have no common factor.} \end{cases}$$

We shall show that f is continuous at all irrational points and not continuous at any rational point of $(0, 1)$.

Let $a \in \mathbb{Q} \cap (0, 1)$. We need to estimate $|f(x) - f(a)|$. If x is irrational, then $|f(x) - f(a)| = |f(a)| = 1/q$ where $x = p/q$ in lowest terms. If $\varepsilon < 1/q$, then there is no hope of making $|f(x) - f(a)| < \varepsilon$ if x is irrational. By the density of irrationals, we can always find an irrational $x \in (a - \delta, a + \delta)$. Hence f is not continuous at a.

Let us write down a textbook proof. Let $a = p/q \in (0, 1)$ in lowest terms. We claim f is not continuous at a. We shall prove this by contradiction. Let us choose $0 < \varepsilon < 1/q$. Suppose there exists $\delta > 0$ such that for $x \in (a - \delta, a + \delta) \cap (0, 1)$, we have $|f(x) - f(a)| < \varepsilon$. Now $(a - \delta, a + \delta) \cap (0, 1)$ is an open interval. There exists an irrational $x \in (a - \delta, a + \delta) \cap (0, 1)$, by the density of irrationals. For this irrational x, we obtain

$$|f(x) - f(a)| = |0 - f(a)| = \frac{1}{q} > \varepsilon,$$

a contradiction. This establishes the claim.

Now let $a \in (0, 1)$ be irrational. We need to estimate $|f(x) - f(a)|$. If x is also irrational, then $|f(x) - f(a)| = 0$. So, we need to concentrate on rational $x = p/q$, in lowest terms. In such a case, $|f(x) - f(a)| = 1/q$. If we want $1/q < \varepsilon$, then we want $q > 1/\varepsilon$. There are only finitely many such rational numbers **in** $(0, 1)$. Since a is irrational, we can choose a $\delta > 0$ such that $(a - \delta, a + \delta)$ avoids these finitely many rational numbers. For any rational $r = p/q$ in $(a - \delta, a + \delta)$, we have $q > 1/\varepsilon$. We now work out a proof based on this argument.

Let $\varepsilon > 0$ be given. If $r = p/q$ (with $p, q \in \mathbb{N}$ and $\gcd(p, q) = 1$) is a rational number in $(0, 1)$, then $f(r) = 1/q$ is less than ε iff $q > 1/\varepsilon$. The number of such positive integers is at most $[1/\varepsilon]$, the integral part of $1/\varepsilon$. Now if $r = p/q \in (0, 1)$ with $1/q \geq \varepsilon$, then $1 \leq p < q$. Hence the set of all rational numbers $p/q \in (0, 1)$ with $1/q \geq \varepsilon$ is

$$\{p/q : 1 \leq q < [1/\varepsilon], 1 \leq p < q\}.$$

This is a finite set, say, $\{r_k = p_k/q_k : 1 \leq k \leq N\}$. Since a is irrational, $a \neq r_k$ for $1 \leq k \leq N$. Therefore, $|a - r_k| > 0$ for $1 \leq k \leq N$. It follows that $\delta := \min\{|a - r_k| : 1 \leq k \leq n\} > 0$. If $|x - a| < \delta$ and $x = p/q$, then $q \geq 1/\varepsilon$. Hence we obtain

$$|f(x) - f(a)| = \begin{cases} 0, & \text{if } x \text{ is irrational} \\ 1/q < \varepsilon, & \text{if } x \text{ is rational.} \end{cases}$$

That is, f is continuous at irrational numbers in $(0,1)$.

(10) Consider $J = \{\frac{1}{n} : n \in \mathbb{N}\}$. Let $f \colon J \to \mathbb{R}$ be any function. Let $a = \frac{1}{k} \in J$ be fixed. We claim that f is continuous at a.

Consider the interval $J_k :- (\frac{1}{k+1}, \frac{1}{k-1})$, if $k \geq 2$ and $(\frac{1}{2}, 1]$ if $k = 1$, the only element of J in J_k is $1/k$.

Note that if we choose $0 < \delta < \frac{1}{k} - \frac{1}{k+1} = \min\{\frac{1}{k} - \frac{1}{k+1}, \frac{1}{k-1} - \frac{1}{k}\}$, if $k \geq 2$, then we conclude if $x \in (a - \alpha, a + \delta) \cap J$ then $x = a$. If $\varepsilon > 0$ is any positive number, this choice of δ will do the job.

Note that $\min\{|x - y| : x, y \in J, \ x \neq y\} = 0$. Compare this with a remark in Example 3.1.7.

(11) Consider $J = \{\frac{1}{n} : n \in \mathbb{N}\} \cup \{0\}$. As seen in the last example, we can show that any $f \colon J \to \mathbb{R}$ is continuous at *any* $a = 1/k$. But, the continuity of f at 0 leads to an interesting connection with convergent sequences.

To give you some indication of this, consider a sequence $y_n := f(1/n)$ and $y := f(0)$. Do you see the connection?

We claim that f is continuous at 0 iff $y_n \to y$.

Assume that f is continuous at 0. Since $1/n \to 0$, by continuity of f at 0, it follows that $y_n \equiv f(1/n) \to f(0) = y$.

Conversely, if $y_n \to 0$, we prove that f is continuous at 0. Let $\varepsilon > 0$ be given. Since $y_n \to 0$,

$$\exists N \in \mathbb{N} \text{ such that } n \geq N \implies |y_n - y| < \varepsilon. \tag{3.3}$$

Let $0 < \delta < 1/N$. Now if $x \in J$, then $x = 1/k$ for some $k \in \mathbb{N}$. So, $|x - 0| < \delta$ is the same as saying that $1/k < \delta$ and hence $k > N$. Hence

$$|f(x) - f(0)| = |f(1/k) - f(0)| < \varepsilon, \text{ by (3.3)}.$$

Exercise 3.2.4. We say that a function $f \colon J \to \mathbb{R}$ is *Lipschitz* on J if there exists $L > 0$ such that $|f(x) - f(y)| \leq L\,|x - y|$ for all $x, y \in J$. Show that any Lipschitz function is continuous.

The slope of the chord joining the points $(x, f(x))$ and $(y, f(y))$ on the graph of f is $\frac{f(x) - f(y)}{x - y}$. Hence f is Lipschitz iff the set of slopes of all possible chords on the graph of f is bounded in $\mathbb{R}$.

Example 3.2.5. We show that the function d_A of Example 3.1.7 is Lipschitz on $\mathbb{R}$. Let $x, y \in \mathbb{R}$. For any $a \in A$, we have $|x - a| \leq |x - y| + |y - a|$. Let $r_a := |x - a|$ and $s_a := |x - y| + |y - a|$. Then the sets $R := \{r_a : a \in A\}$ and $S := \{s_a : a \in A\}$ are indexed by the same set A and we have $r_a \leq s_a$ for each $a \in A$. We have already seen (Item 4 of Exercise 1.3.27 on page 18) that glb $R \leq$ glb S. Clearly, glb $R = d_A(x)$ and glb $S = |x - y| + d_A(y)$. We therefore obtain

$$d_A(x) \leq |x - y| + d_A(y) \text{ and hence } d_A(x) - d_A(y) \leq |x - y|.$$

Interchanging x and y in this inequality, we get $d_A(y) - d_A(x) \leq |x - y|$. Thus, $\pm (d_A(x) - d_A(y)) \leq |x - y|$. We have thus proved $|d_A(x) - d_A(y)| \leq |x - y|$.

Exercise 3.2.6. Let $f \colon J \to \mathbb{R}$ be continuous at c with $f(c) \neq 0$. Use ε-δ definition to show that there exists $\delta > 0$ such that $|f(x)| > |f(c)|/2$ for all $x \in (c - \delta, c + \delta) \cap J$.

In particular, if $f(c) > 0$, then there exists $\delta > 0$ such that $f(x) > f(c)/2$ for all $x \in (c - \delta, c + \delta) \cap J$.

What is the analogue of this when $f(c) < 0$? Conclude that if f is continuous at c and if $f(c) \neq 0$, then f retains the sign of $f(c)$ in an open interval containing c.

Exercise 3.2.7. Let $f \colon J \to \mathbb{R}$ be continuous at $a \in J$. Show that f is *locally bounded*, that is, there exist $\delta > 0$ and $M > 0$ such that

$$\forall \, x \in (a - \delta, a + \delta) \cap J \implies |f(x)| \leq M.$$

Remark 3.2.8. The results of the last two exercises are about the *local* properties of continuous functions. We shall apply them to get some *global* results. See Theorem 3.3.1 and Theorem 3.4.2.

Remark 3.2.9. To verify whether or not a function is continuous at a point $a \in J$, we do not have to know the values of f at each and every point of J; we need only to know the values $f(x)$ for $x \in J$ "near" a or "close to" a, that is, for all $x \in (a - \delta, a + \delta) \cap J$ for some $\delta > 0$.

3.3 Intermediate Value Theorem

Let $f \colon [a, b] \to \mathbb{R}$ be a continuous function such that $f(a)$ and $f(b)$ are of opposite signs. Draw the graphs of a few such functions, say, at least one for which $f(a) > 0$ (necessarily $f(b) < 0$) and another one for which $f(a) < 0$. Do you notice anything in each of the cases? Does the graph meet the x-axis? Can you formulate a result based on your observations?

Theorem 3.3.1, Theorem 3.4.1, and Theorem 3.4.6 are the most important *global* results on continuity. See Remark 3.4.4.

Theorem 3.3.1 (Intermediate Value Theorem). *Let $f \colon [a, b] \to \mathbb{R}$ be a continuous function such that $f(a) < 0 < f(b)$. Then there exists $c \in (a, b)$ such that $f(c) = 0$.*

Strategy: See Figures 3.7, and 3.8. We wish to locate the "first" c from a such that $f(c) = 0$. For all $a \leq x < c$, we observe that $f(t) < 0$ on $[a, x]$. The required point c is the lub of all such x. This suggests us to consider

$$E := \{x \in [a, b] : f(y) \leq 0 \text{ for } y \in [a, x]\},$$

By the local property of f, the set $E \neq \emptyset$ and bounded above by b. Using again the local property, we can show that $f(c) < 0$ and $f(c) > 0$ are not possible.

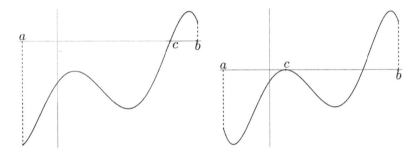

Figure 3.7: IVT, Figure 1.   Figure 3.8: IVT, Figure 2.

Proof. We define $E := \{x \in [a, b] : f(y) \leq 0 \text{ for } y \in [a, x]\}$.

Using the continuity of f at a for $\varepsilon = -f(a)/2$, we can find a $\delta > 0$ such that $f(x) \in (3f(a)/2, f(a)/2)$ for all $x \in [a, a + \delta)$. This shows that $a + \delta/2 \in E$. (See Figure 3.9.)

Since E is bounded by b, there is $c \in \mathbb{R}$ such that $c = \sup E$. Clearly we have $a + \delta/2 \leq c \leq b$ and hence $c \in (a, b]$. We claim that $c \in E$ and that $f(c) = 0$.
Case 1. Assume that $f(c) > 0$. Then by Exercise 3.2.6, there exists $\delta > 0$ such that for $x \in (c-\delta, c+\delta) \cap [a, b]$, we have $f(x) > 0$. Since $c-\delta < c$, there exists $x \in E$ such that $c - \delta < x$. Since $x \in E$, we have $f(t) \leq 0$, for $t \in [a, x] = [a, c - \delta] \cup (c - \delta, x]$, a contradiction. See Figure 3.10.

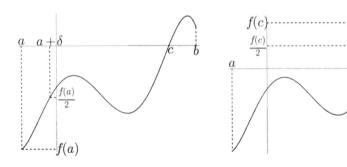

Figure 3.9: IVT, Figure 3.   Figure 3.10: IVT, Figure 4.

Case 2. Assume that $f(c) < 0$. Then by Exercise 3.2.6, there exists $\delta > 0$ such that for $x \in (c - \delta, c + \delta) \cap [a, b]$, we have $f(x) < 0$. Since $c - \delta < c$, there exists $x \in E$

such that $f(t) \leq 0$, for $t \in [a, x]$. Hence $f(t) \leq 0$ for all $t \in [a, x] \cup (c - \delta, c + \delta/2] = [a, c + \delta/2]$. That is, $c + \delta/2 \in E$, contradicting $c = \text{lub } E$. See Figure 3.11.

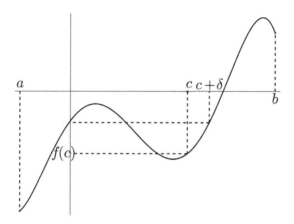

Figure 3.11: IVT, Figure 5.

Hence we are forced to conclude that $f(c) = 0$. □

2nd Proof. Let $J_0 := [a, b]$. Let c_1 be the mid–point of $[a, b]$. Now there are three possibilities for $f(c_1)$. It is zero, negative, or positive. If $f(c_1) = 0$, then the proof is over. If not, we choose one of the intervals $[a, c_1]$ or $[c_1, b]$ so that f assumes values with opposite signs at the end points. To spell it out, if $f(c_1) < 0$, then we take the subinterval $[c_1, b]$. If $f(c_1) > 0$, then we take the subinterval $[a, c_1]$. The chosen subinterval will be called J_1 and we write it as $[a_1, b_1]$.

We now bisect the interval J_1 and choose one of the two subintervals as $J_2 := [a_2, b_2]$ so that f takes values with opposite signs at the end points. We continue this process recursively. We thus obtain a sequence (J_n) of intervals with the following properties:
  (i) If $J_n = [a_n, b_n]$, then $f(a_n) \leq 0$ and $f(b_n) \geq 0$.
  (ii) $J_{n+1} \subset J_n$.
  (iii) $\ell(J_n) = 2^{-n}\ell(J_0) = 2^{-n}(b - a)$.
By nested interval theorem (Theorem 2.4.4), there exists a unique $c \in \cap J_n$. Since $a_n, b_n, c \in J_n$, we have

$$|c - a_n| \leq \ell(J_n) = 2^{-n}(b - a) \text{ and } |c - b_n| \leq \ell(J_n) = 2^{-n}(b - a).$$

Hence it follows that $\lim a_n = c = \lim b_n$. Since $c \in J$ and f is continuous on J, we have

$$f(a_n) \to f(c) \text{ and } f(b_n) \to f(c).$$

Since $f(a_n) \leq 0$ for all n, it follows that $\lim_n f(a_n) \leq 0$, that is, $f(c) \leq 0$, (by the analogue of Item 2 in Exercise 2.1.28). In an analogous way, $f(c) = \lim f(b_n) \geq 0$. We are forced to conclude that $f(c) = 0$. The proof is complete. □

Theorem 3.3.2 (Intermediate Value Theorem – Standard Version). *Let $g\colon [a,b] \to \mathbb{R}$ be a continuous function. Let λ be a real number between $g(a)$ and $g(b)$. Then there exists $c \in (a,b)$ such that $g(c) = \lambda$.*

Proof. **Strategy:** Apply the previous version to the function $f(x) = g(x) - \lambda$.

Assume without loss of generality, $g(a) < \lambda < g(b)$. Then $f(a) < 0$ and $f(b) > 0$. Also g is continuous on $[a,b]$. Hence, by Theorem 3.3.1, there exists a $c \in (a,b)$ such that $f(c) = 0$. That is, $g(c) = \lambda$. □

Remark 3.3.3. We made a crucial use of the LUB property of $\mathbb{R}$ in the proofs of the theorems above. They are not true, for example, in $\mathbb{Q}$. We shall be brief. Consider the interval $[0,2] \cap \mathbb{Q}$ and the continuous function $f(x) = x^2 - 2$. Then $f(0) < 0$ while $f(2) > 0$. We know that there exists no rational number α whose square is 2. Recall also that we have shown that $\mathbb{Q}$ does not enjoy the LUB property (Remark 1.3.20).

Exercise Set 3.3.4. Three typical applications of the intermediate value theorem.

(1) Let $J \subset \mathbb{R}$ be an interval. Let $f\colon J \to \mathbb{R}$ be continuous and $f(x) \neq 0$ for any $x \in J$. Show that either $f > 0$ on J or $f < 0$ on J.

(2) Let $f\colon \mathbb{R} \to \mathbb{R}$ be continuous taking values in $\mathbb{Z}$ or in $\mathbb{Q}$. Then show that f is a constant function.

(3) Let $f\colon [a,b] \to \mathbb{R}$ be a non-constant continuous function. Show that $f([a,b])$ is uncountable.

Remark 3.3.5. The intermediate value theorem says that the image of an interval under a continuous function is again an interval. Let $J \subset \mathbb{R}$ be an interval. Let $f\colon J \to \mathbb{R}$ be continuous. We claim that $f(J)$ is an interval. Let $y_1, y_2 \in f(J)$. Assume that $y_1 < y < y_2$. We need to show that $y \in f(J)$. Let $y_j = f(x_j)$, $j = 1,2$ with $x_j \in J$. Then by the intermediate value theorem, there exists x between x_1 and x_2 such that $f(x) = y$. Hence the claim follows.

Theorem 3.3.6 (Existence of n-th Roots). *Let $\alpha \geq 0$ and $n \in \mathbb{N}$. Then there exists $\beta \geq 0$ such that $\beta^n = \alpha$.*

Strategy: If we draw the graph of $y = x^n$ on $[0,\infty)$, it is obvious that the graph intersects the line $y = \alpha$. (See Figure 3.12.) The point of intersection is of the form $(x, x^n = \alpha)$.

We wish to appeal to the intermediate value theorem. In view of the picture above, the obvious choice for f is $f(x) = x^n$. We then need to find a and b such that $a^n < \alpha < b^n$. Since $\alpha \geq 0$, we may take $a = 0$. Since we want $b^n > \alpha$, we may choose an $N \in \mathbb{N}$ such that $N^n > \alpha$. This suggests that we use the Archimedean property. Choose $N > \alpha$, then $N^n > \alpha$.

Proof. Choose $N \in \mathbb{N}$ such that $N > \alpha$. Then $N^n \geq N > \alpha$. Consider $f\colon [0,\infty) \to [0,\infty)$ defined by $f(x) = x^n - \alpha$. Then $f(0) \leq 0$ and $f(N) > 0$. Intermediate value theorem applied to the pair $(f, [0,N])$ yields a $c \in [0,N]$ such that $f(c) = 0$, that is, $c^n = \alpha$. □

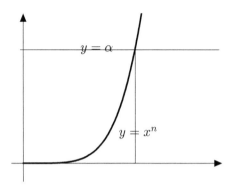

Figure 3.12: Existence of nth root.

Theorem 3.3.7. *Any polynomial of odd degree with real coefficients has a real zero.*

Strategy: It is enough to prove that a monic polynomial

$$P(X) = X^n + a_{n-1}X^{n-1} + \cdots + a_1 X + a_0, (a_j \in \mathbb{R}, 0 \le j \le n-1),$$

of odd degree has a real zero. (Why?)

The basic idea is that as $X \to \infty$, X^n dominates all other terms of the polynomial. For instance, consider $P(X) := X^3 - 10^3 X^2 - 2012$. If $\alpha > 0$ is very large, we see that the sign of $P(\alpha)$ is positive, though there are very large negative terms. How do we show that X^n dominates other terms? How do we show that the sequence (n^2) goes to ∞ much faster than the sequence (n)? It is done by considering the sequence (n^2/n). Similarly, we write $P(X) = X^n Q(1/X)$, where Q is a polynomial in $1/X$. In the example above, $P(X) = X^3 \left(1 - \frac{10^3}{X} - \frac{100}{X^3}\right)$. Now each of the terms in the expression of $Q(1/X)$ is very close to zero if $|X|$ is very large. Hence the term in the brackets is positive. Hence the sign of $P(X)$ for large values of $|X|$ will depend on the sign of X^n. Since n is odd, it follows that for large negative α, $P(\alpha) < 0$ and for large positive α, $P(\alpha) > 0$. Now we can apply the intermediate value theorem.

Proof. We write $P(X) = X^n \left(1 + \frac{a_{n-1}}{X} + \cdots + \frac{a_0}{X^n}\right)$. Note that if $N \in \mathbb{N}$, then $\left|\frac{a_j}{N^{n-j}}\right| \le \left|\frac{a_j}{N}\right|$ for any $1 \le j \le n$. Let $C := \sum_{j=0}^{n-1} |a_j|$. We then have

$$
\left|\frac{a_{n-1}}{N} + \cdots + \frac{a_0}{N^n}\right| \le \left|\frac{a_{n-1}}{N}\right| + \left|\frac{a_{n-2}}{N^2}\right| + \cdots + \left|\frac{a_0}{N^n}\right|
$$

$$
\le \left|\frac{a_{n-1}}{N}\right| + \left|\frac{a_{n-2}}{N}\right| + \cdots + \left|\frac{a_0}{N}\right|
$$

$$
= \frac{C}{N}.
$$

We can choose $N \in \mathbb{N}$ such that $\frac{C}{N} < 1/2$. If $|X| > N$, we have the estimate $\left|\frac{a_{n-1}}{N} + \cdots + \frac{a_0}{N^n}\right| < 1/2$. That is, we have

$$1/2 \le 1 + \frac{a_{n-1}}{X} + \cdots + \frac{a_0}{X^n} \le 3/2. \tag{3.4}$$

Consequently, $P(X) \le X^n/2 < 0$ if $X < -N$ and $P(X) \ge X^n/2 > 0$ if $X > N$. Now the intermediate value theorem asserts the existence of a zero of P in $(-2N, 2N)$. $\square$

Remark 3.3.8. The inequality (3.4) is true for all $n \in \mathbb{N}$, not necessarily for odd integers. This is a very basic inequality about polynomial functions. It says something about the behavior of the polynomial function as $|x| \to \infty$.

Theorem 3.3.9 (Fixed Point Theorem). *Let $f \colon [a, b] \to [a, b]$ be continuous. Then there exists $c \in [a, b]$ such that $f(c) = c$. (Such a c is called a fixed point of f.)*

Proof. Consider $g(x) := f(x) - x$. Then $g(a) \ge 0$ and $g(b) \le 0$. Apply intermediate value theorem to g. $\square$

To appreciate this, try to draw the graph of the continuous functions stated in the theorem. Mark the points $(a, f(a))$ and $(b, f(b))$. Note that $a \le f(a), f(b) \le b$. Try to join them by a curve without lifting your hands off the paper. Draw the diagonal line $y = x$ in the square. Do you see that they intersect? See Figure 3.13.

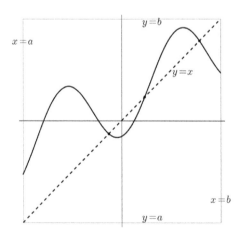

Figure 3.13: Fixed point theorem.

Exercise Set 3.3.10.

(1) Prove that $x = \cos x$ for some $x \in (0, \pi/2)$.

(2) Prove that $xe^x = 1$ for some $x \in (0, 1)$.

(3) Are there continuous functions $f \colon \mathbb{R} \to \mathbb{R}$ such that $f(x) \notin \mathbb{Q}$ for $x \in \mathbb{Q}$ and $f(x) \in \mathbb{Q}$ for $x \notin \mathbb{Q}$?

(4) Let $f \colon [0, 1] \to \mathbb{R}$ be continuous. Assume that the image of f lies in $[1, 2] \cup (5, 10)$ and that $f(1/2) \in [0, 1]$. What can you conclude about the image of f?

(5) Let $f \colon [0, 2\pi] \to [0, 2\pi]$ be continuous such that $f(0) = f(2\pi)$. Show that there exists $x \in [0, 2\pi]$ such that $f(x) = f(x + \pi)$.

(6) Let p be a real polynomial function of odd degree. Show that $p \colon \mathbb{R} \to \mathbb{R}$ is an onto function.

(7) Show that $x^4 + 5x^3 - 7$ has at least two real roots.

(8) Let $p(X) := a_0 + a_1 X + \cdots + a_n X^n$. If $a_0 a_n < 0$, show that p has at least two real roots.

(9) Let J be an interval and $f \colon J \to \mathbb{R}$ be continuous and 1-1. Then f is strictly monotone.

(10) Let I be an interval and $f \colon I \to \mathbb{R}$ be strictly monotone. If $f(I)$ is an interval, show that f is continuous.

(11) Use the last item to conclude that the function $x \mapsto x^{1/n}$ from $[0, \infty) \to [0, \infty)$ is continuous.

3.4   Extreme Value Theorem

Draw the graphs of continuous functions on closed and bounded intervals. Is it possible for you to make them unbounded? For example, can you draw the graph of a continuous function on $[0, 1]$ which takes all positive real numbers as its values?

Theorem 3.4.1 (Weierstrass Theorem). *Let $f \colon [a, b] \to \mathbb{R}$ be a continuous function. Then f is bounded.*

Strategy: The strategy for this proof is very similar to that of the Intermediate Value Theorem. Here we use the local boundedness result, Exercise 3.2.7. Since f is bounded around a, the set of points x such that f is bounded on $[a, x]$ is nonempty. If c is the LUB of this set, then c must be b. For, otherwise, f is bounded around c and hence there exists $c_1 > c$ such that f is bounded on $[a, c_1]$.

Proof. Let $E := \{x \in J := [a, b] : f$ is bounded on $[a, x]\}$. The conclusion of the theorem is that $b \in E$.

Since f is continuous at a, using Exercise 3.2.7, we see that f is bounded on $[a, a + \delta)$ for some $\delta > 0$. Hence $a + \delta/2 \in E$. Obviously, E is bounded by b.

Let $c = \sup E$. Since $a + \delta/2 \in E$ we have $a \le c$. Since b is an upper bound for E, $c \le b$. Thus $a \le c \le b$. We intend to show that $c \in E$ and $c = b$. This will complete the proof.

Since f is continuous at c, it is locally bounded, say, on $(c - \delta, c + \delta) \cap J$. Let $x \in E$ be such that $c - \delta < x \le c$. Then clearly, f is bounded on $[a, c] \subset [a, x] \cup ((c - \delta, c + \delta) \cap J)$. In particular, $c \in E$. If $c < b$, choose δ_1 such that (i) $0 < \delta_1 < \delta$ and (ii) $c + \delta_1 < b$. The above argument shows that $c + \delta_1 \in E$ if $c \ne b$. This contradicts the fact that $c = \sup E$. Hence $c = b$. This proves the result. $\square$

Strategy for second proof: If the result is false, we shall have a sequence (x_n) such that $|f(x_n)| > n$. By Bolzano–Weierstrass theorem (2.7.5), there exists a subsequence converging to an element $x \in [a, b]$. But then f is locally bounded around x whereas x_n's, for arbitrarily large n, lie in the interval around x.

Second Proof. If false, there exists a sequence (x_n) in $[a, b]$ such that $|f(x_n)| > n$ for each $n \in \mathbb{N}$. Since $[a, b]$ is compact, there exists a subsequence, say, (x_{n_k}), which converges to $x \in [a, b]$. Since $|f|$ is continuous, we must have $|f(x_{n_k})| \to |f(x)|$, in particular, the sequence $(|f(x_{n_k})|)$ is bounded, a contradiction. $\square$

Strategy for third proof: The idea for the third proof is to use nested interval theorem. If f is not bounded on $[a, b]$, it is not bounded on at least one of the bisecting subintervals. Inductively, we have a nested sequence of intervals. The function f is bounded around the common point of this nested sequence. But for large values of n, all subintervals lie in this interval around x on which f is bounded.

Third Proof. Assume that f is not bounded on $[a, b]$. Then f is not bounded on one of the subintervals, $[a, (a + b)/2]$ and $[(a + b)/2, b]$. (Why?)

For, if $|f(x)| \le M_1$ for $x \in [a, (a+b)/2]$ and $|f(x)| \le M_2$ for $x \in [(a+b)/2, b]$, then $|f(x)| \le M := \max\{M_1, M_2\}$, for $x \in [a, b]$.

Choose such an interval and call it $J_1 = [a_1, b_1]$. Note that the length of J_1 is half that of $J = [a, b]$. Bisect J_1 into two intervals of equal length. The function f is not bounded on at least one of the subintervals, call it $J_2 = [a_2, b_2]$. Note that $b_2 - a_2 = (b_1 - a_1)/2 = (b - a)/2^2$. Assume by induction, for $n \ge 2$, we have chosen $J_n \subset J_{n-1}$ such that the length of J_n is $2^{-n}(b - a)$ and such that f is unbounded on J_n. Bisect J_n into subintervals of equal length. Then f is not bounded on at least one of them, call it J_{n+1}. Then $J_{n+1} \subset J_n$ and the length of J_{n+1} is half of the length of J_n and hence is $2^{-(n+1)}(b - a)$. Thus we obtain a sequence of intervals (J_n) with the following properties: (1) f is not bounded on each of the J_n's, (2) $J_{n+1} \subset J_n$ for each n, and (3) the length of J_n is $2^{-n}(b - a)$. By the nested interval theorem (Theorem 2.4.4), there exists a unique c common to all these intervals. Using Exercise 3.2.7, we see that f is bounded on $(c - \delta, c + \delta)$ for some $\delta > 0$.

We claim that there exists N such that if $n \ge N$, we have $J_n \subset (c - \delta, c + \delta)$. Let $x \in J_n$. Then we estimate the distance $|x - c|$: $|x - c| < 2^{-n}(b - a)$. Since we want to show that $J_n \subset (c - \delta, c + \delta)$ for all large n, we need to show that for any $x \in J_n$, we have $|x - c| < \delta$. But, both x and c lie in J_n and hence $|x - c| \le 2^{-n}(b - a)$. Hence we need to ensure that $2^{-n}(b - a) < \delta$. Since $2^{-n}(b - a) \to 0$, there exists

$N \in \mathbb{N}$ such that for all $n \geq N$, we have $|2^{-n}(b-a) - 0| = 2^{-n}(b-a) < \delta$. Hence the claim is proved.

Since f is bounded on $(c - \delta, c + \delta)$, it is bounded on $J_n \subset (c - \delta, c + \delta)$. This leads to a contradiction, since f is not bounded on J_n's. $\qquad\square$

Remark 3.4.2. Note that the first proofs of the intermediate value theorem and Weierstrass theorem are quite similar. We defined appropriate subsets of $[a, b]$ and applied the corresponding *local* result (Exercise 3.2.6 and Exercise 3.2.7 respectively) to get the global result. See also Remark 3.2.8.

Remark 3.4.3. If the domain is not bounded or if the domain is not closed, then Weierstrass theorem is not true. For example, if $J = (0, 1)$ and $f(x) = \frac{1}{x}$, then f is not bounded on J.

If $J = (a, \infty)$ and $f(x) = x$, then f is not bounded on J.

Remark 3.4.4. The last two theorems (Theorems 3.3.1 and 3.4.1) are global results in the following sense.

In the first case, we imposed a restriction on the domain, namely, that it is an interval. If the domain is not an interval, the conclusion does not remain valid.

In the second result, we imposed the condition that the interval be closed and bounded.

Remark 3.4.5. If we carefully examine the second proof, we see that we needed only the following fact about the domain J of f:

If (x_n) is a sequence in J, then it has a convergent subsequence whose limit lies in J.

Subsets of $\mathbb{R}$ with this property are called *compact* subsets. What the second proof yields is the following more general version.

Let $J \subset \mathbb{R}$ be a nonempty compact subset. If $f: J \to \mathbb{R}$ is continuous, then f is bounded on J.

If J is any closed and bounded interval, then J is compact. There are sets which not intervals but are compact. For example, $J = \{1/n : n \in \mathbb{N}\} \cup \{0\}$ is compact.

Theorem 3.4.6 (Extreme Values Theorem). *Let the hypothesis be as in the Weierstrass theorem. Then there exists $x_1, x_2 \in [a, b]$ such that $f(x_1) \leq f(x) \leq f(x_2)$ for all $x \in [a, b]$. (In other words, a continuous function f on a closed and bounded interval is bounded and attains its LUB and GLB.)*

Proof. Let $M := \text{lub} \{f(x) : a \leq x \leq b\}$. If there exists no $x \in [a, b]$ such that $f(x) = M$, then $M - f(x)$ is continuous at each $x \in [a, b]$ and $M - f(x) > 0$ for all $x \in [a, b]$. If we let $g(x) := 1/(M - f(x))$ for $x \in [a, b]$, then g is continuous on $[a, b]$. By Theorem 3.4.1, there exists $A > 0$ such that $g(x) \leq A$ for all $x \in [a, b]$. But then we have, for all $x \in [a, b]$, $g(x) := \frac{1}{M - f(x)} \leq A$ or $M - f(x) \geq \frac{1}{A}$. Thus we conclude that $f(x) \leq M - (1/A)$ for $x \in [a, b]$. It follows that $\text{lub} \{f(x) : x \in [a, b]\} \leq M - \frac{1}{A}$. This contradicts our hypothesis that $M = \text{lub} \{f(x) : x \in [a, b]\}$. We therefore conclude that there exists $x_2 \in [a, b]$ such that $f(x_2) = M$.

Let $m := \text{glb} \{f(x) : a \le x \le b\}$. Arguing similarly, we can find an $x_1 \in [a, b]$ such that $f(x_1) = m$. $\qquad\square$

Second Proof. Since $M - \frac{1}{n} < M = \text{lub} \{f(x) : x \in J\}$, there exists $x_n \in J$ such that $M - \frac{1}{n} < f(x_n) \le M$. Hence $f(x_n) \to M$. By Bolzano–Weierstrass, there exists a subsequence (x_{n_k}) such that x_{n_k} converges to some $x \in J$. By continuity of f at x, $f(x_{n_k}) \to f(x)$. Conclude that $f(x) = M$. $\qquad\square$

Look at the examples: (i) $f: (0, 1] \to \mathbb{R}$ given by $f(x) = 1/x$. This is a continuous unbounded function. The interval here, though bounded, is not closed at the end points. (ii) $f: (-1, 1) \to \mathbb{R}$ defined by $f(x) := \frac{1}{1-|x|}$. (iii) $f: \mathbb{R} \to \mathbb{R}$ defined by $f(x) = x$.

None of these examples contradict Theorem 3.4.1.

Note that the theorem remains true if J is assumed to be a compact subset. See Remark 3.4.5. Which proof can be adapted to prove this more general result?

Exercise Set 3.4.7.

(1) Let $f: [a, b] \to \mathbb{R}$ be continuous. Show that $f([a, b]) = [c, d]$ for some $c, d \in \mathbb{R}$ with $c \le d$. Can you identify c, d?

(2) Does there exist a continuous function $f: [0, 1] \to (0, \infty)$ which is onto?

(3) Does there exist a continuous function $f: [a, b] \to (0, 1)$ which is onto?

(4) Let $f: [a, b] \to \mathbb{R}$ be continuous such that $f(x) > 0$ for all $x \in [a, b]$. Show that there exists δ such that $f(x) > \delta$ for all $x \in [a, b]$.

(5) Construct a continuous bijection $f: [a, b] \to [c, d]$ such that f^{-1} is continuous.

(6) Construct a continuous function from $(0, 1)$ *onto* $[0, 1]$. Can such a function be one-one?

(7) Let $f: \mathbb{R} \to \mathbb{R}$ be continuous. Assume that $f(x) \to 0$ as $|x| \to \infty$. (Do you understand this? If not, you may come back after completing this chapter.) Show that there exists $c \in \mathbb{R}$ such that either $f(x) \le f(c)$ or $f(x) \ge f(c)$ for all $x \in \mathbb{R}$. Give an example of a function in which only one of these happens.

(8) Let $f: \mathbb{R} \to \mathbb{R}$ be a function such that (i) $f(\mathbb{R}) \subset (-2, -1) \cup [1, 5)$ and (ii) $f(0) = e$. Can you give realistic bounds for f?

3.5   Monotone Functions

Definition 3.5.1. We say that a function $f: J \subset \mathbb{R} \to \mathbb{R}$ is *strictly increasing* if for all $x, y \in J$ with $x < y$, we have $f(x) < f(y)$.

One defines strictly decreasing in a similar way. A *monotone function* is either strictly increasing or strictly decreasing.

We shall formulate and prove the results for strictly increasing functions. Analogous results for decreasing functions f can be arrived at in a similar way or by applying the result for the increasing functions to $-f$.

Proposition 3.5.2. *Let $J \subset \mathbb{R}$ be an interval. Let $f \colon J \to \mathbb{R}$ be continuous and one-one. Let $a, c, b \in J$ be such that $a < c < b$. Then $f(c)$ lies between $f(a)$ and $f(b)$, that is either $f(a) < f(c) < f(b)$ or $f(a) > f(c) > f(b)$ holds.*

Proof. Since f is one-one, we assume without loss of generality that $f(a) < f(b)$. If the result is false, either $f(c) < f(a)$ or $f(c) > f(b)$.

Let us look at the first case. (See Figure 3.14.) Since $f(c) < f(a) < f(b)$, $y = f(a)$ lies between the values of f at the end points of $[c, b]$. Hence there exists $x \in (c, b)$ such that $f(x) = y = f(a)$. Since $x > a$, this contradicts the fact that f is one-one.

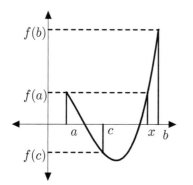
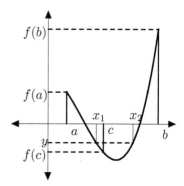

Figure 3.14: Prop. 3.5.2: Figure 1.   Figure 3.15: Prop. 3.5.2: Figure 2.

In case you did not like the way we used y, you may proceed as follows. (See Figure 3.15.) Fix any y such that $f(c) < y < f(a)$. By intermediate value theorem applied to the pair $(f, [a, c])$, there exists $x_1 \in (a, c)$ such that $f(x_1) = y$. Since $f(a) < f(b)$, we also have $f(c) < y < f(b)$. Hence there exists $x_2 \in (c, b)$ such that $f(x_2) = y$. Clearly $x_1 \neq x_2$.

The second case when $f(c) > f(b)$ is similarly dealt with.   □

Theorem 3.5.3. *Let $J \subset \mathbb{R}$ be an interval. Let $f \colon J \to \mathbb{R}$ be continuous and one-one. Then f is monotone.*

Proof. Fix $a, b \in J$, say with $a < b$. We assume without loss of generality that $f(a) < f(b)$. We need to show that for all $x, y \in J$ with $x < y$ we have $f(x) < f(y)$.

(i) If $x < a$, then $x < a < b$ and hence $f(x) < f(a) < f(b)$, by Proposition 3.5.2.

(ii) If $a < x < b$, then $f(a) < f(x) < f(b)$, by Proposition 3.5.2.

(iii) If $b < x$, then $f(a) < f(b) < f(x)$, by Proposition 3.5.2.

In particular, $f(x) < f(a)$ if $x < a$ and $f(x) > f(a)$ if $x > a$.   (3.5)

If $x < a < y$, then $f(x) < f(a) < f(y)$ by (3.5).

If $x < y < a$, then $f(x) < f(a)$ by (3.5) and $f(x) < f(y) < f(a)$ by Proposition 3.5.2.

If $a < x < y$, then $f(a) < f(y)$ by (3.5) and $f(a) < f(x) < f(y)$ by Proposition 3.5.2.

Hence f is strictly increasing.   □

Remark 3.5.4. Recall that the intermediate value theorem says that the image of an interval under a continuous function is an interval. (Remark 3.3.5.)

What is the converse of this statement? The converse is in general not true. (Can you give an example?) A partial converse is found in the next result.

Proposition 3.5.5. *Let J be an interval and $f \colon J \to \mathbb{R}$ be monotone. Assume that $f(J) = I$ is an interval. Then f is continuous.*

Proof. We deal with the case when f is strictly increasing. Let $a \in J$. Assume that a is not an endpoint of J. We prove the continuity of f at a using the ε-δ definition. Look at Figure 3.16.

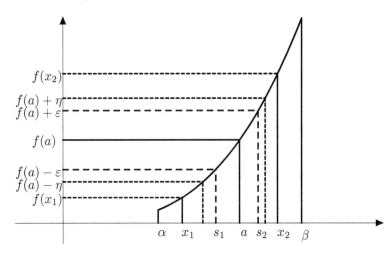

Figure 3.16: Figure for Proposition 3.5.5.

Since a is not an endpoint of J, there exists $x_1, x_2 \in J$ such that $x_1 < a < x_2$ and hence $f(x_1) < f(a) < f(x_2)$. It follows that there exists $\eta > 0$ such that $(f(a) - \eta, f(a) + \eta) \subset (f(x_1), f(x_2)) \subset I$.

Let $\varepsilon > 0$ be given. We may assume $\varepsilon < \eta$. Let $s_1, s_2 \in J$ be such that $f(s_1) = f(a) - \varepsilon$ and $f(s_2) = f(a) + \varepsilon$. Let $\delta := \min\{a - s_1, s_2 - a\}$. If $x \in$

$(a - \delta, a + \delta) \subset (s_1, s_2)$, then $f(a) - \varepsilon = f(s_1) \leq f(x) < f(s_2) = f(a) + \varepsilon$, that is, if $x \in (a - \delta, a + \delta)$, then $f(x) \in (f(a) - \varepsilon, f(a) + \varepsilon)$.

If a is an endpoint of J, an obvious modification of the proof works. $\square$

Corollary 3.5.6 (Inverse Function Theorem). *Let $f \colon J \to \mathbb{R}$ be an increasing continuous function on an interval J. Then $f(J)$ is an interval, $f \colon J \to f(J)$ is a bijection and the inverse $f^{-1} \colon f(J) \to J$ is continuous.* $\square$

Proof. Note that f^{-1} is an increasing function. By Remark 3.3.5, the image $f(J)$ is an interval. Its image under the increasing function f^{-1} is the interval J. Hence, by the last theorem, f^{-1} is continuous. $\square$

Remark 3.5.7. Consider the n-th root function $f \colon [0, \infty) \to [0, \infty)$ given by $f(x) := x^{1/n}$. We can use the last item to conclude that f is continuous, a fact seen by us in Item 7 in Example 3.2.3 on page 75.

3.6   Limits

Definition 3.6.1. Let $J \subset \mathbb{R}$ be an interval and $a \in J$. Let $f \colon J \setminus \{a\} \to R$ be a function. We say that $\lim_{x \to a} f(x)$ exists if there exists $\ell \in \mathbb{R}$ such that for any given $\varepsilon > 0$, there exists $\delta > 0$ such that (see Figure 3.17)

$$x \in J \text{ and } 0 < |x - a| < \delta \implies |f(x) - \ell| < \varepsilon. \tag{3.6}$$

If $\lim_{x \to a} f(x)$ exists, we say that the limit of the function f as x tends to a exists. Note that a need not be in the domain of f. Even if a lies in the domain of f, ℓ need not be $f(a)$. We let $\lim_{x \to a} f(x)$ stand for ℓ and call ℓ as "the" limit of f as $x \to a$.

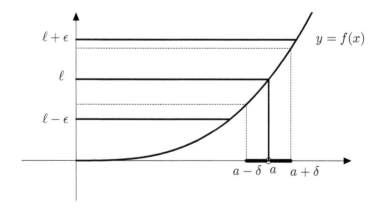

Figure 3.17: $\varepsilon - \delta$–definition of limit.

Standing assumption: Observe that if $\delta > 0$ satisfies (3.6), then any $\delta_1 < \delta$ will also satisfy (3.6). Hence we may as well assume that δ is such that $(a-\delta, a+\delta) \subset J$.

Theorem 3.6.2. *With the notation of the last item, the limit ℓ is unique.*

Proof. Assume that $\lim_{x \to a} f(x) = \ell_1$ and $\lim_{x \to a} f(x) = \ell_2$. We are required to prove that $\ell_1 = \ell_2$. The proof is very similar to that of the uniqueness of the limit of a convergent sequence (Proposition 2.1.15).

Given $\varepsilon > 0$, there exist δ_j, such that $(a - \delta_j, a + \delta_j) \subset J, j = 1, 2$ and

$$0 < |x - a| < \delta_j \implies |f(x) - \ell_j| < \varepsilon, j = 1, 2. \tag{3.7}$$

If $\ell_1 \neq \ell_2$, we can choose $\varepsilon > 0$ such that the sets $(\ell_1 - \varepsilon, \ell_1 + \varepsilon) \cap (\ell_2 - \varepsilon, \ell_2 + \varepsilon) = \emptyset$. Now if we choose x such that $0 < |x - a| < \min\{\delta_1, \delta_2\}$, then (3.7) says that $f(x) \in (\ell_1 - \varepsilon, \ell_1 + \varepsilon) \cap (\ell_2 - \varepsilon, \ell_2 + \varepsilon)$, a contradiction. $\square$

Remark 3.6.3. There is a subtle point in the proof above. To arrive at a contradiction, we choose an x such that $0 < |x - a| < \min\{\delta_1, \delta_2\}$. How do we know that such an x exists?

We used the fact that the domain of f is an interval minus a point. For if we let $\delta = \min\{\delta_1, \delta_2\}$, then $x := a + \delta/2 \in J$. If J is any subset, though we can define $\lim_{x \to a} f(x) = \ell$, ℓ need not be unique.

Consider $J = \mathbb{Z}$ and any function $f \colon J \to \mathbb{R}$. We claim $\lim_{x \to 0} f(x) = \ell$ for any $\ell \in \mathbb{R}$. (Why?) Hence the limit is not unique.

The crucial fact that we needed to prove uniqueness is the observation that in any interval of the form $(a - \delta, a + \delta)$, we can find points (other than a) belonging to the domain of f.

If you do not completely understand this remark, review this again after some time.

Exercise 3.6.4. Let $f \colon \mathbb{R} \to \mathbb{R}$ be given by $f(x) = \frac{x^2 - 4}{x - 2}$ for $x \neq 2$ and $f(2) = e$. Then $\lim_{x \to 2} f(x) = 4$. Prove this using the $\varepsilon - \delta$ definition.

Exercise Set 3.6.5. Find the limits using the $\varepsilon - \delta$ definition.

(1) $\lim_{x \to a} \frac{x^3 - a^3}{x - a}$.

(2) $\lim_{x \to 0} x \sin(1/x)$.

(3) $\lim_{x \to 2} \frac{x^2 - 4}{x^2 - 2x}$.

(4) $\lim_{x \to 6} \sqrt{x + 3}$.

Theorem 3.6.6. *Let $J \subset \mathbb{R}$ be an interval. Let $a \in J$. Let $f \colon J \setminus \{a\} \to R$ be given. Then $\lim_{x \to a} f(x) = \ell$ iff for **every** sequence (x_n) with $x_n \in J \setminus \{a\}$ with the property that $x_n \to a$, we have $f(x_n) \to \ell$.*

Proof. The proof is quite similar to that of Theorem 3.2.2. The reader is encouraged to review the proof of Theorem 3.2.2 and write a proof of the present theorem on his own.

Let $\lim_{x \to a} f(x) = \ell$. Let (x_n) be a sequence in $J \setminus \{a\}$ such that $x_n \to a$. We are required to show that $f(x_n) \to \ell$. Let $\varepsilon > 0$ be given. Since $\lim_{x \to a} f(x) = l$, for this ε, there exists a $\delta > 0$ such that $0 < |x - a| < \delta$ implies that $|f(x) - \ell| < \varepsilon$. Since $x_n \to a$, for this δ, there exists $N \in \mathbb{N}$ such that $n \geq N$ implies that $|x_n - a| < \delta$. Hence it follows that for $n \geq N$, we have $|f(x_n) - \ell| < \varepsilon$, that is, $f(x_n) \to \ell$.

If the converse is not true, then there exists $\varepsilon > 0$ such that for any given $\delta > 0$, there exists $x_\delta \in (a - \alpha, a + \alpha)$, such that $x_\delta \neq a$ and such that $|f(x_\delta) - \ell| \geq \varepsilon$. Apply this to each $\delta = 1/n$, $n \in \mathbb{N}$, to get $x_n \in (a - 1/n, a + 1/n) \cap J$ with $x_n \neq a$ and $|f(x_n) - \ell| \geq \varepsilon$. By the sandwich lemma, $x_n \to a$. Note that (x_n) is a sequence in $J \setminus \{a\}$, but $f(x_n)$ does not converge to ℓ. □

Is there an analogue of the sandwich lemma for the limits of functions?

Lemma 3.6.7. *Let f, g, h be defined on $J \setminus \{a\}$. Assume that*
(1) $f(x) \leq h(x) \leq g(x)$, *for $x \in J$, $x \neq a$.*
(2) $\lim_{x \to a} f(x) = \ell = \lim_{x \to a} g(x)$.
Then $\lim_{x \to a} h(x) = \ell$.

Proof. This is an easy consequence of the last result and the sandwich lemma for sequences. The reader is encouraged to write a proof. □

Let $f \colon J \to \mathbb{R}$ and $c \in J$ be given. Is there any relation between $\lim_{x \to c} f(x)$ and the continuity of f at c?

Theorem 3.6.8. *Let $J \subset \mathbb{R}$ be an interval and $a \in J$. Assume that $f \colon J \to \mathbb{R}$ is a function. Then f is continuous at a iff $\lim_{x \to a} f(x)$ exists and the limit is $f(a)$.*

Proof. Assume that f is continuous at a. Now, by continuity at a, for any sequence (x_n) in J with $x_n \to a$, we have $f(x_n) \to f(a)$. In view of the last theorem, we deduce that $\lim_{x \to a} f(x)$ exists and the limit is $f(a)$.

Conversely, assume that f is defined on J, in particular at a, and that $\lim_{x \to a} f(x) = f(a)$. We need to show that f is continuous at a. We shall prove this by the ε–δ definition of continuity at a and the limit concept. Let $\varepsilon > 0$ be given. Since $\lim_{x \to a} f(x) = f(a)$, for the given ε, there exists $\delta > 0$ such that

$$0 < |x - a| < \delta \implies |f(x) - f(a)| < \varepsilon.$$

Observe that the inequality above holds even for $x = a$. Thus f is continuous at a. □

Remark 3.6.9. In old textbooks, continuity is defined using the limit concept. Later it became standard to define continuity using the ε–δ definition.

The drawback of using limits to define continuity is easily understood if one goes through Remark 3.6.3. While we can define continuity on an arbitrary subset of $\mathbb{R}$, to define $\lim_{x \to a} f(x)$, we need to put a restriction on a so that the limit is unique. We leave it to the reader to ponder over this.

Formulate results analogous to algebra of convergent sequences and algebra of continuous functions. Do you see their proofs in your mind? State the theorem and give a sketch of how it reduces to sequences in view of Theorem 3.6.6.

Theorem 3.6.10 (Algebra of Finite Limits). *Let* $J \subset \mathbb{R}$ *be an interval. Let* $f_k \colon J \setminus \{a\}$, $k = 1, 2$ *be given. Assume that* $\lim_{x \to a} f_k(x) = \ell_k \in \mathbb{R}$. *Let* $\alpha \in \mathbb{R}$. *Then:*
  (i) $\lim_{x \to a} (f_1 + f_2)(x) = \ell_1 + \ell_2$.
  (ii) $\lim_{x \to a} (\alpha f_k)(x) = \alpha \ell_k$.
  (iii) $\lim_{x \to a} (f g)(x) = \ell_1 \ell_2$.
  (iv) *Assume that* $\ell_1 \neq 0$. *Then* $f_1(x) \neq 0$ *for all* $x \in (a - \delta, a + \delta)$ *for some* $\delta > 0$, $x \neq a$. *We also have* $\lim_{x \to a} \frac{1}{f_1}(x) = 1/\ell_1$.

Proof. The proof is easy in view of Theorem 3.6.6 and the algebra of convergent sequences. We leave it to the reader. □

Can you think of a result on the existence of a limit for a composition of functions?

Theorem 3.6.11. *If* $\lim_{x \to a} f(x) = \alpha$ *and if* g *is defined in an interval containing* α *and is continuous at* α, *then* $\lim_{x \to a} (g \circ f)(x)$ *exists and it is* $g(\alpha)$.

Proof. This proof may be compared with that of Item 8 of Theorem 3.1.5.
Assume that g is defined on $(\alpha - r, \alpha + r)$. Let $\varepsilon > 0$ be given. We need to estimate $|(g \circ f)(x) - g(\alpha)|$. Let (x_n) be a sequence in $J \setminus \{a\}$ such that $x_n \to a$. It follows that $f(x_n) \to \alpha$. Hence there exists N such that for all $n \geq N$, $f(x_n) \in (\alpha - r, \alpha + r)$. Now the sequence $(f(x_n))_{n \geq N}$ converges to α. Since g is continuous at α, we have $g(f(x_n)) \to g(\alpha)$. Thus, we have shown that $(g \circ f)(x_n) \to g(\alpha)$. In view of Theorem 3.6.6, the result follows. □

How do we define one-sided limits such as $\lim_{x \to a+} f(x)$? Easy. In the definition of $\lim_{x \to a} f(x)$, we need to restrict x to those $x > a$. That is, $\lim_{x \to a+} f(x) = \ell$ if for $\varepsilon > 0$ there exists $\delta > 0$ such that

$$x > a, x \in J \text{ and } 0 < |x - a| < \delta \implies |f(x) - \ell| < \varepsilon.$$

The definition $\lim_{x \to a-} f(x)$ is left to the reader.
What is the relation between the one-sided limits $\lim_{x \to a+} f(x)$, $\lim_{x \to a-} f(x)$ and the limit $\lim_{x \to a} f(x)$?

Theorem 3.6.12. *Let* $J \subset \mathbb{R}$ *be an interval,* $a \in J$. *Let* $f \colon J \setminus \{a\} \to \mathbb{R}$ *be a function. Then* $\lim_{x \to a} f(x)$ *exists iff the one-sided limits* $\lim_{x \to a+} f(x)$ *and* $\lim_{x \to a-} f(x)$ *exist and are equal.*

Proof. Easy. If $\lim_{x \to a} f(x) = \ell$ exists, then for a given $\varepsilon > 0$, we can find a $\delta > 0$ such that $x \in J$ and $|x - a| < \delta \implies |f(x) - \ell| < \varepsilon$. Clearly, if $x > a$, $x \in J$ and $0 < |x - a| < \delta$. we have $|f(x) - \ell| < \varepsilon$. Thus, $\lim_{x \to a+} f(x)$ exists and the limit is ℓ. Similarly, $\lim_{x \to a-} f(x) = \ell$.

Conversely, assume that the one-sided limits exists and their common value is ℓ. Now given $\varepsilon > 0$, there exists $\delta_1 > 0$ and $\delta_2 > 0$ such that

$$x > a, x \in J, \ 0 < |x - a| < \delta_1 \quad \implies \quad |f(x) - \ell| < \varepsilon,$$
$$x < a, x \in J, \ 0 < |x - a| < \delta_2 \quad \implies \quad |f(x) - \ell| < \varepsilon.$$

Let $\delta := \min\{\delta_1, \delta_2\}$. It is clear that we have

$$x \in J, \ 0 < |x - a| < \delta \implies |f(x) - \ell| < \varepsilon.$$

This proves the theorem. $\square$

How do we assign a meaning to the *symbol* $\lim_{x \to \infty} f(x) = \ell$ for a function $f \colon (a, \infty) \to \mathbb{R}$?

Given $\varepsilon > 0$, we need to find an interval around ∞ such that for x in the interval, we have $|f(x) - \ell| < \varepsilon$. A little thought leads us to consider the intervals of the form (R, ∞), $R > a$.

$\lim_{x \to \infty} f(x) = \ell$ if for a given $\varepsilon > 0$, there exists R such that for $x > R$ we have $|f(x) - \ell| < \varepsilon$.

How do we assign a meaning to $\lim_{x \to -\infty} f(x) = \ell$ for a function $f \colon (-\infty, a) \to \mathbb{R}$? Here the interval around $-\infty$ is an interval of the form $(-\infty, R)$, where $R < a$. Hence the definition is as follows.

$\lim_{x \to -\infty} f(x) = \ell$ if for a given $\varepsilon > 0$, there exists $R > 0$ such that for $x < -R$ we have $|f(x) - \ell| < \varepsilon$. (We require $-R < a$.)

How do we assign a meaning to the *symbol* $\lim_{x \to a} f(x) = \infty$? *Hint:* Recall how we defined a sequence diverging to infinity (page 52).

We say that $\lim_{x \to a} f(x) = \infty$ if for any given $M > 0$ there exists $\delta > 0$ such that $0 < |x - a| < \delta$ implies that $f(x) > M$. (See Figure 3.18.)

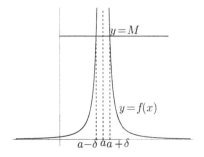

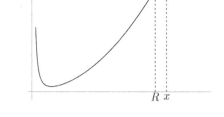

Figure 3.18: $f(x) \to \infty$ as $x \to \alpha$.   Figure 3.19: $f(x) \to \infty$ as $x \to \infty$.

How do we assign a meaning to the *symbol* $\lim_{x \to \infty} f(x) = \infty$? And so on!

We say that $\lim_{x \to \infty} f(x) = \infty$ if for any given $M > 0$ there exists $R > 0$ such that for $x > R$, we have $f(x) > M$. (See Figure 3.19.)

Example 3.6.13. We now give some examples for you to practice the definitions of limits.

(1) Let $f \colon \mathbb{R}^* \to \mathbb{R}$ be given by $f(x) := x/|x|$. Then $\lim_{x \to 0+} f(x) = 1$ and $\lim_{x \to 0-} f(x) = -1$. We need to estimate $|f(x) - 1|$ for $x > 0$. Since $f(x) = 1$ for $x > 0$, for any given $\varepsilon > 0$ we may choose any $\delta > 0$.

(2) $\lim_{x \to 0} f(x) = 0$ where $f(x) = |x|$ if $x \neq 0$ and $f(0) = 23$.

(3) $\lim_{x \to 0} \frac{1}{x^2} = 0$.

(4) $\lim_{x \to 0+} \frac{1}{x} = \infty$ and $\lim_{x \to 0-} \frac{1}{x} = -\infty$.

   We shall prove the first. Let $M > 0$ be given. Now if $x > 0$, we have $f(x) > M$ if $1/x > M$, that is, if $0 < x < 1/M$. Hence we choose $0 < \delta < 1/M$. For $0 < x < \delta$, we have $f(x) = 1/x > 1/\delta > M$.

(5) Let $f \colon \mathbb{R} \to \mathbb{R}$ be defined by $f(x) = \frac{(-1)^n}{n} \sin(\pi x)$ for $x \in [n, n+1)$. Then $\lim_{x \to \pm\infty} f(x) = 0$.

Remark 3.6.14. Using the Equation (3.4) in Theorem 3.3.7, we see that for an odd-degree polynomial $P(X)$ with leading coefficient 1

$$\lim_{x \to \infty} P(X) = \infty \quad \text{and} \quad \lim_{x \to -\infty} P(X) = -\infty.$$

If $P(X)$ is of even degree with leading coefficient 1, what are $\lim_{x \to \infty} P(X)$ and $\lim_{x \to -\infty} P(X)$?

Exercise 3.6.15. Let $J \subset \mathbb{R}$ be an interval. Assume that $a \in J$ and that $f \colon J \setminus \{a\} \to \mathbb{R}$ is such that $\lim_{x \to a} f(x) = \ell$. If we define $f(a) = \ell$, then f is continuous at a.

A typical and standard example is $f \colon \mathbb{R}^* \to \mathbb{R}$ given by $f(x) := \frac{\sin x}{x}$. It is well known that, $\lim_{x \to 0} f(x) = 1$. Hence if we define $g(x) := f(x)$ for $x \neq 0$ and $g(0) = 1$, then $g \colon \mathbb{R} \to \mathbb{R}$ is continuous.

Limit as a Tree Diagram

We exhibit the definitions of various limits as a tree diagram. See Figure 3.20. It lists all the possibilities and shows us a unified way of defining all possible limits.

We may formulate the various definitions above in a uniform fashion as follows. We say that the limit of $\lim_{x \to a} f(x)$ at $x = a$ exists and is ℓ if

$$\forall J_\ell \left(\exists J_a \left(\forall x \in J_a \left(f(x) \in J_\ell \right) \right) \right),$$

where J_a is an interval around a but not containing a and J_ℓ is an interval around ℓ. We tabulate J_a and J_ℓ in various cases.

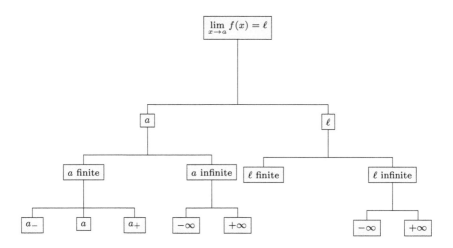

Figure 3.20: Limits: tree diagram.

Table for J_a					
a	a_-	a	a_+	$a = -\infty$	$a = +\infty$
J_a	$(a - \delta, a)$	$(a - \delta, a + \delta) \setminus \{a\}$	$(a, a + \delta)$	$(-\infty, r)$	(r, ∞)

Table for J_ℓ			
ℓ	ℓ finite	$\ell = -\infty$	$\ell = +\infty$
J_ℓ	$(\ell - \epsilon, \ell + \epsilon)$	$(-\infty, R)$	(R, ∞)

Remark 3.6.16 (Indeterminate Forms). Go through the proof of Theorem 3.6.10. Do we have an analogue if the limits are also allowed to be infinite? Look at the following examples:

(i) $f(x) = x$ and $g(x) = x - c$, $h(x) = -x + c$, $c \in \mathbb{R}$. We have $\lim_{x \to \infty} f(x) = \infty$, $\lim_{x \to \infty} g(x) = \infty$, and $\lim_{x \to \infty} h(x) = -\infty$. We have $\lim_{x \to \infty} (f - g)(x) = c = \lim_{x \to \infty} (f + h)(x)$. Thus, the limit of the sum of functions need not be the sum of limits. Note that these examples are instances of the so-called indeterminate form $\infty - \infty$.

(ii) Let $f(x) = cx^2$ and $g(x) = \frac{1}{1+x^2}$. Then $\lim_{x \to \infty} f(x) = \infty$, $\lim_{x \to \infty} g(x) = 0$, but $\lim_{x \to \infty} (fg)(x) = c$. This example gives rise to the indeterminate form of the type $\infty \times 0$.

(iii) Let $f(x) = cx^2$ and $g(x) = \frac{1}{x^2}$. Then $\lim_{x \to 0} f(x) = 0$ and $\lim_{x \to 0} g(x) = \infty$, but $\lim_{x \to 0} (fg)(x) = c$.

(iv) Let $f(x) = x$, $g(x) = \frac{1}{x^2}$, $h(x) = \frac{1}{x^3}$. Then we have $\lim_{x \to 0} f(x) = 0$, $\lim_{x \to 0} g(x) = \infty$, and $\lim_{x \to 0} h(x)$ does not exist. But, $\lim_{x \to 0} (fg)(x)$ does not exist, $\lim_{x \to 0}(fh)(x) = 0$. On the other hand, if we let $f(x) = x^2$ and $g(x) = x^4$, then $\lim_{x \to 0} f(x) = 0 = \lim_{x \to 0} g(x)$, but $\lim_{x \to 0}(\frac{f}{g})(x) = \infty$ while $\lim_{x \to 0}(\frac{g}{f})(x) = 0$.

If we define $f(x) - cx$ and $g(x) = x$, what is $\lim_{x \to 0} \frac{f}{g}(x)$? Note that all these give rise to an indeterminate of the form $\frac{0}{0}$.

We hope that the reader understands how the so-called indeterminate forms arise. These examples should convince you that we cannot blindly apply the algebra of limits rule in these examples as the limits may not exist or if they exist, the result may depend on the relative behavior of the functions near the point where the limit is taken.

L'Hospital's rules, which you learned in calculus courses, give us some contexts in which these limits can be found. Refer to Section 4.3.

We shall have a closer look at the relation between the existence of one-sided limits and the continuity in the case of an increasing function.

Look at the graphs of the following increasing functions. Do you see what happens at the points of discontinuity?

Example 3.6.17. The floor function $f : \mathbb{R} \to \mathbb{R}$ defined by $f(x) := [x]$. See Figure 3.21.

Example 3.6.18. $f : \mathbb{R} \to \mathbb{R}$ defined by $f(x) := [x]$ for $x \notin \mathbb{Z}$ and $f(x) = x + (1/2)$ if $x \in \mathbb{Z}$. See Figure 3.22.

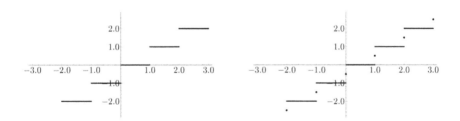

Figure 3.21: Example 3.6.17.   Figure 3.22: Example 3.6.18.

Theorem 3.6.19. *Let $J \subset \mathbb{R}$ be an interval and $f : J \to \mathbb{R}$ be increasing. Assume that $c \in J$ is not an endpoint of J. Then:*
(i) $\lim_{x \to c_-} f = \text{lub } \{f(x) : x \in J; x < c\}$.
(ii) $\lim_{x \to c_+} f = \text{glb } \{f(x) : x \in J; x > c\}$.

Proof. We shall prove (i). Let $\ell := \text{lub } \{f(x) : x \in J; x < c\}$. We wish to prove that $\lim_{x \to c_-} f = \ell$. Let $\varepsilon > 0$ be given. Therefore, there exists $x_0 \in J$, $x_0 < c$ such that $f(x_0) > \ell - \varepsilon$. Now for $x \in J$, $x > x_0$, that is, for $x \in J \cap (x_0, c)$, we have $f(x) > f(x_0) > \ell - \varepsilon$. If we choose $\delta > 0$ such that $x_0 < c - \delta$, the result follows.

(ii) is proved in a similar way.

Compare this proof with that of the convergence of an increasing sequence bounded above. (See Theorem 2.3.2.) □

What is the analogue of the last item for decreasing functions?

Theorem 3.6.20. *Let the hypothesis be as in Theorem 3.6.19. Then the following are equivalent:*

(i) f *is continuous at* c.

(ii) $\lim_{x \to c_-} f = f(c) = \lim_{x \to c_+} f$.

(iii) lub $\{f(x) : x \in J; x < c\} = f(c) = $ glb $\{f(x) : x \in J; x > c\}$.

Proof. In view of Theorem 3.6.19, (ii) is equivalent to (iii). So, it is enough to prove that (i) is equivalent to (iii).

Let $E_1 := \{f(x) : x \in J; x < c\}$ and $E_2 := \{f(x) : x \in J; x > c\}$.

(i) $\implies$ (iii): Let $\ell_- := $ lub $\{f(x) : x \in J; x < c\}$ and $\ell_+ := $ glb $\{f(x) : x \in J; x > c\}$. Since f is increasing, it is clear that $\ell_- \leq f(c) \leq \ell_+$. It suffices to show that $\ell_- = \ell_+$. We shall prove this by contradiction. Let $\varepsilon > 0$ be such that $\ell_- < f(c) - \varepsilon < f(c) + \varepsilon < \ell_+$. (See Figure 3.23.)

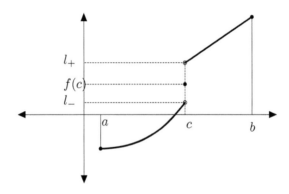

Figure 3.23: Figure for Theorem 3.6.20.

By continuity at c, there exists $\delta > 0$ such that

$$x \in J \cap (c - \delta, c + \delta) \implies |f(x) - f(c)| < \varepsilon.$$

We deduce that if $c - \delta < x < c$, we have $f(x) > f(c) - \varepsilon > \ell_-$. Hence lub $\{f(x) : x \in J; x < c\} \leq \ell_- - \varepsilon$, a contradiction. □

Formulate an analogous result for decreasing functions.
What is the formulation if c is an endpoint of J?

Definition 3.6.21. Let $J \subset \mathbb{R}$ be an interval and $f \colon J \to \mathbb{R}$ be increasing. Assume that $c \in J$ is not an endpoint of J. The *jump* at c is defined as

$$j_f(c) := \lim_{x \to c_+} f - \lim_{x \to c_-} f \equiv \text{glb } \{f(x) : x \in J; x > c\} - \text{lub } \{f(x) : x \in J; x < c\}.$$

How is the jump $j_f(c)$ defined if c is an endpoint?

Proposition 3.6.22. *Let $J \subset \mathbb{R}$ be an interval and $f: J \to \mathbb{R}$ be increasing. Then f is continuous at $c \in J$ iff $j_f(c) = 0$.*

Proof. This is immediate from Theorem 3.6.20. □

Definition 3.6.23. A set E is said to be *countable* if there exists a one-one map of E into $\mathbb{Q}$.

Theorem 3.6.24. *Let $J \subset \mathbb{R}$ be an interval and $f: J \to \mathbb{R}$ be monotone. Then the set D of points of J at which f is discontinuous is countable.*

Proof. Assume that f is increasing. Then $c \in J$ belongs to D iff the interval $J_c := (f(c_-), f(c_+))$ is nonempty. For, $c, d \in D$, with, say, $c < d$, the intervals J_c and J_d are disjoint. (Why?) Look at Figure 3.24.

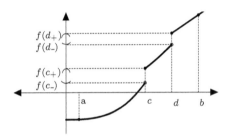

Figure 3.24: $(f(c_-), f(c_+)) \cap (f(d_-), f(d_+)) = \emptyset$.

Let $c < t < d$. Since $f(c_+) \equiv \text{glb} \{f(x) : x > c\}$, and since $t > c$, we see that $f(c_+) \le f(t)$. Similarly, $f(d_-) \equiv \text{lub} \{f(y) : y < d\}$. Since $t < d$, we see that $f(t) \le f(d-)$. Thus, $f(c_+) \le f(t) \le f(d_-)$. In particular, $f(c_+) \le f(d_-)$. Hence if $y \in J_c \cap J_d$, then $y < f(c_+)$ and $y > f(d-)$, that is, $f(d-) < y < f(c_+)$, in particular, $f(d_-) < f(c_+)$, a contradiction.

Thus the collection $\{J_c : c \in D\}$ is a pairwise disjoint family of open intervals. Such a collection is countable. For, choose $r_c \in J_c \cap \mathbb{Q}$. Since $\{J_c : c \in D\}$ is pairwise disjoint, the map $c \mapsto r_c$ from D to $\mathbb{Q}$ is one-one. □

3.7 Uniform Continuity

Definition 3.7.1. Let $J \subset \mathbb{R}$ be any subset. A function $f: J \to \mathbb{R}$ is *uniformly continuous* **on** J if for each $\varepsilon > 0$ there exists $\delta > 0$ such that

$$x_1, x_2 \in J \text{ with } |x_1 - x_2| < \delta \implies |f(x_1) - f(x_2)| < \varepsilon.$$

Note that if we wish to establish uniform continuity of a function on a domain, we need to estimate $|f(x_1) - f(x_2)|$ for any two arbitrary points in the domain of f.

In particular, if f is uniformly continuous on J, then f is continuous on J. Let $a \in J$ and $\varepsilon > 0$ be given. Since f is uniformly continuous on J, for the given $\varepsilon > 0$, there exists a $\delta > 0$ such that for all $x_1, x_2 \in J$ with $|x_1 - x_2| < \delta$, we have $|f(x_1) - f(x_2)| < \varepsilon$. Clearly, if $x \in J$ and if $|x - a| < \delta$, then $|f(x) - f(a)| < \varepsilon$. Observe that δ is the same for all $a \in J$.

A note of caution: The observation in the last paragraph is a corollary of the definition of uniform continuity. Many beginners place undue emphasis on this with the result that they end up estimating $|f(x) - f(a)|$ and look for a uniform δ. It is advisable to start estimating $|f(x_1) - f(x_2)|$ for x_1, x_2 in the domain. See Item 3 of Example 3.2.3.

Remark 3.7.2. Every uniformly continuous function is continuous, but the converse is not true.

Remark 3.7.3. Unlike continuity, uniform continuity is a *global* concept.

The notion of uniform continuity of f gives us control on the variation of the images of a pair of points which are close to each other independent of where they lie in domain of f. Go through Example 3.7.4 and Example 3.7.5. In these two examples, it is instructive to draw the graphs of the functions under discussion and try to understand the remark above.

Example 3.7.4. Let $\alpha > 0$. Let $f \colon (0, \infty) \to \mathbb{R}$ be given by $f(x) = 1/x$. Then f is uniformly continuous on (α, ∞). (See Figure 3.25.)

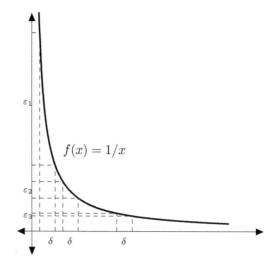

Figure 3.25: Uniform continuity of $1/x$.

$$|f(x) - f(y)| = \frac{|x-y|}{xy} \le \frac{|x-y|}{\alpha^2}.$$

This suggests that if $\varepsilon > 0$ is given, then we choose $\delta \le \alpha^2 \varepsilon$. Now write a textbook proof.

The function $g \colon (0, \infty) \to \mathbb{R}$ given by $g(x) = 1/x$ is not uniformly continuous. Assume the contrary. Look at the graph of f near $x = 0$. You will notice that if x and y are very close to each other and are also very near to 0 (which is not in the domain of f, though), their values vary very much. This suggests to us a method of attack. If f is uniformly continuous on $(0, \infty)$, then for $\varepsilon = 1$, there exists $\delta > 0$. How do we choose points which are δ-close and are close to 0? Points of the form $1/n$ for $n \gg 0$ are close to zero. This suggests that we choose $x = 1/N$ and $y = 1/2N$. To make them δ-close, we choose $\frac{1}{N} < \delta$. Then $|f(x) - f(y)| \ge 1$. Now you can write a textbook proof.

Example 3.7.5. (See Figure 3.26.)

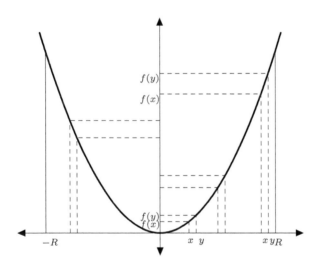

Figure 3.26: $f(x) = x^2$ on $[-R, R]$.

Let $f \colon \mathbb{R} \to \mathbb{R}$ be given by $f(x) = x^2$. Let A be any bounded subset of $\mathbb{R}$, say, $A = [-R, R]$. Then f is uniformly continuous on A but not on $\mathbb{R}$! If you look at the graph of f, you will notice that if x is very large (that is, near to ∞), and if y is very near to x, the variations in $f(x)$ and $f(y)$ become large. If f were uniformly continuous on $\mathbb{R}$, then for $\varepsilon = 1$ we can find a δ as in the definition. Choose N so that $N > 1/\delta$. Take $x = N$ and $y = N + 1/N$. Then $|f(x) - f(y)| \ge 2$.

If $x, y \in [-R, R]$, then

$$|f(x) - f(y)| = |x + y|\,|x - y| \le 2R\,|x - y|,$$

which establishes the uniform continuity of f on $[-R, R]$.

Exercise 3.7.6. Any Lipschitz function is uniformly continuous. (See Exercise 3.2.4 on page 77.)

Exercise 3.7.7. Let $J \subset$ be an interval. Let $f: J \to R$ be differentiable with bounded derivative, that is, $|f'(x)| \leq L$ for some $L > 0$. Then f is Lipschitz with Lipschitz constant L. In particular, f is uniformly continuous on J. *Hint:* Recall the mean value theorem (Theorem 4.2.5).

Specific examples: $f(x) = \sin x$, $g(x) = \cos x$ are Lipschitz on $\mathbb{R}$. The inverse of tan, $\tan^{-1}: (-\pi/2, \pi/2) \to \mathbb{R}$ is Lipschitz.

Exercise Set 3.7.8.

(1) Let $f: J \to \mathbb{R}$ be uniformly continuous. Then f maps Cauchy sequences in J to Cauchy sequences in $\mathbb{R}$.

(2) The converse of the last exercise is not true.

Theorem 3.7.9. *Let J be a closed and bounded interval. Then any continuous function $f: J \to \mathbb{R}$ is uniformly continuous.*

Proof. If f is not uniformly continuous, then there exists $\varepsilon > 0$ such that for all $1/n$ we can find $a_n, b_n \in J$ such that $|a_n - b_n| < 1/n$ but $|f(a_n) - f(b_n)| \geq \varepsilon$. By Bolzano-Weierstrass theorem, there exists a subsequence (a_{n_k}) such that $a_{n_k} \to a \in J$. Observe that

$$|b_{n_k} - a| \leq |b_{n_k} - a_{n_k}| + |a_{n_k} - a|.$$

It follows that $b_{n_k} \to a$. By continuity, $f(a_{n_k}) \to f(a)$ and also $f(b_{n_k}) \to f(a)$. In particular, for all sufficiently large k, we must have $|f(a_{n_k}) - f(b_{n_k})| < \varepsilon$, a contradiction.   □

Remark 3.7.10. The theorem remains true of we assume that J is a compact subset of $\mathbb{R}$. See Remark 3.4.5.

Example 3.7.11. If $\emptyset \neq A \subseteq \mathbb{R}$, show that $f = d_A$ is uniformly continuous where $d_A(x) := \text{glb} \{|x - a| : a \in A\}$. (See Example 3.2.5.)

Exercise Set 3.7.12.

(1) Let $f: \mathbb{R} \to \mathbb{R}$ be continuous and periodic with period $p > 0$. That is, for all $x \in \mathbb{R}$, we have $f(x + p) = f(x)$. Show that f is uniformly continuous. (Examples are the sine and cosine functions sin and cos with period $p = 2\pi$.)

(2) Let $f(x) = \frac{1}{x+1} \cos x^2$ on $[0, \infty)$. Show f is uniformly continuous.

(3) Let f be uniformly continuous on $[a, c]$ and also on $[c, b]$. Show that it is uniformly continuous on $[a, b]$.

(4) Let $f\colon [0, \infty)$ be continuous. Assume that f is uniformly continuous on $[R, \infty)$ for some $R > 0$. Show that f is uniformly continuous on $[0, \infty)$.

(5) Let $f(x) = x^{1/2}$ on $[0, \infty)$. Is f uniformly continuous?

(6) Show that $f(x) = \frac{|\sin x|}{x}$ is uniformly continuous on $(-1, 0)$ and $(0, 1)$ but not on $(-1, 0) \cup (0, 1)$. Compare this exercise with Exercise 3 above.

(7) A function $f\colon \mathbb{R} \to \mathbb{R}$ is uniformly continuous iff whenever (x_n) and (y_n) are sequences of $\mathbb{R}$ such that $|x_n - y_n| \to 0$, then we have $|f(x_n) - f(y_n)| \to 0$.

(8) Let $f\colon B \subset \mathbb{R} \to \mathbb{R}$ be uniformly continuous on a bounded set B. Show that $f(B)$ is bounded.

(9) Let $f\colon J \subset \mathbb{R} \to \mathbb{R}$ be uniformly continuous with $|f(x)| \geq \eta > 0$ for all $x \in X$. Then $1/f$ is uniformly continuous on J.

(10) Let $f(x) := \sqrt{x}$ for $x \in [0, 1]$. Then f is uniformly continuous but not Lipschitz on $[0, 1]$.

Hence the converse of Exercise 3.7.6 is not true.

(11) Check for uniform continuity of the functions on their domains:
   (a) $f(x) := \sin(1/x)$, $x \in (0, 1]$.
   (b) $g(x) := x \sin(1/x)$, $x \in (0, 1]$.

3.8 Continuous Extensions

This section is optional. The material in this section is not used in the rest of the book. However, a keen reader is advised to go through this section as it raises some standard questions and solves them employing useful tools of analysis.

Let X and Y be sets and $A \subset B \subset X$. Let $f\colon A \to Y$ be given. We say a function $g\colon B \to Y$ is an *extension* of f (to B) if $g(a) = f(a)$ for $a \in A$.

Example 3.8.1. Let X, Y, A, and f be as above. Fix $z \in Y$. Define

$$g(x) = \begin{cases} f(x), & x \in A \\ z, & x \notin A. \end{cases}$$

Clearly g is an extension of f. Our construction also shows that such an extension need not be unique. (Why?)

Example 3.8.2. Let $X = \mathbb{R} = Y$ and $A = \mathbb{R} \setminus \{2\}$. Let $f(x) = \frac{x^2 - 4}{x - 2}$, $x \in A$. Note that f is continuous on A. Consider $g(x) = f(x)$ for $x \in A$ and $g(2) = 4$. Also, $h(x) = f(x)$ for $x \in A$ and $h(2) = 0$. Then both g and h are extensions of f. But we prefer g since g is continuous. The function g is called a *continuous extension* of f.

Exercise 3.8.3. Keep the notation of the last exercise. Can we have two distinct continuous extensions of f?

Exercise 3.8.4. This is a generalization of the last exercise. Let $A \subset B \subset \mathbb{R}$. Assume that for any given $b \in B$, there exists a sequence (a_n) in A such that $a_n \to b$. Let $f: A \to \mathbb{R}$ be continuous. Show that if g and h are continuous extensions of f to B, then $g = h$. That is, under the stated assumptions, continuous extensions are unique.

Example 3.8.5. The most standard example is the function $f: \mathbb{R}^* \to \mathbb{R}$ defined by $f(x) = \frac{\sin x}{x}$ for $x \neq 0$. This is continuous on $\mathbb{R}^*$. If we define $g(0) = 1$ and $g(x) = f(x)$ for $x \neq 0$, then g is a continuous extension of f.

   Another standard example is the function $f: \mathbb{R}^* \to \mathbb{R}$ defined by $f(x) = x \sin(1/x)$ for $x \neq 0$. This is continuous on $\mathbb{R}^*$. If we define $g(0) = 0$ and $g(x) = f(x)$ for $x \neq 0$, then g is a continuous extension of f.

   Do continuous extensions always exist?

Exercise Set 3.8.6. Let $A = \{x \in \mathbb{R} : x > 0\}$. Define $f(x) = 1/x$ for $x \in A$. Consider $B := \{x \in \mathbb{R} : x \geq 0\}$. Show that there exists no continuous extension of f to B.

   We are now ready to state a positive and a most useful result in this direction.

Theorem 3.8.7. *Let $A \subset B \subset \mathbb{R}$. Assume that for any given $b \in B$, there exists a sequence (a_n) in A such that $a_n \to b$. Let $f: A \to \mathbb{R}$ be uniformly continuous. Then there exists a unique continuous extension $g: B \to \mathbb{R}$ of f. Furthermore, g is also uniformly continuous on B.*

> **Strategy:** Given $b \in B$, we choose (a_n) in A such that $a_n \to b$. By the uniform continuity of f, the sequence $(f(a_n))$ is Cauchy in $\mathbb{R}$ (Exercise 3.7.8.1) and hence is convergent, say, to c. We would like to define $g(b) = c$. (Why would we like it this way?)
>
> There is an ambiguity here. What happens if we got another sequence (x_n) in A such that $x_n \to b$? The sequence $(f(x_n))$ is Cauchy and it may converge to $d \in \mathbb{R}$. Do we let $g(b) = c$ or $g(b) = d$? We shall show that if (x_n) in A converges to b, the sequences $(f(x_n))$ converge to the same limit $c := \lim f(a_n)$.
>
> Now that g is defined without ambiguity, we show that such a g is uniformly continuous on B.
>
> Let $x, y \in B$. We need to estimate $|g(x) - g(y)|$. If (x_n) and (y_n) are sequences in A such that $x_n \to x$ and $y_n \to y$, we then have the following estimate:
>
> $$|g(x) - g(y)| \leq |g(x) - g(x_n)| + |g(x_n) - g(y_n)| + |g(y_n) - g(y)|$$
> $$\leq |g(x) - f(x_n)| + |f(x_n) - f(y_n)| + |f(y_n) - g(y)|. \quad (3.8)$$

It is now easy to estimate the three terms on the right side. Give $\varepsilon > 0$, we choose $N \in \mathbb{N}$ such that

$$n \geq N \implies |f(x_n) - g(x)| < \varepsilon/3 \ \& \ |f(y_n) - g(y)| < \varepsilon/3.$$

(How can we do this?) By the uniform continuity of f, for a give $\varepsilon > 0$, we can find $\delta > 0$ so that if $s, t \in A$ with $|s - t| < \delta$, we then have $|f(s) - f(t)| < \varepsilon/3$. As seen earlier, we can find $N_2 \in \mathbb{N}$ such that for $n \geq N_2$, we have

$$|x_n - y_n| \leq |x_n - x| + |x - y| + |y_n - y| < 3 \times \delta/3.$$

This allows us to estimate the middle term in (3.8):

$$|f(x_n) - f(y_n)| < \varepsilon/3.$$

See below how we turn these ideas into a precise proof.

Proof. Let $b \in B$. Let (a_n) be a sequence in A such that $a_n \to b$. By Exercise 3.7.8.1, the sequence $(f(a_n))$ is Cauchy in $\mathbb{R}$ and hence is convergent, say, to c. Let (x_n) be any sequence in A such that $x_n \to b$. We shall show that $f(x_n) \to c$. Let $\varepsilon > 0$ be given. By the uniform continuity of f on A, for the given ε, there exists $\delta > 0$ such that for

$$x, y \in A \ \text{and} \ |x - y| < \delta \implies |f(x) - f(y)| < \varepsilon/2. \tag{3.9}$$

Since $a_n \to b$ and $x_n \to b$, for δ as above, there exists N_1 such that for all $n \geq N_1$, we have $|x_n - b| < \delta/2$ and $|a_n - a| < \delta/2$. We obtain

$$n \geq N_1 \implies |x_n - a_n| \leq |x_n - b| + |a_n - b| < \delta.$$

We deduce the following:

$$\text{If } n \geq N_1, \text{ then } |f(x_n) - f(a_n)| < \varepsilon/2. \tag{3.10}$$

Since $f(a_n) \to c$, for the given $\varepsilon > 0$, there exists N_2 such that

$$n \geq N_2 \implies |f(a_n) - c| < \varepsilon/2. \tag{3.11}$$

We observe, for $n \geq \max\{N_1, N_2\}$, that

$$|f(x_n) - c| \leq |f(x_n) - f(a_n)| + |f(a_n) - c|$$
$$< \varepsilon/2 + \varepsilon/2, \ \text{using} \ (3.10), (3.11).$$

This proves that $f(x_n) \to c$.

What we have shown tells us that we can define $g(b) = \lim f(a_n)$ without ambiguity, provided that (a_n) is a sequence in A converging to b.[1]

---

[1] The correct technical jargon is that g is well-defined.

Observe that this also applies to if $b \in A$! (Why? We may take the constant sequence (b).) Hence $g(a) = f(a)$ for $a \in A$. Thus we have an extension g of f to B.

We now show that g so defined is uniformly continuous on B. Let $\varepsilon > 0$ be given. Let $\delta > 0$ be chosen so that

$$s, t \in A \quad \text{and} \quad |s - t| < \delta \implies |f(s) - f(t)| < \varepsilon/3. \qquad (3.12)$$

Let $x, y \in B$ such that $|x - y| < \delta/3$. Since $x_n \to x$ and $y_n \to y$, we can find $N_1 \in \mathbb{N}$ such that for $n \geq N_1$, we have

$$|x_n - x| < \delta/3 \quad \text{and} \quad |y_n - y| < \delta/3.$$

It follows that for $n \geq N_1$, we have

$$|x_n - y_n| \leq |x_n - x| + |x - y| + |y - y_n| < \delta. \qquad (3.13)$$

Since $f(x_n) \to g(x)$ and $f(y_n) \to g(y)$, for ε as above, there exists $N_2 \in \mathbb{N}$ such that

$$n \geq N_2 \implies |f(x_n) - g(x)| < \varepsilon/3 \ \& \ |f(y_n) - g(y)| < \varepsilon/3. \qquad (3.14)$$

We are now ready to estimate $|g(x) - g(y)|$ for $x, y \in B$ with $|x - y| < \delta/3$. Choose an $n \geq \max\{N_1, N_2\}$. We then have

$$\begin{aligned} |g(x) - g(y)| &\leq |g(x) - g(x_n)| + |g(x_n) - g(y_n)| + |g(y_n) - g(y)| \\ &\leq |g(x) - f(x_n)| + |f(x_n) - f(y_n)| + |f(y_n) - g(y)|. \end{aligned}$$

The first and the third terms on the right is less than $\varepsilon/3$ thanks to (3.14). The middle term is less than $\varepsilon/3$ by (3.13) and (3.12).   □

Did you observe that we used the curry-leaves trick in the proof of (3.8)?

We now apply this result to define a^x for $a > 0$ and $x \in \mathbb{R}$. We make some preliminary observations.

For $n \in \mathbb{N}$, a^n makes sense. Since $a > 0$, a^{-1} makes sense, namely, the reciprocal $1/a$ of a. Hence we define $a^{-n} := (1/a)^n$. If we define $a^0 = 1$, then the law of exponents holds:

$$a^{m+n} = a^m a^n \text{ for all } m, n \in \mathbb{Z}.$$

We wish to extend this to rational exponents, that is, we wish to define a^r for $r \in \mathbb{Q}$. Let us write $r = m/n$ with $n \in \mathbb{N}$ and with m and n relatively prime. (This means that they do not have any nontrivial common divisors.) There are two ways to define a^r:

(1) the n-th positive root of a^m and
(2) the m-th power of n-th positive root of a.

It is easy to see that both the definitions lead to the same positive real number. We need to show that $(a^{1/n})^m = (a^m)^{1/n}$.

If we let $b = a^{1/n}$ be the unique positive n-th root of a, we need to show that b is the unique n-th root of a^m. We have $b^n = a$ so that $(b^n)^m = (b^m)^n = a^m$. That is, b^m is the unique n-th root of a_m.

The definition is made in such a way that the law of exponents holds: $a^{r+s} = a^r a^s$ for $r, s \in \mathbb{Q}$. Also, if $r < s$, then $a^r < a^s$.

Let $R > 0$ be fixed. We show that the map $f : r \mapsto a^r$ is uniformly continuous on $A_R := [-R, R] \cap \mathbb{Q}$. We start the proof by establishing the continuity of f at $0 \in A_R$.

Given $\varepsilon > 0$, choose $n > a$ such that $1 < n^{\frac{1}{n}} < 1 + \varepsilon$. Let $0 < r < \frac{1}{n}$. Then observe that $a^r < a^{\frac{1}{n}} < n^{\frac{1}{n}} < 1 + \varepsilon$. Thus, if $x_n > 0$, $\lim x_n = 0$, then $\lim a^{x_n} = 1$. Also, $a^{-x_n} = \frac{1}{a^{x_n}} \to 1$.

Let $\varepsilon > 0$ be given. Let $\delta > 0$ correspond to the continuity of f at 0 for $a^{-R}\varepsilon$. Now let $x, y \in A_R$ be such that $|x - y| < \delta$. . Then, we have

$$|a^x - a^y| = a^x |a^{x-y} - 1| < a^R a^{-R}\varepsilon.$$

We conclude that f is uniformly continuous on A_R. Therefore, there exists a unique continuous extension g of f to $[-R, R]$. We denote $g(x) = a^x$ for $|x| \le R$. Note that if $R_2 > R_1 > 0$ is given and the continuous extensions are denoted by g_1 and g_2 respectively, we have $g_1 = g_2$ on $[-R_1, R_1]$ by the uniqueness of the continuous extension. We thus have a unique continuous function $g : \mathbb{R} \to \mathbb{R}$ denoted by $g(x) = a^x$ which agrees with our definition of $a^{m/n}$ if $x = m/n$.

Exercise 3.8.8. Show that for all $x, y \in \mathbb{R}$, we have $a^{x+y} = a^x a^y$.

Exercise 3.8.9. If f is continuous, not identically zero, and satisfies the functional equation $f(x + y) = f(x)f(y)$ for all $x, y \in \mathbb{R}$, then $f(x) = a^x$ for some $a > 0$.

Exercise 3.8.10. Define the function $\exp(x) := \sum_{n=0}^{\infty} \frac{x^n}{n!}$. Then $\exp(x) = e^x$.
    This exercise is best done at the end of Section 7.5.

Exercise 3.8.11. Let $a > 1$. Let $f(x) = a^x$. Define $\log_a(x) = u$ if $a^u = x$. Then $\log_a$ is well-defined on the set of positive reals. Since it is the inverse of f, it is continuous, one-one and strictly increasing.

Exercise 3.8.12. Prove the following:

(1) $\log_a(xy) = \log_a(x) + \log_a(y)$ for all $x, y \in \mathbb{R}^+$.

(2) $\log_a(1) = 0$.

(3) $\log_a(x^y) = y \log_a(x)$ for all $x > 0$ and $y \in \mathbb{R}$.

(4) $\log_x y \log_y z = \log_x z$ whenever both sides make sense.

Chapter 4

Differentiation

Contents

The basic idea of differential calculus (as perceived by modern mathematics) is to approximate (at a point) a given function by an affine (linear) function (or a first-degree polynomial).

Let J be an interval and $c \in J$. Let $f : J \to \mathbb{R}$ be given. We wish to approximate $f(x)$ for x near c by a polynomial of the form $a + b(x - c)$. To keep the notation simple, let us assume $c = 0$. What is meant by approximation? If $E(x) := f(x) - a - bx$ is the error by taking the value of $f(x)$ as $a + bx$ near 0, what we want is that the error goes to zero much faster than x goes to zero. As we have seen earlier this means that $\lim_{x \to 0} \frac{f(x) - a - bx}{x} = 0$.

If this happens, then $\lim_{x \to 0}(f(x) - a - bx) = 0$. We thus arrive at $a = \lim_{x \to 0} f(x)$. In particular, if f is continuous at 0, then $a = f(0)$. Hence the requirement for a function f to be approximable at 0 is that there exists a real number b such that $\lim_{x \to 0} \dfrac{f(x) - f(0)}{x} = b$.

If such is the case, we say that f is *differentiable* at $c = 0$ and denote the (unique) real number b by $f'(0)$. It is called the derivative of f at 0.

In general, f is said to be differentiable at c, if there exists a real number α such that

$$\lim_{x \to c} \frac{f(x) - f(c)}{x - c} = \alpha.$$

We say that f is differentiable at $c \in J$ if we can approximate the increment $f(x) - f(c) \equiv f(c + h) - f(c)$ in the dependent variable by a linear polynomial $\alpha(x - c) = \alpha h$ in the increment of the independent variable. Approximation here means that the error should go to zero much faster than the increment going to zero.

4.1   Differentiability of Functions

Definition 4.1.1. Let J be an interval and $c \in J$. Let $f \colon J \to \mathbb{R}$. Then f is said to be *differentiable* at c, if there exists a real number α such that

$$\lim_{x \to c} \frac{f(x) - f(c)}{x - c} = \alpha. \tag{4.1}$$

It is sometimes useful to use the variable h for the increment $x - c$ and reformulate (4.1) as follows:
f is said to be differentiable at c, if there exists a real number α such that

$$\lim_{h \to 0} \frac{f(c + h) - f(c)}{h} = \alpha. \tag{4.2}$$

In view of the definition of limit, the differentiability condition in (4.1) can be defined using ε-δ as follows.

We say that f is *differentiable* at c if there exists $\alpha \in \mathbb{R}$ such that for any given $\varepsilon > 0$, there exists a $\delta > 0$ such that

$$x \in J \text{ and } 0 < |x - c| < \delta \implies |f(x) - f(c) - \alpha(x - c)| < \varepsilon |x - c|. \tag{4.3}$$

We say that f is differentiable on J if it is differentiable at each $c \in J$.

To check if a given function is differentiable at $x = c$, the basic idea is write the term $f(x) - f(c)$ as $\alpha(x - c) + $ 'something' and check whether something goes to zero much faster than $x - c$ goes to zero.

Example 4.1.2.

(1) Let $f \colon J \to \mathbb{R}$ be a constant, say, C. Then f is differentiable at $c \in J$ with $f'(c) = 0$.

   In this case, we have $f(x) - f(c) = 0$ for all c. Therefore, an obvious choice for α is 0.

   Let c be any real number and $\varepsilon > 0$ be given. Let $\alpha = 0$. Let us estimate the error term:

$$|f(x) - f(c) - \alpha(x - c)| = |C - C - \alpha(x - c)| = |\alpha|\,|(x - c)| = 0.$$

   This suggests that we can choose any $\delta > 0$. Thus if f is a constant function, then it is differentiable on $\mathbb{R}$ with $f'(c) = 0$ for $c \in \mathbb{R}$.

(2) Let $f\colon J \to \mathbb{R}$ be given by $f(x) = ax + b$.

Let c be an arbitrary real number. Consider the expression $f(x) - f(c) = a(x - c)$. This suggests, $\alpha = f'(c) = a$.

Let $\varepsilon > 0$ be given. Now, let us try to estimate the error term:

$$|f(x) - f(c) - \alpha(x - c)| = |a(x - c) - a(x - c)| = 0.$$

This suggests that we can choose any $\delta > 0$ for any $\varepsilon > 0$.

Let $\varepsilon > 0$ be given. Let $\delta > 0$ be arbitrary. We estimate the error term:

$$|f(x) - f(c) - \alpha(x - c)| = 0 < \varepsilon |x - c|.$$

Since c is an arbitrary real number, f is differentiable on $\mathbb{R}$, and $f'(c) = a$.

(3) If $f\colon J \to \mathbb{R}$ is given by $f(x) = x^n$, $n \in \mathbb{N}$, then $f'(c) = nc^{n-1}$.

Let c be an arbitrary real number. Let us look at

$$f(c + h) - f(c) = (c + h)^n - c^n$$

$$= c^n + nc^{n-1}h + \binom{n}{2}c^{n-2}h^2 + \cdots + h^n - c^n$$

$$= nc^{n-1}h + \text{ terms involving higher powers of } h.$$

This suggests that we take $\alpha = nc^{n-1}$.

With this choice of α, we estimate the error.

$$\left|f(c + h) - f(c) - nc^{n-1}h\right| = \left|\sum_{k=2}^{n} \binom{n}{k} c^{n-k} h^k\right|$$

$$\leq \sum_{k=2}^{n} \binom{n}{k} |c|^{n-k} |h|^k$$

$$\leq \sum_{k=2}^{n} \binom{n}{k} |c|^{n-k} |h|^2 \qquad \text{for } |h| < 1$$

$$= |h|^2 \left[\sum_{k=2}^{n} \binom{n}{k} |c|^{n-k}\right]$$

$$= |h|^2 M,$$

where $M := \sum_{k=2}^{n} \binom{n}{k} |c|^{n-k}$.

The above approximation suggests that we may take $\delta = \varepsilon/M$.

Now let us look at a formal proof.

For given $\varepsilon > 0$, choose $\delta = \min\{1, \frac{\varepsilon}{M}\}$. For x such that $0 < |h| < \delta$, using the above estimates, we have

$$\left|f(c + h) - f(c) - nc^{n-1}h\right| \leq |h^2| M < \varepsilon |h|.$$

Hence f is differentiable at c and $f'(c) = nc^{n-1}$. Since c is arbitrary, f is differentiable on $\mathbb{R}$.

The next result gives a powerful, and at the same time a very simple characterization of differentiability of a function at a point.

Theorem 4.1.3. *Let $f\colon J \to \mathbb{R}$ be given. Then f is differentiable at $c \in J$ iff there exists a function $f_1\colon J \to \mathbb{R}$ satisfying the following two conditions:*
1. *We have*
$$f(x) = f(c) + f_1(x)(x - c) \text{ for } x \in J. \tag{4.4}$$
2. *f_1 is continuous at c.*
In such a case, $f'(c) = f_1(c)$.

Proof. Let us assume that there exists a function $f_1\colon J \to \mathbb{R}$ continuous at c such that
$$f(x) = f(c) + f_1(x)(x - c) \text{ for } x \in J.$$
We need to prove that f is differentiable at $x = c$ and $f'(c) = f_1(c)$. It is easy to see that under this assumption we have
$$|f(x) - f(c) - f_1(c)(x - c)| = |x - c|\,|f_1(x) - f_1(c)|\,.$$

Let $\varepsilon > 0$ be given. Since f_1 is continuous at c, there exists $\delta > 0$ such that for all x with $|x - c| < \delta$, we have $|f_1(x) - f_1(c)| < \varepsilon$. Hence, for all x with $0 < |x - c| < \delta$, we get
$$|f(x) - f(c) - f_1(c)(x - c)| = |x - c|\,|f_1(x) - f_1(c)| < \varepsilon\,|x - c|\,.$$
Thus f is differentiable at $x = c$.

Assume that f is differentiable at c. We have to find an f_1 satisfying the two conditions. The first condition that $f(x) = f(c) + f_1(x)(x - c)$ for all $x \in J$ leads us to arrive at
$$f_1(x) = \frac{f(x) - f(x)}{x - c}, \text{ for } x \in J, x \neq c.$$
We need to define $f_1(c)$. Since f_1 should be continuous at c, we must have $\lim_{x \to c} f_1(x) = f_1(c)$. By substituting the expression for $f_1(x)$, $x \neq c$, we see that $f_1(c) = \lim_{x \to c} \frac{f(x) - f(c)}{x - c}$. Since by our hypothesis, f is differentiable at c, this suggests that we define $f_1(c) = f'(c)$. Thus we have defined f_1 on J. We now turn to a formal proof.

Define
$$f_1(x) := \begin{cases} \frac{f(x) - f(c)}{x - c}, & \text{for } x \in J \text{ and } x \neq c \\ f'(c), & \text{if } x = c. \end{cases}$$

From the very definition, it follows that for $x \in J$, we have $f(x) = f(c) + f_1(x)(x - c)$. Since
$$\lim_{x \to c} f_1(x) = \lim_{x \to c} \frac{f(x) - f(c)}{x - c} = f'(c) = f_1(c),$$
the continuity of f_1 at c follows. $\qquad\square$

In spite of its simplicity, this is a very powerful characterization of differentiability at a point. We illustrate its use in the next few examples. Recall how our definition of continuity using sequences reduced the proofs of algebra of continuous functions to those of algebra of convergent sequences. Our characterization above will reduce the problem of proving the algebra of differentiable functions to the algebra of continuous functions.

Example 4.1.4.

(1) Let $f(x) = \frac{1}{x}$ for $x \neq 0$. Then f is differentiable at $c \neq 0$ and $f'(c) = -\frac{1}{c^2}$ for $c \neq 0$.

We have

$$f(x) - f(c) = \frac{1}{x} - \frac{1}{c} = \frac{c-x}{cx} = \left[\frac{-1}{cx}\right](x - c).$$

This suggests that we define $f_1(x) := \frac{-1}{cx}$ for $x \neq 0$. We then have $f(x) = f(c) + f_1(x)(x - c)$. Clearly f_1 is continuous for $x \neq 0$. This proves that $f(x) = \frac{1}{x}$ is differentiable at any $c \neq 0$ and that $f'(c) = f_1(c) = -1/c^2$.

(2) Let $f(x) = x^n$ for $x \in \mathbb{R}$. We have

$$f(x) - f(c) = (x - c)[x^{n-1} + x^{n-2}c + \cdots + xc^{n-2} + c^{n-1}].$$

Thus we are led to define $f_1(x) := x^{n-1} + x^{n-2}c + \cdots + xc^{n-2} + c^{n-1}$.

It is clear that f_1 is continuous at $x = c$ and that $f_1(c) = nc^{n-1}$. Hence f is differentiable at c and that $f'(c) = nc^{n-1}$.

(3) Let $n \in \mathbb{N}$. We now show that $f: (0, \infty) \to (0, \infty)$ defined by $f(x) := x^{1/n}$ is differentiable. We shall adopt the notation of Example 7 on page 75. It would be a good idea to revisit the example now. Let $a > 0$ be fixed.

We would like to guess f_1 by looking at $x^{1/n} - a^{1/n}$. We set $t := x^{1/n}$ and $s := a^{1/n}$ and arrive at

$$t - s = (x - a)\frac{1}{t^{n-1} + \cdots + s^{n-1}}.$$

This suggests that we define $f_1(x) = \frac{1}{x^{\frac{n-1}{n}} + \cdots + a^{\frac{n-1}{n}}}$. Clearly, f_1 is as required and we have $f_1(a) = \frac{1}{na^{\frac{n-1}{n}}} = \frac{1}{n}a^{\frac{1}{n}-1}$.

(4) Let $f(x) = e^x$, $x \in \mathbb{R}$. Using the standard facts about the exponential function, we show that $f'(c) = e^c$. As we have not rigorously defined the exponential function and also we are going to use infinite series expansion, the argument below is not satisfactory at this level. However, the purpose of this is to show the reader how to guess f_1 in a formal way and justify the

steps rigorously later. If the reader is not happy with this, he may skip this example.

$$f(x) - f(c) = e^c(e^x e^{-c} - 1) = e^c(e^{x-c} - 1)$$

$$= e^c(x - c) \sum_{n=1}^{\infty} \frac{(x - c)^{n-1}}{n!}$$

$$= \left[e^c \sum_{n=1}^{\infty} \frac{(x - c)^{n-1}}{n!} \right] (x - c)$$

$$= f_1(x)(x - c).$$

The steps above can be justified. We can show that f_1 is continuous and that $f_1(c) = e^c$ and it is continuous at c.

Exercise 4.1.5. Let m, n be positive integers. Define

$$f(x) = \begin{cases} x^n & \text{for } x \geq 0 \\ x^m & \text{for } x < 0. \end{cases}$$

Discuss the differentiability of f at $x = 0$.

Proposition 4.1.6. *If f is differentiable at c, then f is continuous at c.*

Proof. We use the notation of Theorem 4.1.3. Since f_1 is continuous at $x = c$, $f_1(x)(x - c)$ is also continuous at c. Hence $f(x) = f(c) + f_1(x)(x - c)$ continuous at c. □

Remark 4.1.7. Like continuity, differentiability is also a local concept. That is, to check whether a function $f\colon J \to \mathbb{R}$ is differentiable at c or not, we need to know f only on a small interval $(c - \delta, c + \delta)$ around c. This is evident from the definition of derivative as the limit of difference quotient.

Theorem 4.1.8 (Algebra of Differentiable Functions). *Let $f, g\colon J \to \mathbb{R}$ be differentiable at $c \in J$. Then the following hold:*
(a) *$f + g$ is differentiable at c with $(f + g)'(c) = f'(c) + g'(c)$.*
(b) *αf is differentiable at c with $(\alpha f)'(c) = \alpha f'(c)$.*
(c) *fg is differentiable at c with $(fg)'(c) = f(c)g'(c) + f'(c)g(c)$.*
(d) *If f is differentiable at c with $f(c) \neq 0$, then $\varphi := 1/f$ is differentiable at c with $\varphi'(c) = -\frac{f'(c)}{(f(c))^2}$.*

Strategy: In each of the proofs, we exploit the existence of f_1 as stipulated in Theorem 4.1.3 and manipulate the expressions to find the required auxiliary function f_1.

Proof. We first prove (a). Since f and g are differentiable at c, using Theorem 4.1.3, there exist continuous functions f_1 and g_1, respectively, such that

$f(x) = f(c) + f_1(x)(x - c)$ and $g(x) = g(c) + g_1(x)(x - c)$. Define $h(x) := f(x) + g(x)$. Then,

$$h(x) = f(x) + g(x) = [f(c) + g(c)] + [f_1((x) + g_1(x)](x - c).$$

We define $h_1(x) = f_1(x) + g_1(x)$. Then h_1 is continuous at c. Hence $f + g$ is differentiable at c and $(f + g)'(c) = f'(c) + g'(c)$.

We now prove (b). Since f is differentiable at c, using Theorem 4.1.3, there exists a continuous function f_1 such that $f(x) = f(c) + f_1(x)(x - c)$. Hence $\alpha f(x) = \alpha f(c) + \alpha f_1(x)(x - c)$. Note that αf is continuous at c. Therefore, αf is differentiable at c and its derivative is $\alpha f'(c)$.

To prove (c), let us write $\varphi(x) := (fg)(x) = f(x)g(x)$. We again make use of the existences of continuous functions f_1 and g_1 such that $f(x) = f(c) + f_1(x)(x - c)$ and $g(x) = g(c) + g_1(x)(x - c)$. We multiply the expressions for $f(x)$ and $g(x)$ to get an expression for $\varphi(x)$ and try to guess the auxiliary φ_1 below.

$$\varphi(x) = [f(c) + f_1(x)(x - c)][g(c) + g_1(x)](x - c)$$
$$= \varphi(c) + f(c)g_1(x)(x - c) + g(c)f_1(x)(x - c) + f_1(x)(g_1(x)(x - c)^2$$
$$= \varphi(c) + [f(c)g_1(x) + g(c)f_1(x) + f_1(x)g_1(x)(x - c)](x - c).$$

This suggests that we define

$$\varphi_1(x) := [f(c)g_1(x) + g(c)f_1(x) + f_1(x)g_1(x)(x - c)].$$

Then φ is as required and its value at c is $f(c)g_1(c) + g(c)f_1(c)$. This proves that fg is differentiable at c and its derivative at c is given by $f(c)g_1(c) + g(c)f_1(c) = f(c)g'(c) + g(c)f'(c)$.

Before we prove (d), let us make some preliminary remarks. Since we assume f is differentiable at c, it is continuous at c. Since $f(c) \neq 0$, there exists $\delta > 0$ such that for $x \in (c - \delta, c + \delta) \subset J$, we have $f(x) \neq 0$. (See Theorem 3.1.5.) Hence the function $1/f$ is defined on the interval $(c - \delta, c + \delta)$.

Given that f is differentiable at c and $f(c) \neq 0$. Using Theorem 4.1.3, there exists a continuous function f_1 such that $f(x) = f(c) + f_1(x)(x - c)$. Let $g = 1/f$. We need to find a continuous function g_1 such that $g(x) = g(c) + g_1(x)(x - c)$. Let us look at

$$g(x) - g(c) = \frac{1}{f(c) + f_1(x)(x - c)} - \frac{1}{f(c)}$$
$$= \frac{-f_1(x)}{f(c)[f(c) + f_1(x)(x - c)]}(x - c).$$

Note that $f(x) = f(c) + f_1(x)(x - c) \neq 0$ in the open interval $(c - \delta, c + \delta)$. Define $g_1(x) := \frac{-f_1(x)}{f(c)[f(c) + f_1(x)(x - c)]}$. Note that $g_1(x)$ makes sense, as the denominator does not vanish on $(c - \delta, c + \delta)$. Also g_1 is continuous at c and $g_1(c) = \frac{-f'(c)}{f(c)^2}$. Hence $g = 1/f$ is differentiable at c and its derivative is $-\frac{f'(c)}{f(c)^2}$. $\qquad\square$

Corollary 4.1.9. *Let $D_a(J)$ (respectively $C_a(J)$) denote the set of functions on J differentiable (respectively continuous) at a. Then $D_a(J)$ is a vector subspace of $C_a(J)$.*

Proof. This follows from Theorem 4.1.8.   □

Exercise Set 4.1.10.

(1) Show that $f \colon \mathbb{R} \to \mathbb{R}$ given by $f(x) = |x|$ is not differentiable at $x = 0$.

(2) Let $f \colon \mathbb{R} \to \mathbb{R}$ be given by $f(x) = x^2$ if $x \in \mathbb{Q}$ and $f(x) = 0$ if $x \notin \mathbb{Q}$. Show that f is differentiable at $x = 0$. Find $f'(0)$.

(3) Show that $f(x) = x^{1/3}$ is not differentiable at $x = 0$.

(4) Let $n \in \mathbb{N}$. Define $f \colon \mathbb{R} \to \mathbb{R}$ by $f(x) = x^n$ for $x \geq 0$ and $f(x) = 0$ if $x < 0$. For which values of n,

    (a) is f continuous at 0?

    (b) is f differentiable at 0?

    (c) is f' continuous at 0?

    (d) is f' differentiable at 0?

(5) Let $f \colon \mathbb{R} \to \mathbb{R}$ be differentiable. Let $n \in \mathbb{N}$. Fix $a \in \mathbb{R}$. Find

$$\lim_{x \to a} \frac{a^n f(x) - x^n f(a)}{x - a}.$$

(6) Can you generalize the last exercise?

(7) Let $f \colon J \to \mathbb{R}$ be differentiable. Let $x_n < c < y_n$ be such that $y_n - x_n \to 0$. Show that

$$\lim_{n \to \infty} \frac{f(y_n) - f(x_n)}{y_n - x_n} = f'(c).$$

(8) Use the identity $1 + x + \cdots + x^n = \frac{1 - x^{n+1}}{1 - x}$ for $x \neq 1$ to arrive at a formula for the sum $1 + x + 2x^2 + \cdots + nx^n$.

(9) Let $f \colon \mathbb{R} \to \mathbb{R}$ be an even function, that is, $f(-x) = f(x)$ for all $x \in \mathbb{R}$. Assume that f is differentiable. Show that f' is odd.

(10) Let $f \colon (a, b) \to \mathbb{R}$ be differentiable at $c \in (a, b)$. Assume that $f'(x) \neq 0$. Show that there exists $\delta > 0$ such that for $x \in (c - \delta, c + \delta) \cap (a, b)$, we have $f(x) \neq f(c)$.

Theorem 4.1.11 (Chain Rule). *Let $f: J \to \mathbb{R}$ be differentiable and $f(J) \subset J_1$, an interval and if $g: J_1 \to \mathbb{R}$ is differentiable at $f(c)$, then $g \circ f$ is differentiable at c with $(g \circ f)'(c) = g'(f(c)) \cdot f'(c)$.*

Proof. Since f is differentiable at $x = c$, there exists $f_1: J \to \mathbb{R}$ which is continuous at c such that

$$f(x) = f(c) + f_1(x)(x - c) \text{ and } f'(c) = f_1(c).$$

Similarly, there exists $g_1: J_1 \to \mathbb{R}$, continuous at $d = f(c) \in J_1$ such that

$$g(y) = g(d) + g_1(y)(y - d), \text{ and } g'(d) = g_1(d) \quad \text{for } y \in J_1. \tag{4.5}$$

Let $h = g \circ f$. We have to find a function $h_1: J \to \mathbb{R}$ which continuous at c such that

$$h(x) = h(c) + h_1(x)(x - c) = g(f(c)) + h_1(x)(x - c).$$

Consider the composition

$$\begin{aligned} (g \circ f)(x) = g(f(x)) &= g(f(c) + f_1(x)(x - c)) \\ &= g(d + f_1(x)(x - c)) \\ &= g(y), \quad \text{where } y = d + f_1(x)(x - c). \end{aligned}$$

Using (4.5), the last expression on the right can be written as

$$\begin{aligned} g(y) &= g(d) + g_1(y)(y - d) \\ &= g(d) + g_1[f(c) + f_1(x)(x - c)][f(c) + f_1(x)(x - c) - f(c)] \\ &= g(f(c)) + g_1(f(c) + f_1(x)(x - c))f_1(x)(x - c). \end{aligned}$$

The right-hand side of the above equation suggests that we may choose

$$h_1(x) = g_1(f(c) + f_1(x)(x - c))f_1(x).$$

It follows from the continuity of composition of continuous functions and algebra of continuous functions that h_1 is continuous at c. Also, we have

$$h_1(c) = g_1(f(c))f_1(c) = g'(f(c))f'(c).$$

This completes the proof. $\square$

Example 4.1.12. We shall assume that the properties of the following functions and their derivatives are known.

(1) $f(x) = e^x$, $f'(x) = e^x$.

(2) $f(x) = \log x$, $x > 0$, $f'(x) = 1/x$.

(3) $f(x) = \sin x$, $f'(x) = \cos x$.

(4) $f(x) = \cos x$, $f'(x) = -\sin x$.

Example 4.1.13. Let $\alpha \in \mathbb{R}$ and $x > 0$. Recall the definition of x^α from your calculus course. We define $x^\alpha := e^{\alpha \log x}$. Thus, we have a function $\varphi \colon (0, \infty) \to \mathbb{R}$ defined by $\varphi(x) := x^\alpha$. Note that φ is the composition of $f \colon x \mapsto \alpha \log x$ followed by $g \colon t \mapsto e^t$. Hence by the Chain rule, φ is differentiable and we have

$$\varphi'(x) = g'(f(x))f'(x) = \left(e^{\alpha \log x}\right)\left(\frac{\alpha}{x}\right) = (\alpha x^\alpha)\left(\frac{1}{x}\right) = \alpha x^{\alpha-1}.$$

Exercise Set 4.1.14.

(1) Let $f \colon \mathbb{R} \to \mathbb{R}$ be differentiable at $x = 0$. Define $g(x) = f(x^2)$. Show that g is differentiable at 0 (i) by Theorem 4.1.3 and (ii) by the chain rule.

(2) Let $r \in \mathbb{Q}$. Define $f(x) = x^r \sin(1/x)$ for $x \neq 0$ and $f(0) = 0$. For what values of r, is f differentiable at 0?

There are two important ways of looking at derivatives. One interpretation is physical and the other is geometric.

Let $f \colon [a, b] \to \mathbb{R}$ be a function. One thinks of the domain as the time interval and $f(t) - f(c)$ as the distance traveled by a particle in $t - c$ units of time and so that the velocity is $\frac{f(t)-f(c)}{t-c}$. The derivative which is the limit of these velocities as $t \to c$ is called the *instantaneous velocity* of the motion of the particle at the instant $t = c$. This is a very useful way of looking at the derivative.

The geometric interpretation of the derivative $f'(c)$ is that it is the slope of the tangent line at $(c, f(c))$ to the graph $\{(t, f(t)) : t \in [a, b]\}$. Figure 4.1 shows that the tangent line is the limiting position of the chords.

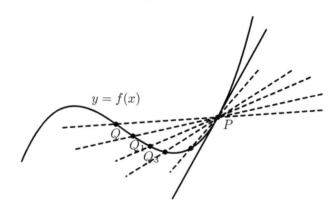

Figure 4.1: Tangent line is the limiting position of the chords.

What does Exercise 4.1.10.7 say? The slope of the chord joining $(x_n, f(x_n))$ and $(y_n, f(y_n))$ is $\frac{f(y_n)-f(x_n)}{y_n-x_n}$. Thus if $x_n < c < y_n$, $x_n \to c$ and $y_n \to c$, then the sequence $\left(\frac{f(y_n)-f(x_n)}{y_n-x_n} \right)$ of the slopes of the chords converge to the slope of the tangent at $(c, f(c))$ to the graph.

Look at Figure 3.3 on page 70. Are these graphs of differentiable functions?

Students should notice that if the graph of a function has sharp corners, then it is not differentiable at such points. (See Figure 4.2). You may recall the graph of $|x|$. Do you think $\min\{\sin x, \cos x\}$ is differentiable on $\mathbb{R}$? Figure 3.4 on page 70 may help you decide.

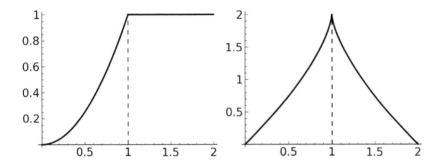

Figure 4.2: Non-differentiable functions at $x = 1$.

Exercise Set 4.1.15.

1. Construct a continuous function $f \colon \mathbb{R} \to \mathbb{R}$ which is not differentiable at $x = 2$.

2. Construct a continuous function $f \colon \mathbb{R} \to \mathbb{R}$ which is not differentiable at integers.

Weierstrass has constructed a function $f \colon \mathbb{R} \to \mathbb{R}$ which is continuous on $\mathbb{R}$ but differentiable nowhere. See Appendix E in [2].

4.2   Mean Value Theorems

As the reader has already learned from calculus courses, differential calculus is a powerful tool in problems of maxima–minima. In this section we shall establish the so-called first derivative test. There are certain misconceptions here, so we shall start with precise definitions.

Definition 4.2.1. Let $J \subset \mathbb{R}$ be an interval and $f \colon J \to \mathbb{R}$ be a function. We say that a point $c \in J$ is a point of *local maximum* if there exists $\delta > 0$ such that $(c - \delta, c + \delta) \subset J$ and $f(x) \leq f(c)$ for all $x \in (c - \delta, c + \delta)$. (Look at Figure 4.3.)

A *local minimum* is defined similarly. A point c is said to be a *local extremum* if it is either a local maximum or a local minimum.

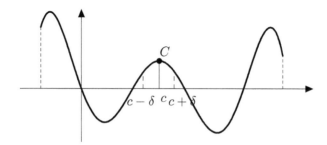

Figure 4.3: Local maximum.

A point $x_0 \in J$ is said to be a point of *(global) maximum* on J if $f(x) \le f(x_0)$ for all $x \in J$. *Global minimum* is defined similarly.

How do we define global extremum?

Note that a local extremum need not be a global extremum. Similarly, a global extremum need not be a local extremum.

The points of local minimum and local maximum should be "interior points" in the domain. In Figure 4.4, the points, c_1, c_3, and c_5 are local maximum whereas c_2 and c_4 are points of local minimum. The point c_1 is the global maximum and c_2 is the global minimum.

Do you believe that there is any relation between a global maximum and a local maximum? For example, is any global maximum necessarily a local maximum? or is any local maximum necessarily a global maximum?

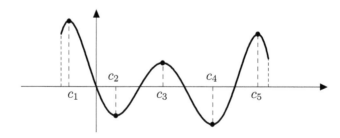

Figure 4.4: Local maximum and minimum.

Example 4.2.2. Look at $f \colon [a, b] \to \mathbb{R}$ where $f(x) = x$. Then b is a point of global maximum but not a local maximum. What can you say about a?

On the other hand, look at $g\colon [-2\pi, 2\pi] \to \mathbb{R}$ defined by $g(x) = \cos x$. The point $x = 0$ is a local maximum as well as a global maximum. What can you say about the points $x = \pm 2\pi$?

Theorem 4.2.3. *Let $J \subset \mathbb{R}$ be an interval. Let $f\colon J \to \mathbb{R}$ be differentiable on J and $c \in J$. If c is a local extremum of f, then $f'(c) = 0$.*

Strategy: Let c be a local maximum of f. Look at the difference quotient $\frac{f(c+h)-f(c)}{h}$. It is ≤ 0 if $h > 0$ and is ≥ 0 if $h < 0$. So if you take one-sided limits, the right-side limit should be ≤ 0 and the left-side limit should be ≥ 0. Hence the limit must be 0.

Proof. Let f have local maximum at c. Then there exists a $\delta > 0$ such that $f(x) \leq f(c)$ for all $x \in (c-\delta, c+\delta)$. That is, $f(c+h) \leq f(c)$ and $f(c-h) \leq f(c)$ for all $h \in (-\delta, \delta)$. Since f is differentiable at c, we have $f'(c) = \lim\limits_{h \to 0} \dfrac{f(c+h) - f(c)}{h}$.

$$f'(c) = \lim_{h \to 0+} \frac{f(c+h) - f(c)}{h} \leq 0 \quad \text{and} \quad f'(c) = \lim_{h \to 0-} \frac{f(c+h) - f(c)}{h} \geq 0.$$

Thus from the above two inequalities, we deduce $f'(c) = 0$.

The proof when c is a local minimum is similar. $\qquad\square$

Question: Where did we use the fact that c is a local maximum in the proof? Compare the result with the function of f of Example 4.2.2. What are $f'(b)$, $f'(a)$?

Theorem 4.2.4 (Rolle's Theorem). *Let $f\colon [a, b] \to \mathbb{R}$ be such that (i) f is continuous on $[a, b]$, (ii) f is differentiable on (a, b), and (iii) $f(a) = f(b)$. Then there exists $c \in (a, b)$ such that $f'(c) = 0$.*

The geometric interpretation is that there exists $c \in (a, b)$ such that the slope of the tangent to the graph of f at c equals zero. That is, the tangent at $(c, f(c))$ is parallel to the x-axis. (See Figure 4.5.)

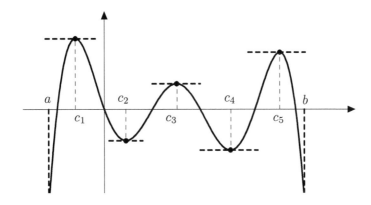

Figure 4.5: Rolle's theorem.

Strategy: Draw a few pictures of differentiable functions satisfying $f(a) = f(b)$. You will observe that the likely choices of c are points of local extremum. Since f is continuous on the closed and bounded interval, what we are assured of is the existence of global extrema! If the maximum of f and the minimum of f are different, then one of them is assumed by f at point in (a, b). We show that such a point is a local extremum. What happens if the maximum and minimum of f coincide?

Proof. By the extreme value theorem (Theorem 3.4.6), there exist $x_1, x_2 \in [a, b]$ such that $f(x_1) \leq f(x) \leq f(x_2)$ for all $x \in [a, b]$. If $f(x_1) = f(x_2)$, then the result is trivial. (Why? For any $x \in [a, b]$ we have $f(x_1) \leq f(x) \leq f(x_2)$. Hence f is a constant. So, for any $c \in (a, b)$, we obtain $f'(c) = 0$.)

If $f(x_1) \neq f(x_2)$, then at least one of x_1, x_2 is different from a and b. (Why? Since $f(a) = f(b)$, $\{x_1, x_2\} = \{a, b\}$ would imply $f(x_1) = f(x_2)$, a contradiction.)

Suppose x_2 is different from a and b. Hence $a < x_2 < b$. Let $c = x_2$. Let $\delta := \min\{c-a, b-c\}$. It follows that $(c-\delta, c+\delta) \subset (a, b)$. Since $c = x_2$ is a global maximum, we have $f(x) \leq f(c)$ for $x \in (c-\delta, c+\delta)$. Therefore we conclude that c is local maximum of f. By Theorem 4.2.3 we have $f'(c) = 0$.   □

Next we look at the most important result in differentiation.

Theorem 4.2.5 (Mean Value Theorem). *Let $f: [a, b] \to \mathbb{R}$ be such that (i) f is continuous on $[a, b]$ and (ii) f is differentiable on (a, b). Then there exists $c \in (a, b)$ such that*

$$f(b) - f(a) = f'(c)(b - a). \tag{4.6}$$

We rewrite (4.6) as $\frac{f(b)-f(a)}{b-a} = f'(c)$. Now the left side is the slope of the chord joining $(a, f(a))$ and $(b, f(b))$. As mentioned earlier, we may consider $f'(c)$ as the slope of the tangent of the tangent line at $(c, f(c))$ to the graph of f. Thus we arrive at the geometric interpretation: Under the given conditions, there exists c such that the slope of the tangent to the graph of f at c equals that of the chord joining the two points $(a, f(a))$ and $(b, f(b))$. Look at Figure 4.6.

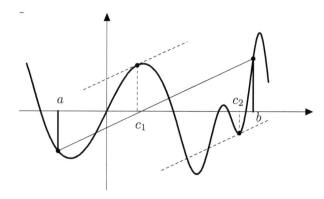

Figure 4.6: Mean value theorem.

Strategy: The basic idea is to apply Rolle's theorem. That means, we need to get a function $\varphi(x)$ such that $\varphi(a) = \varphi(b)$. One natural way to get this is to look at the difference between the graph of $f(x)$ and the chord joining the points $(a, f(a))$ and $(b, f(b))$. Note that the equation of the chord is, $\frac{y-f(a)}{f(a)-f(b)} = \frac{x-a}{a-b}$. Therefore, the equation of the chord is $\ell(x) = f(a) + \frac{f(b)-f(a)}{b-a}(x-a)$. That is, the function we would like to consider is $\varphi(x) := f(x) - \ell(x)$.

Proof. Consider $g(x) = f(x) - \ell(x)$, where $\ell(x) := f(a) + \frac{f(b)-f(a)}{b-a}(x-a)$. Clearly, $g(a) = g(b) = 0$ and g satisfies all conditions of Rolle's theorem. Hence there exists $c \in (a, b)$ such that $g'(c) = 0$. This implies,

$$f'(c) - \ell'(c) = 0 \implies f(b) - f(a) = f'(c)(b - a).$$

This completes the proof. □

The mean value theorem is also known as Lagrange's mean value theorem. It is the single most important result in the theory of differentiation. Below are some typical applications. Many beginners try to prove the first two applications below ab initio, that is, they try to deduce them from the definition of the derivative. Please keep it in mind that perhaps the only way to derive them is to use the mean value theorem.

Example 4.2.6 (Applications of MVT).

(1) Let J be an interval and $f\colon J \to \mathbb{R}$ be differentiable with $f'(x) = 0$ for all $x \in J$. Then f is a constant on J.

For any $x, y \in J$, by MVT, we have $f(y) - f(x) = f'(z)(y - x)$ for some z between x and y. Since $f'(z) = 0$, we get $f(y) = f(x)$ for all $x, y \in J$. This implies that f is a constant function. (Why? Fix $x_0 \in J$. Then for each $x \in J$, $f(x) = f(x_0)$.)

Is this result true if J is not an interval?

Note that the concept of differentiability of a function f can be defined on a set U, which is also the union of open intervals. However, it may happen that $f\colon U \to \mathbb{R}$ is differentiable with $f' = 0$ on U, but f is not a constant. For instance, consider $U = (-1, 0) \cup (0, 1)$ and $f(x) = -1$ on $(-1, 0)$ and $f(x) = 1$ on $(0, 1)$. Then $f'(x) = 0$ for all x, but f is not a constant function.

(2) Let J be an interval and $f\colon J \to \mathbb{R}$ be differentiable with $f'(x) > 0$ for $x \in J$. Then f is increasing on J.

For any $x, y \in J$ with $x < y$ by MVT, we have $f(y) - f(x) = f'(z)(y - x)$ for some z between x and y. Since $f'(z) > 0$ and $y - x > 0$, the right side is positive. This implies, if $x < y$, then $f(x) < f(y)$. Thus f is strictly increasing.

What is the corresponding result when $f'(x) < 0$ for $x \in J$? Formulate the result and prove it.

(3) The following application was already seen while discussing uniform continuity. Look at Exercise 3.7.7 on page 102.

Let $f\colon J \to \mathbb{R}$ be differentiable. Assume that there exists $M > 0$ such that $|f'(x)| \le M$. Then f is uniformly continuous on J.

We need to estimate $|f(x_1) - f(x_2)|$ for $x_1, x_2 \in J$. Since J is an interval by the MVT, there exists c in between x_1 and x_2 such that $f(x_1) - f(x_2) = f'(c)(x_1 - x_2)$. Hence

$$|f(x_1) - f(x_2)| = |f'(c)|\,|(x_1 - x_2)| \le M\,|x_1 - x_2|.$$

We conclude that f is Lipschitz and hence uniformly continuous. See Exercise 3.7.6.

Note that in each of the three applications we used the hypothesis that the domain of f is an interval J. If $x, y \in J$, then the interval $[x, y] \subset J$ so that the point c given by the mean value theorem lies in J. Hence our hypothesis on the derivative of f can be applied to $f'(c)$.

Remark 4.2.7. It should be noted that to conclude that f is increasing in Item 2 in Example 4.2.6 above, we required that $f' > 0$ *on J*.

A common mistake is that some beginners believe that if $f'(c) > 0$ then f is increasing in an interval of the form $(c - \delta, c + \delta)$. This is false. A counterexample is given in Item 11 in Exercise 4.2.16.

Exercise 4.2.8. Let $f\colon \mathbb{R} \to \mathbb{R}$ be such that $|f(x) - f(y)| \le (x - y)^2$ for all x, y. Show that f is differentiable, and the derivative is zero. Hence conclude that f is a constant.

The mean value theorem is quite useful in proving certain inequalities. Here are some samples.

Example 4.2.9. We have $e^x > 1 + x$ for all $x \in \mathbb{R}$.
Suppose $x > 0$. Consider the function $f(x) = e^x$ on the interval $[0, x]$. Since e^x is a differentiable on $\mathbb{R}$, we can apply mean value theorem to f on the interval $[0, x]$. Hence there exists $c \in (0, x)$ such that

$$e^x - e^0 = f'(c)(x - 0) = e^c x.$$

Note that $f'(x) = e^x > 1$ for $x > 0$. So the displayed equation yields $e^x - 1 = e^c x > x$.

If $x < 0$, then consider the interval $[x, 0]$.

Example 4.2.10. We have $\frac{y-x}{y} < \log \frac{y}{x} < \frac{y-x}{x}$, $0 < x < y$.
Let $0 < x < y$ and $f(x) = \log x$ on $[x, y]$. We know that $\log x$ is a differentiable function on $x > 0$. Hence using the MVT, there exists $c \in (x, y)$ such that

$$\log y - \log x = \frac{1}{c}(y - x) \implies \log \frac{y}{x} = (y - x)\frac{1}{c}.$$

Since $0 < x < c < y$, we have $\frac{1}{y} < \frac{1}{c} < \frac{1}{x}$. Hence we get

$$\frac{y-x}{y} < \log \frac{y}{x} = \frac{1}{c}(y-x) < \frac{y-x}{x}. \tag{4.7}$$

Exercise Set 4.2.11. Some standard applications.
(1) Prove that $e^x > ex$ for $x \in \mathbb{R}$.
(2) Prove that $\frac{x}{1+x} < \log(1+x) < x$, $x > 0$.
(3) Prove that $n(b-a)a^{n-1} < b^n - a^n < n(b-a)b^{n-1}$, $0 < a < b$.
(4) Show that $\sin x \le x$ for $x > 0$.
(5) Show that $0 < \frac{1}{x}\log\left(\frac{e^x-1}{x}\right) < 1$ for $x > 0$.
(6) Prove that $\frac{\sin x}{x}$ is strictly increasing on $(0, \pi/2)$.

Which is greater, e^π or π^e? We prove a more general inequality which answers this question:

Example 4.2.12. If $e \le a < b$, then $a^b > b^a$.
Using (4.7), we have

$$\frac{b-a}{b} < \log\left(\frac{b}{a}\right) < \frac{b-a}{a}.$$

Since $a \log\left(\frac{b}{a}\right) < b - a$, we have $\frac{b^a}{a^a} = e^{a\log(b/a)} < e^{b-a}$. That is, $b^a < e^{b-a}a^a$. If $e \le a$, then $e^t \le a^t$ for $t \ge 0$ and hence we conclude that $b^a < e^{b-a}a^a \le a^{b-a}a^a = a^b$.

Theorem 4.2.13 (Inverse Function Theorem). *Let $f: I := (a,b) \to \mathbb{R}$ be continuously differentiable with $f'(x) \ne 0$ for all x. Then (i) f is strictly monotone. (ii) $f(I) = J$ is an interval and (ii) $g := f^{-1}$ is (continuous and) differentiable on the interval $J := f((a,b))$ and we have*

$$g'(f(x)) = \frac{1}{f'(x)} = \frac{1}{f'(g(y))} \text{ for all } x = g(y) \in [a,b].$$

Proof. Since f' is continuous on I, by the intermediate value theorem exactly one of the following holds: either $f' > 0$ or $f' < 0$ on I. (Why?) Hence f is strictly monotone on I. By Proposition 3.5.4, J is an interval. Note that by Corollary 3.5.6, the inverse function g is continuous on J. Fix $c \in (a,b)$. Let $d := f(c)$. Then

$$\frac{g(y)-g(d)}{y-d} = \frac{x-c}{f(x)-f(c)} = \frac{1}{\frac{f(x)-f(c)}{x-c}}. \tag{4.8}$$

Since g is continuous, if (y_n) is a sequence in J converging to d, then $x_n := g(y_n) \to c$. Hence

$$\lim_{n\to\infty} \frac{g(y_n)-g(d)}{y_n-d} = \lim_{n\to\infty} \frac{1}{\frac{f(x_n)-f(c)}{x_n-c}} = \frac{1}{f'(c)}.$$

Since this is true for any sequence (y_n) in J converging to d, we conclude that

$$\lim_{y \to d} \frac{g(y) - g(d)}{y - d} = \frac{1}{f'(c)}.$$

One may also take $\lim_{y \to d}$ in (4.8) and observe that $y \to d$ iff $x \to c$, thanks to the continuity of f and g.   □

Exercise 4.2.14. Let $f \colon (0, \infty) \to (0, \infty)$ be defined by $f(x) = x^{1/n}$. Use the inverse function theorem to compute the derivative of f.

Theorem 4.2.15 (Cauchy's Form of MVT). *Let* $f, g \colon [a, b] \to \mathbb{R}$ *be differentiable. Assume that* $g'(x) \neq 0$ *for any* $x \in (a, b)$. *Then there exists* $c \in (a, b)$ *such that*

$$\frac{f(b) - f(a)}{g(b) - g(a)} = \frac{f'(c)}{g'(c)}. \tag{4.9}$$

Geometrically, Cauchy's form of MVT means the following:

We look at the map $t \mapsto (g(t), f(t))$ from J to $\mathbb{R}^2$ as a parameterized curve in the plane. For example, $t \mapsto (\cos t, \sin t)$, $t \in [0, 2\pi]$ is a parameterization of a circle. If $f \colon [a, b] \to \mathbb{R}$ is a function, then $t \mapsto (t, f(t))$ is a parameterization of the graph of the function.

Then the slope of the chord joining the points $(g(a), f(a))$ and $(g(b), f(b))$ is $\frac{f(b) - f(a)}{g(b) - g(a)}$. The tangent "vector" to the parameterized curve at a point $(g(c), f(c))$ is $(g'(c), f'(c))$ and hence the tangent line at c has the slope $f'(c)/g'(c)$. Thus Cauchy's mean value theorem says that there exists a point $t_0 \in (a, b)$ such that slope $f'(t_0)/g'(t_0)$ of the tangent to the curve at t_0 is equal to the slope of the chord joining the end points of the curve. Look at Figure 4.7.

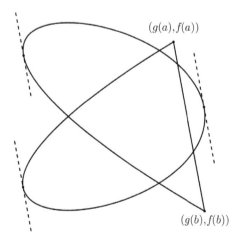

Figure 4.7: Cauchy mean value theorem.

Strategy: The basic idea is to use Rolle's theorem. That is, we wish to get a function $h(x)$ satisfying conditions of Rolle's theorem and that $h'(c) = 0$ gives $f'(c)/g'(c)$. If we look at $h(x) = f(x) - \lambda g(x)$ and find λ so that $h(a) = h(b)$.

Proof. Note that $g(a) \neq g(b)$. (Why?) Otherwise, by Rolle's theorem there exists $c \in (a, b)$ such that $g'(c) = 0$. This is a contradiction.

Let $h(x) := f(x) - \lambda g(x)$ where $\lambda \in \mathbb{R}$ is chosen so that $h(b) = h(a)$. It is easy to see that $\lambda = \frac{f(b)-f(a)}{g(b)-g(a)}$. It is clear that h satisfies the condition of Rolle's theorem on $[a, b]$. By Rolle's theorem, there exists $c \in (a, b)$ such that $h'(c) = 0$. This implies $\lambda = \frac{f'(c)}{g'(c)}$. Hence there exists $c \in (a, b)$ such that there exists $c \in (a, b)$ such that

$$\frac{f(b) - f(a)}{g(b) - g(a)} = \frac{f'(c)}{g'(c)}. \qquad \square$$

As observed in the remark on the geometric interpretation, if $g(x) = x$ in the Cauchy mean value theorem, it reduces to Lagrange mean value theorem.

Exercise Set 4.2.16.

(1) Let $P(X) := \sum_{k=0}^{n} a_k X^k$, $n \geq 2$ be a real polynomial. Assume that all the roots of P lie in $\mathbb{R}$. Show that all the roots of its derivative $P'(X)$ also are real.

(2) Does there exist a *differentiable* function $f \colon \mathbb{R} \to \mathbb{R}$ such that $f'(x) = 0$ if $x < 0$ and $f'(x) = 1$ if $x > 0$?

(3) Let $f \colon \mathbb{R} \to \mathbb{R}$ be differentiable such that $|f'(x)| \leq M$ for some $M > 0$ for all $x \in \mathbb{R}$.

   (a) Show that f is uniformly continuous on $\mathbb{R}$.

   (b) If $\varepsilon > 0$ is sufficiently small, then show that the function $g_\varepsilon(x) := x + \varepsilon f(x)$ is one-one.

(4) Let $f \colon (a, b) \to \mathbb{R}$ be differentiable at $x \in (a, b)$. Prove that

$$\lim_{h \to 0} \frac{f(x + h) - f(x - h)}{2h} = f'(x).$$

Give an example of a function where the limit exists but the function is not differentiable.

(5) Let $f \colon [0, 2] \to \mathbb{R}$ be given by $f(x) := \sqrt{2x - x^2}$. Show that f satisfies the conditions of Rolle's theorem. Find a c such that $f'(c) = 0$.

(6) Use MVT to establish the following inequalities:

   (a) Let $b > a > 0$. Show that $b^{1/n} - a^{1/n} < (b - a)^{1/n}$.

(b) Show that $|\sin x - \sin y| \leq |x - y|$.

(c) Show that

$$nx^{n-1}(y - x) \leq y^n - x^n \leq ny^{n-1}(y - x) \text{ for } 0 \leq x \leq y.$$

(d) **Bernoulli's Inequality.** Let $\alpha > 0$ and $h \geq -1$. Then

$$(1 + h)^\alpha \leq 1 + \alpha h, \qquad\qquad \text{for } 0 < \alpha \leq 1, \qquad (4.10)$$
$$(1 + h)^\alpha \geq 1 + \alpha h, \qquad\qquad \text{for } \alpha \geq 1. \qquad (4.11)$$

(7) Assume that $f \colon (a, b) \to \mathbb{R}$ is differentiable on (a, b) except possibly at $c \in (a, b)$. Assume that $\lim_{x \to c} f'(x)$ exists. Prove that $f'(c)$ exists and f' is continuous at c.

(8) Show that the function $f(x) = x^3 - 3x^2 + 17$ is not one-one on the interval $[-1, 1]$.

(9) Prove that the equation $x^3 - 3x^2 + b = 0$ has at most one root in the interval $[0, 1]$.

(10) Show that $\cos x = x^3 + x^2 + 4x$ has exactly one root in $[0, \pi/2]$.

(11) Let $f(x) = x + 2x^2 \sin(1/x)$ for $x \neq 0$ and $f(0) = 0$. Show that $f'(0) = 1$ but f is not monotonic in any interval around 0.

(12) Let J be an open interval and $f, g \colon J \to \mathbb{R}$ be differentiable. Assume that $f(a) = 0 = f(b)$ for $a, b \in J$ with $a < b$. Show that $f'(c) + f(c)g'(c) = 0$ for some $c \in (a, b)$.

(13) Let $f, g \colon \mathbb{R} \to \mathbb{R}$ be differentiable. Assume that $f(0) = g(0)$ and $f'(x) \leq g'(x)$ for all $x \in \mathbb{R}$. Show that $f(x) \leq g(x)$ for $x \geq 0$.

(14) Let $f \colon \mathbb{R} \to \mathbb{R}$ be differentiable. Assume that $1 \leq f'(x) \leq 2$ for $x \in \mathbb{R}$ and $f(0) = 0$. Prove that $x \leq f(x) \leq 2x$ for $x \geq 0$.

(15) Let $f, g \colon \mathbb{R} \to \mathbb{R}$ be differentiable. Let $a \in \mathbb{R}$. Define $h(x) = f(x)$ for $x < a$ and $h(x) = g(x)$ for $x \geq a$. Find necessary and sufficient conditions which will ensure that h is differentiable at a. (This is a gluing lemma for differentiable functions.)

(16) Let $f \colon [2, 5] \to \mathbb{R}$ be continuous and be differentiable on $(2,5)$. Assume that $f'(x) = (f(x))^2 + \pi$ for all $x \in (2, 5)$. True or false: $f(5) - f(2) = 3$.

(17) Let $f \colon (0, \infty) \to \mathbb{R}$ be differentiable. If $f'(x) \to \ell$ as $x \to \infty$, then show that $f(x)/x \to \ell$ as $x \to \infty$.

(18) Let $f \colon (a, b) \to \mathbb{R}$ be differentiable. Assume that $\lim_{x \to a+} f(x) = \lim_{x \to b-} f(x)$. Show that there exists $c \in (a, b)$ such that $f'(c) = 0$.

(19) Let $f: (0,1] \to \mathbb{R}$ be differentiable with $|f'(x)| < 1$. Define $a_n := f(1/n)$. Show that (a_n) converges.

(20) Let $f: [a,b] \to \mathbb{R}$ be continuous and differentiable on (a,b). Assume further that $f(a) = f(b) = 0$. Prove that for any given $\lambda \in \mathbb{R}$, there exists $c \in (a,b)$ such that $f'(c) = \lambda f(c)$.

(21) Let $f, g: [a,b] \to \mathbb{R}$ be continuous and differentiable on (a,b). Assume further that $f(a) = f(b) = 0$. Prove that for any given $\lambda \in \mathbb{R}$, there exists $c \in (a,b)$ such that $f'(c) + g'(c)f(c) = 0$.

(22) Show that $f(x) := x\,|x|$ is differentiable for all $x \in \mathbb{R}$. What is $f'(x)$? Is f' continuous? Does f'' exist?

(23) Let $f: \mathbb{R} \to \mathbb{R}$ be differentiable with $f(0) = -3$. Assume that $f'(x) \le 5$ for $x \in \mathbb{R}$. How large can $f(2)$ possibly be?

(24) Let $f: \mathbb{R} \to \mathbb{R}$ be differentiable with $f(1) = 10$ and $f'(x) \ge 2$ for $1 \le x \le 4$. How small can $f(4)$ possibly be?

(25) Let $f(x) = 1/x$ for $x \ne 0$ and $g(x) = \begin{cases} 1/x, & \text{if } x > 0 \\ 1 + (1/x), & \text{if } x < 0. \end{cases}$
Let $h = f - g$. Then $h' = 0$ but h is not a constant. Explain.

(26) Let $f: [0,1] \to \mathbb{R}$ be continuous. Assume that $f'(x) \ne 0$ for $x \in (0,1)$. Show that $f(0) \ne f(1)$.

(27) Show that on the graph of any quadratic polynomial f the chord joining the points $(a, f(a))$ and $(b, f(b))$ is parallel to the tangent line at the midpoint of a and b.

(28) Let $n = 2k - 1 \in \mathbb{N}$. Let $f(x) = x^n$ for $x \in \mathbb{R}$. Show that f maps $\mathbb{R}$ bijectively onto itself.

(29) Let $f(x) = x^{2k} + ax + b$, $k \in \mathbb{N}$, $a, b \in \mathbb{R}$. Show that f has at most two zeros in $\mathbb{R}$.

(30) Let $f: (0, \infty) \to \mathbb{R}$ be differentiable. Assume that $f(0) = 0$ and f' is increasing. Prove that $f(x)/x$ is increasing.

(31) Let $f: [a,b] \to \mathbb{R}$ be differentiable. Assume that there exists no $x \in [a,b]$ such that $f(x) = 0 = f'(x)$. Prove that the set $\{t \in [a,b] : f(t) = 0\}$ of zeros of f is finite.

(32) Let $f: [a,b] \to [a,b]$ be differentiable. Assume that $f'(x) \ne 1$ for $x \in [a,b]$. Prove that f has a unique fixed point in $[a,b]$.

4.3   L'Hospital's Rules

As an application to the Cauchy mean value theorem, we look at L'Hospital's Rules.

Theorem 4.3.1 (L'Hospital's Rule). *Let J be an open interval. Let either $a \in J$ or a is an endpoint of J. (Note that it may happen that $a = \pm\infty$!) Assume that*
(i) *$f, g \colon J \setminus \{a\} \to \mathbb{R}$ is differentiable,*
(ii) *$g(x) \neq 0 \neq g'(x)$ for $x \in J \setminus \{a\}$, and*
(iii) *$A := \lim_{x \to a} f(x) = \lim_{x \to a} g(x)$ where A is either 0 or ∞.*
   Assume that $B := \lim_{x \to a} \frac{f'(x)}{g'(x)}$ exists either in $\mathbb{R}$ or $B = \pm\infty$. Then

$$\lim_{x \to a} \frac{f(x)}{g(x)} = \lim_{x \to a} \frac{f'(x)}{g'(x)} \equiv B.$$

Strategy: The trick is to to bring the expression $f(x)/g(x)$ to the form $\frac{f(x)-f(c)}{g(x)-g(c)}$ so that Cauchy's mean value theorem can be applied. The proofs below show a clever way of achieving this. They are worth going through a couple of times to master the tricks of analysis.

Proof. We attend to a simple case first where $A = 0$, $a \in \mathbb{R}$, and $B \in \mathbb{R}$.
   Set $f(a) = 0 = g(a)$. Then f and g are continuous on J. Let (x_n) be a sequence in J such that either $x_n > a$ or $x_n < a$ for all $n \in N$ and $x_n \to a$. By Cauchy's MVT, there exists c_n between a and x_n such that

$$\frac{f(x_n) - f(a)}{g(x_n) - g(a)} = \frac{f'(c_n)}{g'(c_n)}.$$

Since $f(a) = 0 = g(a)$, it follows that

$$\frac{f(x_n)}{g(x_n)} = \frac{f(x_n) - f(a)}{g(x_n) - g(a)} = \frac{f'(c_n)}{g'(c_n)}.$$

Clearly, $c_n \to a$. By hypothesis, the sequence $f'(c_n)/g'(c_n) \to B$ and hence the result.
   Let us now look at the case when $A = \infty$. Write $h(x) = f(x) - Bg(x)$, $x \in J \setminus \{a\}$. Then $h'(x) = f'(x) - Bg'(x)$ so that

$$\lim_{x \to a} \frac{h'(x)}{g'(x)} = 0.$$

We want to show that $\lim_{x \to a} \frac{h(x)}{g(x)} = 0$. Let $\varepsilon > 0$ be given. Then there exists $\delta_1 > 0$ such that

$$g(x) > 0 \quad \text{and} \quad \left| \frac{h'(x)}{g'(x)} \right| < \frac{\varepsilon}{2} \text{ for } x \in (a, a + \delta_1]. \qquad (4.12)$$

If $x \in (a, a + \delta_1)$, then by Cauchy's mean value theorem,

$$\frac{h(x) - h(a + \delta_1)}{g(x) - g(a + \delta_1)} = \frac{h'(c_x)}{g'(c_x)} \text{ for some } c_x \in (x, a + \delta_1). \tag{4.13}$$

From (4.12)–(4.13), we get

$$\left| \frac{h(x) - h(a + \delta_1)}{g(x) - g(a + \delta_1)} \right| < \frac{\varepsilon}{2} \text{ for } x \in (a, a + \delta_1). \tag{4.14}$$

Since $\lim_{x \to a} g(x) = \infty$, there exists $\delta_2 < \delta_1$ such that

$$g(x) > g(a + \delta_1) \text{ for } x \in (a, a + \delta_2). \tag{4.15}$$

From (4.12) and (4.15), we deduce

$$0 < g(x) - g(a + \delta_1) < g(x), \text{ for } x \in (a, a + \delta_2). \tag{4.16}$$

From (4.14) and (4.16), we get

$$\frac{|h(x) - h(a + \delta_1)|}{g(x)} < \frac{|h(x) - h(a + \delta_1)|}{g(x) - g(a + \delta_1)} < \frac{\varepsilon}{2}, \text{ for } x \in (a, a + \delta_2). \tag{4.17}$$

Now choose $\delta_3 < \delta_2$ so that

$$\frac{|h(a + \delta_1)|}{g(x)} < \frac{\varepsilon}{2} \text{ for } x \in (a, a + \delta_3). \tag{4.18}$$

Algebra gives us

$$\frac{h(x)}{g(x)} = \frac{h(x) - h(a + \delta_1)}{g(x)} + \frac{h(a + \delta_1)}{g(x)}.$$

Using this, if $x \in (a, a + \delta_3)$, we have

$$\left| \frac{h(x)}{g(x)} \right| \leq \frac{|h(x) - h(a + \delta_1)|}{g(x)} + \frac{|h(a + \delta_1)|}{g(x)}. \tag{4.19}$$

Hence by (4.17) and (4.18)

$$\left| \frac{h(x)}{g(x)} \right| < \varepsilon, \text{ for } x \in (a, a + \delta_3). \tag{4.20}$$

(4.20) says that $\lim_{x \to a} \frac{h(x)}{g(x)} = 0$. Since

$$\frac{f(x)}{g(x)} = \frac{h(x)}{g(x)} + B,$$

the result follows.

The other cases are left to the reader as instructive exercises. $\square$

Our proof above follows the exposition in [4].

Example 4.3.2. We now give a few typical applications. It is important that the reader should keep in mind that the conclusions say something about the behavior of functions such as which of the two goes to zero or to infinity faster/slower. Analysis most often deals with the comparison of the behavior of functions.

(1) $f(x) = \log x$ and $g(x) = x$ for $x > 0$. We know that $\lim_{x \to \infty} f(x) = \infty$ and $\lim_{x \to \infty} g(x) = \infty$. Also,

$$\lim_{x \to \infty} \frac{f'(x)}{g'(x)} = 0 \text{ so that } \lim_{x \to \infty} \frac{\log x}{x} = 0.$$

(2) Repeated application of L'Hospital's rule yields

$$\lim_{x \to \infty} \frac{x^n}{e^x} = \lim_{x \to \infty} \frac{nx^{n-1}}{e^x} = \cdots = \lim_{x \to \infty} \frac{n!}{e^x} = 0.$$

(3) Recall that $\lim_{x \to 0+} f(1/x) = \lim_{x \to \infty} f(x)$. (Why?) From the last item we conclude

$$\lim_{x \to 0+} x^{-n} e^{-1/x} = 0.$$

(4) The last example gives rise to an interesting example of a function. Consider

$$f(x) := \begin{cases} e^{-1/x} & \text{for } x > 0 \\ 0 & \text{for } x \le 0. \end{cases}$$

Then we have from the last item,

$$f'(0) = \lim_{x \to 0+} \frac{g(x) - g(0)}{x} = \lim_{x \to 0+} \frac{e^{-1/x}}{x} = 0.$$

(5)

$$g(x) := \begin{cases} e^{-1/x^2} & \text{for } x \ne 0 \\ 0 & \text{for } x = 0. \end{cases}$$

Then

$$g'(0) = \lim_{x \to 0+} \frac{g(x) - g(0)}{x} = \lim_{x \to 0+} \frac{e^{-1/x^2}}{x} = 0.$$

Theorem 4.3.3 (Darboux Theorem). *Let $f : [a, b] \to \mathbb{R}$ be differentiable. Assume that $f'(a) < \lambda < f'(b)$. Then there exists $c \in (a, b)$ such that $f'(c) = \lambda$. (Thus, though f' need not be continuous, it enjoys the intermediate value property.)*

Proof. Let $f'(a) < \lambda < f'(b)$ and consider $g(x) = f(x) - \lambda x$. Then g is differentiable on $[a, b]$. Note that $g'(a) < 0$ and $g'(b) > 0$.

It attains a global minimum at some $c \in [a, b]$. We claim that c cannot be any of the endpoints. For, if $c = a$, then $g(a + h) - g(a) \geq 0$ for $h > 0$ and hence

$$g'(a) = \lim_{h \to 0+} \frac{g(a + h) - g(a)}{h} \geq 0.$$

This contradicts our observation that $g'(a) < 0$. Similarly, if $c = b$, then $g(b - h) - g(b) \geq 0$ so that the difference quotient $\frac{g(b-h)-g(b)}{-h} \leq 0$. Hence we conclude that $g'(b) = \lim_{h \to 0-} \frac{g(b+h)-g(b)}{h} \leq 0$. This contradicts the fact that $g'(b) > 0$. Hence we conclude that $a < c < b$. Hence c is a local minimum for the differentiable function g. It follows that $g'(c) = 0$, that is, $f'(c) = \lambda$. $\square$

Remark 4.3.4. Why can't we work with a maximum? How does the argument go if we assume $f'(b) < \lambda < f'(a)$?

Exercise 4.3.5. The proof above says something about the derivative of functions at the endpoints, if the endpoints turn out to be points of extrema of f. Can you derive them?

To make sure that the Darboux theorem can be applied to a larger class of functions, we look at some functions which are differentiable and whose derivatives are not continuous.

Example 4.3.6. Define

$$f(x) = \begin{cases} x^2 \sin \frac{1}{x}, & x \neq 0 \\ 0, & x = 0. \end{cases}$$

Then f is differentiable at all points including 0. Then

$$f'(x) = \begin{cases} 2x \sin \frac{1}{x} - \cos \frac{1}{x}, & x \neq 0 \\ 0, & x = 0. \end{cases}$$

It is easy to see that f' is not continuous. Look at Figures 4.8–4.9 for the graphs of f and f'. For example, if we consider $x_n = \frac{1}{n\pi}$, then $x_n \to 0$. But $\cos x_n = \cos n\pi = (-1)^n$ is not convergent. Hence we conclude that f' is not continuous at $x = 0$.

According to Darboux theorem, f' enjoys the intermediate value property, even though it is not continuous.

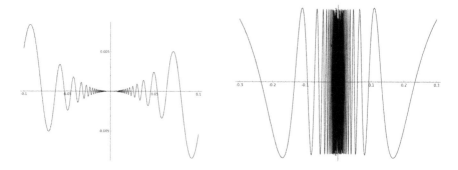

Figure 4.8: Example 4.3.6, graph of f. Figure 4.9: Example 4.3.6, graph of f'.

Exercise Set 4.3.7.

(1) What are all the differentiable functions $f: [0,1] \to \mathbb{R}$ the slopes of the tangents to their graphs are always rational?

(2) Let $f: (a,b) \to \mathbb{R}$ be differentiable. Assume that $f'(x) \neq 0$ for $x \in (a,b)$. Show that f is monotone on (a,b).

4.4   Higher-order Derivatives

Let J be an interval. Let $f: J \to \mathbb{R}$ be differentiable on J. Since the derivative at $c \in J$ is unique, we have a (well-defined) function $f': J \to \mathbb{R}$ given by $x \mapsto f'(x)$. Let $c \in J$. If $f': J \to \mathbb{R}$ is differentiable at c, then we say that f has a second-order derivative at c and it is denoted by $f''(c)$ or by $f^{(2)}(c)$. If $f''(c)$ exists for all $c \in J$, we say that f is twice differentiable on J. The number $f''(c)$ is called the second derivative of f at c.

Note that to talk of f being twice differentiable at c, it is *not* enough if f is differentiable at c. It has to be differentiable in an interval around c.

How do we define thrice differentiable and third derivative of f?

Assume that we have defined $f^{(n-1)}$, $(n-1)$-th order derivative of f on J. We then say f is n-times differentiable on J if the function $f^{(n-1)}$ is differentiable on J. If it exists, we denote the derivative of $f^{(n-1)}$ by $f^{(n)}$. The number $f^{(n)}(c)$ is called the n-th derivative of f at c.

Note that if f'' exists, then f' is continuous. We say f is C^1 on J and write $f \in C^1(J)$ if f is differentiable and f' is continuous. Such a function is said to be *continuously differentiable*. (Do you recall an example of a function which is differentiable but its derivatives are not continuous?)

How do we define n-times continuously differentiable? If $f^{(n)}$ exists on J and if it is continuous on J, we say that f is n-times continuously differentiable on J and denote it by $f \in C^n(J)$. We also say that f is a C^n-function or a function of class C^n.

If f is a function all of whose derivatives exist, it is called an *infinitely differentiable* function, or a *smooth function*. Note that if f is smooth, then it is C^k for all $k \in \mathbb{N}$. Because of this, one also says that f is C^∞ when it is smooth. (Note that we do not define the infinite-th derivative of f!)

Are there C^∞ functions? Yes, constants, more generally polynomials, the transcendental functions such the exponential, logarithmic, trigonometric, and hyperbolic functions in their respective domains.

Let us look at some examples.

Example 4.4.1.

(1) Any polynomial function is a smooth function.

(2) e^x, $\sin x$, and $\cos x$ are smooth functions.

(3) Let $f(x) := \begin{cases} x^n, & x > 0 \\ 0, & x \leq 0. \end{cases}$

Then f is $(n-1)$-times differentiable at $x = 0$. However, $f^{(n)}(0)$ does exist.

(4) Let $f(x) := \begin{cases} x^n \sin(1/x), & x \neq 0 \\ 0, & x = 0. \end{cases}$

Then $f^{(k)}(0)$ exists for all $k < n$, but $f^{(n)}(0)$ does not exist.

(5) Let $f(x) = \begin{cases} x^2, & \text{if } x < 0 \\ x^3, & \text{if } x \geq 0. \end{cases}$

Then $f'(0)$ exists but $f''(0)$ does not.

(6) Let $f(x) := |x|$. Define $g_1(x) := \int_0^x f(t)\, dt$. Then, by the fundamental theorem of calculus, g_1 is differentiable with derivative $g_1'(x) = f(x)$. Define recursively, $g_n(x) := \int_0^x g_{n-1}(t)\, dt$. Then g_n is n-times continuously differentiable, but not $(n+1)$-times differentiable.

(7) The function in item 5 of Example 4.3.2 is infinitely differentiable with $g^{(n)}(0) = 0$. See Example 4.5.3 on page 139.

Exercise 4.4.2. Let $m, n \in \mathbb{N}$. Consider the function

$$f(x) := \begin{cases} x^n, & x > 0 \\ x^m, & x \leq 0. \end{cases}$$

Discuss its higher-order differentiability.

Proposition 4.4.3 (Leibniz Formula). *If $h = fg$ is a product of two functions with derivatives up to order n, then*

$$h^{(n)}(x) = \sum_{k=0}^{n} \binom{n}{k} f^{(k)}(x) g^{(n-k)}(x). \tag{4.21}$$

Proof. We use induction on n, the order of derivative. We know from the product rule of derivative

$$h'(x) = f(x)g'(x) + f'(x)g'(x) = \binom{1}{0}f^{(0)}(x)g^{(1)}(x) + \binom{1}{1}f^{(1)}(x)g^{(0)}(x).$$

This means that the result is true for $n = 1$. Let $n = 2$.

$$h''(x) = (f(x)g'(x) + f'(x)g'(x))' = f(x)g''(x) + 2f'(x)g'(x) + f''(x)g(x)$$

$$= \binom{2}{0}f(x)g''(x) + \binom{2}{1}f'(x)g'(x) + \binom{2}{2}f''(x)g(x).$$

Thus the result is also true for $n = 2$. Let us assume that the Leibniz rule is true for n.

We shall prove it for $n + 1$. Let $h = fg$ with f and g both have derivatives of order $n + 1$.

$$h^{(n+1)}(x) = (h^{(n)})'(x)$$

$$= \left(\sum_{k=0}^{n}\binom{n}{k}f^{(k)}(x)g^{(n-k)}(x)\right)' \quad \text{by induction}$$

$$= \sum_{k=0}^{n}\binom{n}{k}\left(f^{(k)}(x)g^{(n-k)}(x)\right)'$$

$$= \sum_{k=0}^{n}\binom{n}{k}\left[f^{(k+1)}(x)g^{(n-k)}(x) + f^{(k)}(x)g^{(n-k+1)}(x)\right]$$

$$= \sum_{k=0}^{n}\binom{n}{k}\left[f^{(k+1)}(x)g^{(n-k)}(x)\right]$$

$$+ \sum_{k=0}^{n}\binom{n}{k}\left[f^{(k)}(x)g^{(n-k+1)}(x)\right]. \tag{4.22}$$

Here onward, the proof is exactly similar to that of the binomial theorem for an integral exponent. We advise the reader to go through it and complete the proof of the Leibniz formula. □

Example 4.4.4. Let $y = x^2 e^{kx}$. Use Leibniz theorem to find the y_n. For $n > 2$,

$$y_n = \binom{n}{0}\frac{d^n}{dx^n}(e^{kx})x^2 + \binom{n}{1}\frac{d^{n-1}}{dx^{n-1}}(e^{kx})2x + \binom{n}{1}\frac{d^{n-2}}{dx^{n-2}}(e^{kx})2 + 0$$

$$= k^n e^{kx}x^2 + 2nk^{n-1}e^{kx}x + n(n-1)k^{n-2}e^{kx}.$$

4.5 Taylor's Theorem

The simplest functions which we can easily construct and work with are polynomial functions of the form

$$f(x) = a_0 + a_1 x + \cdots a_n x^n, \quad a_i \in \mathbb{R}.$$

The next class of functions are $|x|$ and x^α where $\alpha \in \mathbb{Q}$. The transcendental functions such as exponential, trigonometric, and hyperbolic functions need a lot of analysis even to define them rigorously. We would like to approximate a function $f \colon \mathbb{R} \to \mathbb{R}$ by means of polynomial functions. We already know that if the function is differentiable at a, we can approximate $f(x)$ for x near a by a first-degree polynomial, namely, $f(a) + f'(a)(x - a)$. We also have control over the error. Taylor's theorem extends this result to $(n+1)$-times differentiable functions and approximates them by means of an n-th degree polynomial function.

Theorem 4.5.1 (Taylor's Theorem). *Assume that $f \colon [a, b] \to \mathbb{R}$ is such that $f^{(n)}$ is continuous on $[a, b]$ and $f^{(n+1)}(x)$ exists on (a, b). Fix $x_0 \in [a, b]$. Then for each $x \in [a, b]$ with $x \neq x_0$, there exists c between x and x_0 such that*

$$f(x) = f(x_0) + \sum_{k=1}^{n} \frac{(x - x_0)^k}{k!} f^{(k)}(x_0) + \frac{(x - x_0)^{n+1}}{(n + 1)!} f^{(n+1)}(c). \qquad (4.23)$$

Strategy: Proof of any version of Taylor's theorem involves some trick. The basic idea is to apply Rolle's theorem to a suitably defined function on the interval $[x, x_0]$ or $[x_0, x]$. The function is obtained by replacing x_0 by t on the right-side expression and choosing the coefficient M of $(x - t)^{n+1}$ so that Rolle's theorem can be applied.

Proof. Define

$$F(t) := f(t) + \sum_{k=1}^{n} \frac{(x - t)^k}{k!} f^{(k)}(t) + M(x - t)^{n+1},$$

where M is chosen so that $F(x_0) = F(x) \equiv f(x)$. That is,

$$M = \frac{1}{(x - x_0)^{n+1}} \left(f(x) - f(x_0) - \sum_{k=1}^{n} \frac{(x - x_0)^k}{k!} f^{(k)}(x_0) \right).$$

This is possible since $x \neq x_0$.

Note that the smallest interval, (either $[x, x_0]$ or $[x_0, x]$), containing x and x_0 lies inside $[a, b]$, since x, x_0 are points of an interval. Clearly, F is continuous on $[a, b]$, differentiable on (a, b) and $F(x) = f(x) = F(x_0)$.

Hence by Rolle's theorem applied to the interval $[x, x_0]$ or $[x_0, x]$, as the case may be, defined by x and x_0, there exists $c \in (a, b)$ such that

$$0 = F'(c) = \frac{(x - c)^n}{n!} f^{(n+1)}(c) - (n + 1)M(x - c)^n.$$

Thus, $M = \frac{f^{(n+1)}(c)}{(n+1)!}$. Hence

$$f(x) = F(x_0) = f(x_0) + \sum_{k=1}^{n} \frac{(x - x_0)^k}{k!} f^{(k)}(x_0) + \frac{(x - x_0)^{n+1}}{(n + 1)!} f^{(n+1)}(c).$$

This is what we wanted. $\qquad\square$

The right-hand side of (4.23) is called the n-th order (or n-th degree) Taylor expansion of the function f at x_0. The expression

$$f(x_0) + \sum_{k=1}^{n} \frac{f^{(k)}(x_0)}{k!}(x - x_0)^k$$

is called the n-th degree *Taylor polynomial* of f at x_0.

The term $\frac{f^{(n)}(c)}{n!}(x - x_0)^{n+1}$ is called the remainder term in the Taylor expansion. It is usually denoted by R_n. This form of the remainder is the simplest and is known as the *Lagrange's form*. There are two other forms which are more useful: Cauchy's form and the integral form of the remainder. We shall establish the Cauchy form of remainder at the end of the chapter while the integral form will be derived in Chapter 6.

The remainder term is the "error term" if we wish to approximate f near x_0 by the n-th order Taylor polynomial. If we assume that $f^{(n+1)}$ is bounded, say, by M on (a, b), then R_n goes to zero much faster than $(x - x_0)^n \to 0$, since $\left| \frac{R_n(x)}{(x-x_0)^n} \right| \leq \frac{M}{(n+1)!} |x - x_0|$.

Illustration of Taylor's approximation. Look at Figure 4.10. Notice that if the degree of Taylor's polynomial is higher, the approximation is better.

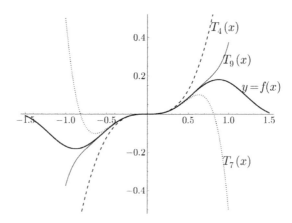

Figure 4.10: Illustration of Taylor's series expansions.

Example 4.5.2. We now find the Taylor polynomials for some of the standard functions.

(1) Let $f(x) = x^n$. Let $a \in \mathbb{R}$. Let us look at the Taylor expansion of $f(x)$ about a. Since it has to be a polynomial in powers of $x - a$, we think of the binomial expansion

$$x^n = (x - a + a)^n = a^n + \sum_{k=1}^{n} \binom{n}{k} a^{n-k}(x - a)^k.$$

Observe that the coefficients of $(x-a)^k$ are precisely $f^{(k)}(a)$. Can the reader write down the m-th degree Taylor's polynomial of $f(x) = x^n$ around a?

(2) Consider $f(x) = e^x$. Let us find the n-th degree Taylor polynomial of $f(x) = e^x$ about $x = 0$.

Note that $f^{(n)}(x) = e^x$ for all $n \in \mathbb{N}$, therefore, $f^{(n)}(0) = 1$ for all n. Hence the n-th degree Taylor's polynomial of $f(x) = e^x$ about $x = 0$ is given by $1 + \frac{x}{1!} + \frac{x^2}{2!} + \cdots + \frac{x^n}{n!}$.

(3) Let $f(x) = \sin x$. It is easy to see that

$$f^{(n)}(0) = \begin{cases} 0, & \text{when } n \text{ is even} \\ 1, & \text{when } n = 1, 5, 9, 13, \ldots, 4k+1 \\ -1, & \text{when } n = 3, 7, 11, 15, \ldots, 4k-1. \end{cases}$$

Hence the n-th degree Taylor's polynomial of $\sin x$ about $x = 0$ is given by

$$T_n(x) = f(0) + \sum_{k=0}^{n} \frac{f^{(n)}(0)}{k!} x^k$$

$$= x - \frac{x^3}{3!} + \frac{x^5}{5!} - \cdots + \varepsilon \frac{x^p}{p!},$$

where p is n if n is odd and is $n-1$ when n is even and where $\varepsilon = +1$ when $n = 4k+1$ and is -1 when $n = 4k-1$.

(4) Let $f(x) := \log(1+x)$ for $x \in (-1, \infty)$. Again, we shall find the n-degree Taylor's polynomial of f about $x = 0$. We have

$$f'(x) = \frac{1}{1+x}, \quad f''(x) = \frac{-1}{(1+x)^2}, \quad f'''(x) = \frac{2}{(1+x)^3}.$$

By induction, it is easy to show that

$$f^{(n)}(x) = \frac{(-1)^{n-1}(n-1)!}{(1+x)^n}.$$

Hence the n-th degree Taylor's polynomial of $\log(1+x)$ about $x = 0$ is given by

$$T_n(x) = f(0) + \sum_{k=0}^{n} \frac{f^{(n)}(0)}{k!} x^k$$

$$= x - \frac{x^2}{2} + \frac{x^3}{3} - \frac{x^4}{4} + \cdots \pm \frac{x^n}{n}.$$

Example 4.5.3. We now construct a very interesting C^∞ function which is very useful in higher aspects of analysis. It is also significant in understanding the subtle issues involved with Taylor expansions.

Consider $f : \mathbb{R} \to \mathbb{R}$ defined by

$$f(t) = \begin{cases} 0, & \text{for } t \leq 0 \\ \exp(-1/t), & \text{for } t > 0. \end{cases}$$

See Figure 4.11. We claim that f is infinitely differentiable on $\mathbb{R}$.

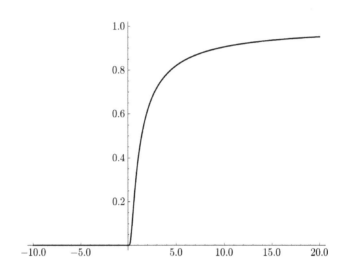

Figure 4.11: Graph of the function in Example 4.5.3.

In view of item 2 of Example 4.5.2, we have

$$e^x = 1 + \frac{x}{1!} + \frac{x^2}{2!} + \cdots + \frac{x^{n+1}}{(n+1)!} + R_{n+1},$$

where $R_{n+1} = \frac{e^c x^{n+2}}{(n+1)!} > 0$, if $x > 0$. Observe that this implies that if $x > 0$, we have $e^x > \frac{x^k}{k!}$ for $k \in \mathbb{N}$. If we replace x by $1/t$ where $t > 0$, we obtain $e^{1/t} > \frac{1}{t^{n+1}(n+1)!}$. This in turn leads us to conclude that $e^{-1/t} < (n+1)! t^{n+1}$. Finally, we get $t^{-n} e^{-1/t} < (n+1)! t$ for $t > 0$. By the sandwich lemma for the limits (on page 92), we conclude that

$$\lim_{t \to 0_+} t^{-n} e^{-1/t} = 0 \text{ for } n \in \mathbb{Z}_+. \tag{4.24}$$

By the algebra of limits, we conclude if $p(1/t) := a_0 + \frac{a_1}{t} + \cdots + \frac{a_n}{t^n}$, then

$$\lim_{t \to 0_+} p(1/t) e^{-1/t} = 0.$$

Now let us show that $f^{(n)}(0) = 0$ for all $n \in \mathbb{Z}_+$. If we take $n = 0$ in (4.24), it follows that $\lim_{t \to 0_+} f(t) = 0$. Since $f(t) = 0$ for $t \leq 0$, the continuity of f at 0 follows.

Let us now show that f is differentiable at 0. We start with the difference quotient $\frac{e^{-1/t}-0}{t} = t^{-1}e^{-1/t}$. By (4.24), the limit $\lim_{t\to 0_+} t^{-1}e^{-1/t} = 0$. The left-sided limit $\lim_{t\to 0_-} \frac{f(t)-f(0)}{t}$ is obviously zero and hence it follows that f is differentiable at 0 with $f'(0) = 0$.

What is f'? We have

$$f'(t) = \begin{cases} \frac{1}{t^2}e^{-1/t}, & t > 0 \\ 0, & t \le 0. \end{cases}$$

Clearly, f' is differentiable at any $t \ne 0$. We show that $f''(0) = 0$. We again need to work with the case when the limit is $t \to 0_+$. We have

$$\lim_{t\to 0_+} \frac{t^{-2}e^{-1/t} - 0}{t} = \lim_{t\to 0_+} t^{-3}e^{-1/t} = 0, \text{ by (4.24)}.$$

Assume that we have proved that $f^{(n)}(0) = 0$. We wish to show that $f^{(n+1)}(0) = 0$. Observe that the n-th derivative of $e^{-1/t}$ will be of the form $p(1/t)$ where the "degree" of p is $n + 1$. Hence $\lim_{t\to 0_+} \frac{p(1/t)e^{-1/t}-0}{t}$ is of the form $\lim_{t\to 0_+} p_1(1/t)e^{-1/t}$ where $p_1(1/t) = (1/t)p(1/t)$. Hence we conclude that $f^{(n+1)}(0) = 0$.

This shows that the function f is C^n for any $n \in \mathbb{N}$ and hence is infinitely differentiable. The remarkable feature of this function is that all its Taylor polynomials at $x_0 = 0$ are 0, as $f^{(k)}(0) = 0$ for any $k \in \mathbb{Z}_+$. What does this say about the behavior of f near 0?

We know that any polynomial function is infinitely differentiable. We cannot find a nonzero polynomial which is zero on a non-degenerate interval. (Why?) We are going to construct an infinitely differentiable function which vanishes on infinite intervals.

Exercise Set 4.5.4. This set extends the example above and leads to the construction of a non-constant C^∞ function which is zero outside an interval.

(1) Let $f : \mathbb{R} \to \mathbb{R}$ defined by

$$f(t) = \begin{cases} e^{-\frac{1}{t^2}}, & t > 0 \\ 0, & t \le 0. \end{cases}$$

Look at Figure 4.12. Show that f is infinitely differentiable.

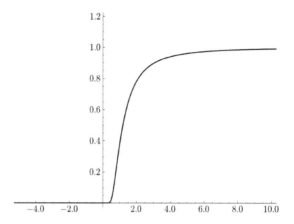

Figure 4.12: Graph of the function in Item 1 of Exercise 4.5.4.

(2) Let f be as in Item 1 of Exercise 4.5.4. Let $\varepsilon \in \mathbb{R}$. Define $g_\varepsilon(t) := f(t)/(f(t) + f(\varepsilon - t))$ for $t \in \mathbb{R}$. Then g_ε is differentiable, $0 \leq g_\varepsilon \leq 1$, $g_\varepsilon(t) = 0$ iff $t \leq 0$ and $g_\varepsilon(t) = 1$ iff $t \geq \varepsilon$. See Figure 4.13.

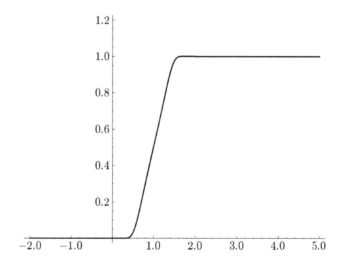

Figure 4.13: Graph of the function in Item 2 of Exercise 4.5.4 for $\varepsilon = 2$.

(3) For $0 < a < b$, define $h_{[a,b]}(x) = g_a(x)g_{-b}(-x)$. Look at the Figure 4.14. Show that if $h_{[a,b]}(x) \neq 0$, then $x \in [a, b]$. We say that this is a C^∞ function whose "support" lies in $[a, b]$.

Definition 4.5.5. Let $f \colon J \to \mathbb{R}$ be a differentiable function on an interval J. A point $c \in J$ is a *point of inflection* of f if $f(x) - f(c) - f'(c)(x - c)$ changes sign

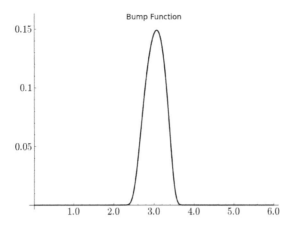

Figure 4.14: Graph of the function in Item 3 of Exercise 4.5.4.

as x increases through c in an interval containing c. Geometrically, this means that the graph as a curve crosses the tangent at the point of inflection. We let $T(x) := f(x) - f(c) - f'(c)(x - c)$ denote the tangent to the graph of f at $(c, f(c))$.

We now give a sufficient condition for the local extrema of a C^2-function.

Theorem 4.5.6. *Let $n \geq 2$, $r > 0$. Let $f^{(n)}$ be continuous on $[a - r, a + r]$. Assume that $f^{(k)}(a) = 0$ for $1 \leq k \leq n - 1$, but $f^{(n)}(a) \neq 0$. If n is even, then a is a local extremum. It is a minimum if $f^{(n)}(a) > 0$ and a local maximum if $f^{(n)}(a) < 0$.*
If n is odd, then a is a point of inflection.

Proof. By Taylor's theorem we have

$$f(x) = f(a) + \frac{(x - a)^n}{n!} f^{(n)}(c), \text{ for some } c \text{ between } a \text{ and } x. \qquad (4.25)$$

Let us first prove it when n is even. In this case $(x - a)^n \geq 0$ for all x. Now, $f^{(n)}(a) < 0$ implies that $f^{(n)}(c) < 0$ for x near a by the continuity of $f^{(n)}$. Hence (4.25) implies that $f(x) \leq f(a)$ for all such x. Thus a is a local maximum. The other case is similar and left as an exercise.
What happens if n is odd?
Let $n \geq 3$ be odd. We have

$$f(x) - f(a) = \frac{(x - a)^n}{n!} f^{(n)}(c).$$

Note that $(x - a)^n$ changes sign from positive to negative as x increases through a. Now whatever the sign of $f^{(n)}(a)$, $f(x) - f(a)$ changes sign as x increases through a. Hence a is a point of inflection. $\qquad \square$

Example 4.5.7. It is possible for f to have derivatives of all orders that all vanish at a strict local maximizer. Let us consider

$$f(x) = \begin{cases} -e^{-\frac{1}{x^2}}, & \text{when } x \neq 0 \\ 0, & \text{when } x = 0. \end{cases}$$

It is easy to check that 0 is the global maximizer of f. Further, f has derivatives of all order at 0 and they vanish.

4.6 Convex Functions

Look at the pictures. In Figure 4.15, if we draw *any* chord joining $(x, f(x))$ and $(y, f(y))$ on the graph, the part of the graph of f on the interval $[x, y]$ (or on $[y, x]$) lies below the chord. Look at the Figure 4.16 in which this phenomenon does not happen.

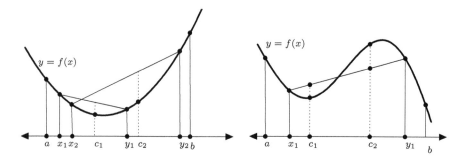

Figure 4.15: Convex function.   Figure 4.16: Non-convex function.

The functions of the first type are called convex functions.

We need a few definitions.

The line segment joining the two points $(x_1, f(x_1))$ and $(x_2, f(x_2))$ is given by $(1 - t)(x_1, f(x_1)) + t(x_2, f(x_2))$ for $t \in [0, 1]$. This is a parametric version of the equation (learned in coordinate/analytic geometry) for the line joining two points in $\mathbb{R}^2$. For, if (x, y) is point on the line, then we can solve for t so that $(x, y) = (1 - t)(x_1, f(x_1)) + t(x_2, f(x_2))$. Note that $x = (1 - t)x_1 + tx_2$ and $y = (1 - t)f(x_1) + tf(x_2)$. That is, we have

$$x - x_1 = t(x_2 - x_1) \text{ and } y - f(x_1) = t(f(x_2) - f(x_1)).$$

We arrive at $t = \frac{x - x_1}{x_2 - x_1} = \frac{y - f(x_1)}{f(x_2) - f(x_1)}$. Also note that $x = (1 - t)x_1 + tx_2$. Hence the condition for the convexity of f is that for each $t \in [0, 1]$, the y-coordinate of $(x, f(x))$ must be less than or equal to the y-coordinate of $(x = (1 - t)x_1 + tx_2, f((1 - t)x_1 + tf(x_2))$. We have thus arrived at the following definition.

Definition 4.6.1. Let J be intervals in $\mathbb{R}$. A function $f\colon J \to \mathbb{R}$ is said to be *convex* if for all $t \in [0,1]$ and for all $x, y \in J$, the following inequality holds:

$$f(tx + (1-t)y) \le tf(x) + (1-t)f(y). \tag{4.26}$$

If for $0 < t < 1$, strict inequality holds in (4.26), then the function is said to be *strictly convex*.

We say that f is *concave* if the reverse inequality holds in (4.26). (See Figure 4.17.)

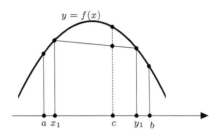

$$y = f(x)$$

Figure 4.17: Graph of a concave function.

Theorem 4.6.2 (Derivative Test for Convexity). *Assume that $f\colon [a,b] \to \mathbb{R}$ is continuous and differentiable on (a,b). If f' is increasing on (a,b), then f is convex on $[a,b]$. In particular, if f'' exists and is non-negative on (a,b), then f is convex.*

Proof. Let $x, y \in J := [a,b]$. Any point between x and y is given by $z = (1-t)x + ty$ for some $0 \le t \le 1$. Write $f(z) = (1-t)f(z) + tf(z)$. We wish to show

$$f(z) \le (1-t)f(x) + tf(y) \implies (1-t)f(z) + tf(z) \le (1-t)f(x) + tf(y).$$

In particular, it suffices to show that

$$(1-t)\,(f(z) - f(x)) \le t\,(f(y) - f(z))\,. \tag{4.27}$$

By the MVT, there exist $x < r < z$ and $z < s < y$ such that

$$f(z) - f(x) = f'(r)(z - x) \quad \text{and} \quad f(y) - f(z) = f'(s)(y - z).$$

Since $(1-t)z + tz = z = (1-t)x + ty$, we have

$$(1-t)(z - x) = t(y - z). \tag{4.28}$$

We have

$$\begin{aligned}
(1-t)[f(z) - f(x)] &= f'(r)(1-t)(z - x)\\
&= f'(r)t(y - z) \quad \text{by (4.28)}\\
&\le tf'(s)(y - z) = t[f(y) - f(z)].
\end{aligned}$$

This establishes (4.27) and hence the theorem. $\qquad\square$

Exercise 4.6.3. State and prove an analogue of Theorem 4.6.2 for concave functions.

Example 4.6.4. Let $f(x) := x^3 - 6x^2 + 9x$. Let us examine the convexity of f.

We have $f'(x) = 3(x - 1)(x - 3)$ and $f''(x) = 6x - 12$. It is easy to see that $f''(x) > 0$ if $x > 2$ and $f''(x) < 0$ if $x < 2$. Hence f is convex for $x > 2$ and concave for $x < 2$.

Example 4.6.5 (Examples of convex functions.).

(1) e^x is (strictly) convex on $\mathbb{R}$.

(2) x^α is convex on $(0, \infty)$ for $\alpha \geq 1$.

(3) $-x^\alpha$ is strictly convex on $(0, \infty)$ for $0 < \alpha < 1$.

(4) $x \log x$ is strictly convex on $(0, \infty)$.

(5) $f(x) = x^4$ is strictly convex but $f''(0) = 0$.

(6) $f(x) = x + (1/x)$ is convex on $(0, \infty)$.

(7) $f(x) = 1/x$ is convex on $(0, \infty)$.

Theorem 4.6.6. *If $f : (a, b) \to \mathbb{R}$ is convex and $c \in (a, b)$ is a local minimum, then c is a minimum for f on (a, b). That is, local minima of convex functions are global minima.*

Remark 4.6.7. This result is one of the reasons why convex functions are very useful in applications especially in optimization problems.

Proof. Let $f : (a, b) \to \mathbb{R}$ be convex and $c \in (a, b)$, a local minimum of f. If c is not a point of global minimum, then there exists a $d \in (a, b)$ ($d \leq c$ or $d \geq c$) with $f(d) < f(c)$. Consider the curve $\gamma(t) = (1 - t)c + td$. Then $\gamma(0) = c$, $\gamma(1) = d$, and

$$\begin{aligned}
f(\gamma(t)) &\leq (1 - t)f(c) + tf(d) \\
&< (1 - t)f(c) + tf(c), \qquad\quad t \neq 0 \\
&= f(c), \qquad\qquad\qquad\qquad \text{for all } t \in (0, 1 \,]. \qquad (4.29)
\end{aligned}$$

But for t sufficiently small, $\gamma(t) \in (c - \varepsilon, c + \varepsilon)$ so that

$$f(\gamma(t)) \geq f(c) \quad \text{for } 0 < t < \varepsilon,$$

which contradicts (4.29). $\qquad\qquad\qquad\qquad\qquad\qquad\qquad\qquad\qquad\qquad\square$

Theorem 4.6.8. *Let $f \colon [a, b] \to \mathbb{R}$ be convex. Then for any $x \in (a, b)$, we have*

$$\frac{f(x) - f(a)}{x - a} \leq \frac{f(b) - f(a)}{b - a} \leq \frac{f(b) - f(x)}{b - x}. \qquad (4.30)$$

Proof. Observe that

$$x = a + \frac{x-a}{b-a}(b-a) = \left(1 - \frac{x-a}{b-a}\right)a + \left(\frac{x-a}{b-a}\right)b.$$

Since f is convex, we conclude that

$$f(x) \le \left(1 - \frac{x-a}{b-a}\right)f(a) + \left(\frac{x-a}{b-a}\right)f(b).$$

We deduce from this

$$(b-a)f(x) \le ((b-a) - (x-a))\,f(a) + (x-a)f(b).$$

Rearranging the terms we obtain

$$(f(x) - f(a))(b-a) \le (f(b) - f(a))(x-a),$$

from which the leftmost inequality in (4.30) follows.

To arrive at the rightmost inequality in (4.30), start with

$$x = b + \frac{x-b}{b-a} = \frac{b-x}{b-a}a + \left(1 - \frac{b-x}{b-a}\right)b.$$

$\square$

Corollary 4.6.9. *Let $f: (a,b) \to \mathbb{R}$ be differentiable. Then f is convex iff*

$$f(y) \ge f(x) + f'(x)(y-x), \qquad \text{for } x, y \in (a,b). \tag{4.31}$$

Proof. Assume that $x < x_1 < y$. From (4.30), we get

$$\frac{f(x_1) - f(x)}{x_1 - x} \le \frac{f(y) - f(x)}{y - x}.$$

In the above, we let $x_1 \to x$ from the right, and we arrive at $f'(x) \le \frac{f(y)-f(x)}{y-x}$.
This is the desired result.

If $y < x$, the proof above may be adapted.

Conversely, assume that $f(y) \ge f'(x)(y-x)$ for $x, y \in (a,b)$. We claim that
f' is increasing. Let $x < y$. We then have

$$f(y) \ge f(x) + f'(x)(y-x) \tag{4.32}$$
$$f(x) \ge f(y) + f'(y)(x-y)$$
$$= f(y) - f'(y)(y-x). \tag{4.33}$$

We obtain

$$f(x) + f'(y)(y-x) \ge f(y), \qquad\qquad \text{from (4.33)}$$
$$\ge f(x) + f'(x)(y-x), \qquad \text{from (4.32)}.$$

The result follows from this.

$\square$

Corollary 4.6.10. *Let $f\colon (a,b) \to \mathbb{R}$ be differentiable. Let $x_0 \in (a,b)$ be such that $f'(x_0) = 0$. Then f has a global minimum at x_0.*

Proof. Easy. For any $x \in (a,b)$, in view of (4.31), we get

$$f(x) \geq f(x_0) + f'(x_0)(x - x_0) = f(x_0).$$

$\square$

Exercise Set 4.6.11.

(1) Let $f\colon (a,b) \to \mathbb{R}$. Prove that f is convex iff the slopes of the chords is an increasing function. (Formulate this rigorously and then prove it.)

(2) Use the last exercise to prove that a differentiable function $f\colon (a,b) \to \mathbb{R}$ is convex iff f' is increasing.

(3) Let $g\colon J \to \mathbb{R}$ be convex. Then show that $e^{g(x)}$ is convex.

(4) Show that $1/g(x)$ is convex if g is concave and positive.

(5) If $f, g\colon J \to \mathbb{R}$ are convex functions, then what can you say about $f + g$? Justify your answer.

(6) If $f, g\colon J \to \mathbb{R}$ are convex functions, then what can you say about $\max\{f, g\}$ and $\min\{f, g\}$?

(7) Let $f\colon (a,b) \to \mathbb{R}$ be convex and differentiable. Let $c \in (a,b)$ be fixed. Prove that for any $x \in (a,b)$,

$$f(x) - f(c) \geq f'(c)(x - c).$$

What does this mean geometrically?

(8) Let $f\colon \mathbb{R} \to \mathbb{R}$ be differentiable and convex. If f is bounded, prove that f is a constant.

Application: AM-GM Inequality

Let $x_1, \ldots, x_n$ be non-negative real numbers. Their arithmetic mean and geometric mean are defined by

$$\mathrm{AM} := \frac{x_1 + \cdots + x_n}{n} \quad \text{and} \quad \mathrm{GM} := (x_1 \cdots x_n)^{1/n}.$$

The inequality of the title says that the arithmetic mean is greater than or equal to the geometric mean and equality holds iff all the x_i's are equal.

We prove this by mathematical induction and calculus. For $n = 1$, the statement holds true with equality.

Assume that the AM-GM statement is true for any set of n non-negative real numbers.

Let $n+1$ non-negative real numbers $x_1, \ldots, x_{n+1}$ be given. We need to prove that

$$\frac{x_1 + \cdots + x_n + x_{n+1}}{n+1} - (x_1 \cdots x_n x_{n+1})^{\frac{1}{n+1}} \geq 0, \qquad (4.34)$$

with equality only if all the $n+1$ numbers are equal.

To avoid trivial cases, we may assume that all $n+1$ numbers are positive. We consider the last number x_{n+1} as a variable and define the function

$$f(t) = \frac{x_1 + \cdots + x_n + t}{n+1} - (x_1 \cdots x_n)^{\frac{1}{n+1}}, \qquad t > 0.$$

It suffices to show that $f(t) \geq 0$ for all $t > 0$, with $f(t) = 0$ only if $x_1, \ldots, x_n$ and t are all equal. We employ the first and second derivative tests of calculus. We have

$$f'(t) = \frac{1}{n+1} - \frac{1}{n+1}(x_1 \cdots x_n)^{\frac{1}{n+1}} t^{-\frac{n}{n+1}}, \qquad t > 0.$$

We are looking for points t_0 such that $f'(t_0) = 0$. Thus we obtain

$$(x_1 \cdots x_n)^{\frac{1}{n+1}} t_0^{-\frac{n}{n+1}} = 1.$$

That is, t_0 satisfies

$$t_0^{\frac{n}{n+1}} = (x_1 \cdots x_n)^{\frac{1}{n+1}}.$$

Or what is the same

$$t_0 = (x_1 \cdots x_n)^{\frac{1}{n}}.$$

That is, the only critical point t_0 of f is the geometric mean of $x_1, \ldots, x_n$. Note that if $t = R^n$ for very large $R \gg 1$, $f(t) \to \infty$ as $R \to \infty$. Hence it follows that f has a strict global minimum at t_0. Note that $f'' > 0$ and hence the function is convex. Hence t_0 must be a point of global minimum. We now compute $f(t_0)$.

$$f(t_0) = \frac{x_1 + \cdots + x_n + (x_1 \cdots x_n)^{1/n}}{n+1} - (x_1 \cdots x_n)^{\frac{1}{n+1}}(x_1 \cdots x_n)^{\frac{1}{n(n+1)}}$$

$$= \frac{x_1 + \cdots + x_n}{n+1} + \frac{1}{n+1}(x_1 \cdots x_n)^{\frac{1}{n}} - (x_1 \cdots x_n)^{\frac{1}{n}}$$

$$= \frac{x_1 + \cdots + x_n}{n+1} - \frac{n}{n+1}(x_1 \cdots x_n)^{\frac{1}{n}}$$

$$= \frac{n}{n+1}\left(\frac{x_1 + \cdots + x_n}{n} - (x_1 \cdots x_n)^{\frac{1}{n}}\right).$$

The term within the brackets in the last step is non-negative in view of the induction hypothesis. The hypothesis also says that we can have equality only when $x_1, \ldots, x_n$ are all equal. In this case, their geometric mean t_0 has the same value. Hence, unless $x_1, \ldots, x_n, x_{n+1}$ are all equal, we have $f(x_{n+1}) > 0$. This completes the proof. $\qquad \square$

Convex functions are used to establish various inequalities. We refer the reader to the article (g) in [8].

4.7   Taylor's Theorem: Cauchy's Form of the Remainder

Readers may omit this section on the first reading.

We establish a version of Taylor's theorem in which the remainder term is expressed in a form which allows us to get subtle estimates. The way to appreciate its power is to see how it is used in the proof of binomial theorem for non-integral indices. This is done in Section 7.7.

Theorem 4.7.1 (Taylor's Theorem with Remainder). *Let n and p be natural numbers. Assume that $f: [a, b] \to \mathbb{R}$ is such that $f^{(n-1)}$ is continuous on $[a, b]$ and $f^{(n)}(x)$ exists on (a, b). Then there exists $c \in (a, b)$ such that*

$$f(b) = f(a) + (b-a)f'(a) + \frac{(b-a)^2}{2!}f''(a) + \cdots + \frac{(b-a)^{n-1}}{(n-1)!}f^{(n-1)}(a) + R_n, \quad (4.35)$$

where

$$R_n = \frac{(b-c)^{n-p}(b-a)^p}{p(n-1)!}f^{(n)}(c). \quad (4.36)$$

In particular, when $p = n$, we get Lagrange's form of the remainder

$$R_n = \frac{(b-a)^n}{n!}f^{(n)}(c), \quad (4.37)$$

and when $p = 1$, we get Cauchy's form of the remainder.

$$R_n = (b-a)\frac{(b-c)^{n-1}}{(n-1)!}f^{(n)}(c), \quad \text{where } c = a + \theta(b-a), \ 0 < \theta < 1 \quad (4.38)$$

$$= \frac{(b-a)^n}{(n-1)!}(1-\theta)^{n-1}f^{(n)}(a + \theta(b-a)), \quad \text{where } 0 < \theta := \frac{(c-a)}{(b-a)} < 1. \quad (4.39)$$

Proof. The proof is an obvious modification of an earlier proof.

Consider

$$F(x) = f(b) - f(x) - (b-x)f'(x) - \frac{(b-x)^2}{2!}f''(x) - \cdots - \frac{(b-x)^{n-1}}{(n-1)!}f^{(n-1)}(x).$$

We have for $x \in (a, b)$,

$$F'(x) = \frac{-(b-x)^{n-1}f^{(n)}(x)}{(n-1)!}. \quad (4.40)$$

We now define

$$g(x) = F(x) - \left(\frac{b-x}{b-a}\right)^p F(a). \quad (4.41)$$

The g is continuous on $[a, b]$, and differentiable on (a, b) and $g(a) = 0 = g(b)$. Hence by Rolle's theorem, there exists $c \in (a, b)$ such that $g'(c) = 0$. Using the definition of g in (4.41) we get

$$0 = g'(c) = F'(c) + \frac{p(b-c)^{p-1}}{(b-a)^p} F(a). \qquad (4.42)$$

Using (4.40) in (4.42) and simplifying we get

$$\frac{(b-c)^{n-1}}{(n-1)!} f^{(n)}(c) = \frac{p(b-c)^{p-1}}{(b-a)^p} F(a). \qquad (4.43)$$

That is, $F(a) = \frac{(b-c)^n - p(b-a)^p}{p(n-1)!} f^{(n)}(c)$. This is what we set out to prove.

Lagrange's form of the remainder (4.37) and Cauchy's form (4.38) are obvious. If we write $c = a + \theta(b - a)$ for some $\theta \in (0, 1)$, Cauchy's form (4.39) of the remainder is obtained from (4.36). $\qquad \square$

We shall see later the power of Cauchy's form of the remainder in Section 7.7 where it is used to prove the convergence of the binomial series.

Chapter 5

Infinite Series

Contents

Since the addition of real numbers is a binary operation, that is, a map $+\colon \mathbb{R} \times \mathbb{R} \to \mathbb{R}$, given an ordered pair (x, y) of real numbers we can add them to get $x+y := +(x, y)$. It is the associativity of the addition that allows us to add a given finite ordered tuple $(x_1, \ldots, x_n)$. This is established in algebra books (especially in group theory). To "add" an infinite tuple, that is, to add a sequence, (x_n) of real numbers, we need analysis to give a sensible meaning to "$\sum_{n=1}^{\infty} x_n$." For instance, if $x_n = (-1)^{n-1}$, and if we want to "add" them based on our earlier experience we may manipulate the terms and end up with "different" answers as follows.

$$1 - 1 + 1 - 1 + \cdots = (1 - 1) + (1 - 1) + \cdots$$
$$= 0 + 0 + \cdots + 0 + \cdots$$
$$= 0,$$

$$1 - 1 + 1 - 1 + \cdots = 1 + (-1 + 1) + (-1 + 1) + \cdots$$
$$= 1 + 0 + 0 + \cdots$$
$$= 1.$$

This absurdity shows that we should define the "the sum" of a sequence (x_n) of real numbers in a rigorous way so as to avoid this.

When we write $\frac{1}{3} = 0.3333\ldots$, what do we mean by it? Recall that the right side is the infinite sum $\frac{3}{10} + \frac{3}{10^2} + \frac{3}{10^3} + \cdots$. We would like to arrive at a definition of the sum of an infinite series using which we should be able to prove rigorously $\frac{1}{3} = 0.33333\ldots$.

5.1   Convergence and Sum of an Infinite Series

Definition 5.1.1. Given a sequence (a_n) of real numbers, a formal sum of the form $\sum_{n=1}^{\infty} a_n$ (or $\sum a_n$, for short) is called an *infinite series*.

For any $n \in \mathbb{N}$, the finite sum $s_n := a_1 + \cdots + a_n$ is called the (*n*-th) *partial sum* of the series $\sum a_n$.

A more formal definition of an infinite series is as follows. By the symbol $\sum_n a_n$ we mean the sequence (s_n) where $s_n := a_1 + \cdots + a_n$.

We say that the infinite series $\sum a_n$ is *convergent* if the sequence (s_n) of partial sums is convergent. In such a case, the limit $s := \lim s_n$ is called the *sum of the series* and we denote this fact by the **symbol** $\sum a_n = s$.

We say that the series $\sum a_n$ is *divergent* if the sequence of its partial sums is divergent.

The series $\sum_n a_n$ is said to be *absolutely convergent* if the infinite series $\sum_n |a_n|$ is convergent. Note that a series $\sum a_n$ of non-negative terms, (that is, $a_n \geq 0$ for all n) is convergent iff it is absolutely convergent.

If a series is convergent but not absolutely convergent, then it is said to be *conditionally convergent*.

Let us look at some examples of series and their convergence.

Example 5.1.2. Let (a_n) be a constant sequence $a_n = c$ for all n. Then the infinite series $\sum a_n$ is convergent iff $c = 0$. For, the partial sum is $s_n = nc$. Thus (s_n) is convergent iff $c = 0$. (Why? Use the Archimedean property.)

Example 5.1.3. Let a_n be non-negative real numbers and assume that $\sum a_n$ is convergent. Since $s_{n+1} = s_n + a_{n+1}$, it follows that the sequence (s_n) is increasing. We have seen (Theorem 2.3.2) that (s_n) is convergent iff it is bounded above. Hence a series of non-negative terms is convergent iff the sequence of partial sums is bounded. Note that if $\sum a_n$ is convergent, then $\sum a_n = \text{lub } \{s_n : n \in \mathbb{N}\}$.

Example 5.1.4 (Geometric Series). This is the most important example. Let $z \in \mathbb{R}$ be such that $|z| < 1$. Consider the infinite series $\sum_{n=0}^{\infty} z^n$. We claim that the series converges to $\alpha := 1/(1 - z)$ for $|z| < 1$. Its n-th partial sum s_n is given by

$$s_n := \sum_{k=0}^{n} z^k = \frac{1 - z^{n+1}}{1 - z}.$$

Now, $|\alpha - s_n| = \frac{z^{n+1}}{1-z}$, which converges to 0. Hence we conclude that $\sum z^n = \frac{1}{1-z}$ if $|z| < 1$.

In particular,

$$0.3333\cdots = \sum_{n=1}^{\infty} \frac{3}{10^n} = \frac{3}{10} \sum_{n=0}^{\infty} \frac{1}{10^n} = \frac{3}{10} \frac{1}{1 - \frac{1}{10}} = \frac{1}{3}.$$

Also note that if $|z| \geq 1$, then the n-term does not go to 0, so the series cannot be convergent in this case. (See Corollary 5.1.11.)

Exercise 5.1.5. What is $0.9999\cdots$?

Example 5.1.6 (Telescoping Series). Let (a_n) and (b_n) be two sequences such that $a_n = b_{n+1} - b_n$, $n \geq 1$. We note that $s_1 = a_1 = b_2 - b_1$, $s_2 = a_1 + a_2 = (b_2 - b_1) + (b_3 - b_2) = b_3 - b_1$ and

$$s_n = a_1 + \cdots + a_n = (b_2 - b_1) + (b_3 - b_2) + \cdots + (b_{n+1} - b_n) = b_{n+1} - b_1.$$

Thus we see that $\sum a_n$ converges iff $\lim b_n$ exists, in which case we have

$$\sum a_n = -b_1 + \lim b_n.$$

A typical example: $\sum \frac{1}{n(n+1)}$. For, observe that $\frac{1}{n(n+1)} = \frac{1}{n} - \frac{1}{n+1}$. Here $b_n = -1/n$ so that the sum is $-b_1 + \lim b_{n+1} = 1 + \lim \frac{-1}{n+1} = 1$.

Example 5.1.7. Consider $\sum_{n=1}^{\infty} \frac{n}{n^4 + n^2 + 1}$. This is one of the series for which we can find the sum! We observe

$$a_n = \frac{n}{n^4 + n^2 + 1}$$
$$= \frac{n}{(n^2 + 1)^2 - n^2}$$
$$= \frac{n}{(n^2 + 1 + n)(n^2 + 1 - n)}$$
$$= \frac{1}{2}\left[\frac{1}{n^2 - n + 1} - \frac{1}{n^2 + n + 1}\right].$$

Note that the sum in the brackets is a telescoping series with $b_n = \frac{1}{2}\left(\frac{1}{n^2 - n + 1}\right)$. Hence we get $s_n = \frac{1}{2} - \frac{1}{2}\left(\frac{1}{n^2 + n + 1}\right) \to \frac{1}{2}$.

Example 5.1.8. Let us look at the series $\sum_n \frac{1}{n^2}$ of positive terms. Observe that $\frac{1}{n^2} < \frac{1}{n(n-1)}$ for $n \geq 2$. If s_n denotes the partial sum of the series $\sum_n \frac{1}{n^2}$ and t_n that of $\sum \frac{1}{n(n-1)}$, it follows that $s_n \leq t_n$. Since (t_n) is bounded above (Example 5.1.6), the sequence (s_n) is bounded above. Hence in view of Example 5.1.3 we see that the series $\sum n^{-2}$ is convergent.

This is a special case of the comparison test to be seen below.

Example 5.1.9 (Harmonic Series). The series $\sum_{n=1}^{\infty} \frac{1}{n}$ is divergent. Observe that the harmonic series is a series of positive terms. Hence it is convergent iff its partial sums are bounded above (see Example 5.1.3). We show that the subsequence (s_{2^k}) is not bounded above. The key observation is that each of the terms in

$$\frac{1}{2^{k-1} + 1} + \cdots + \frac{1}{2^k}$$

is greater than or equal to $\frac{1}{2^k}$ and there are 2^{k-1} terms in the sum. Hence the sum is greater than $2^{k-1} \times \frac{1}{2^k} = \frac{1}{2}$. We have the following estimates:

$$s_1 = 1$$

$$s_2 = 1 + \frac{1}{2} = \frac{3}{2}$$

$$s_{2^2} = 1 + \frac{1}{2} + \left(\frac{1}{3} + \frac{1}{4}\right)$$

$$> 1 + \frac{1}{2} + \left(\frac{1}{4} + \frac{1}{4}\right)$$

$$= 1 + \frac{1}{2} + 2 \cdot \frac{1}{4} = 1 + \frac{1}{2} + \frac{1}{2} = 1 + \frac{2}{2}$$

$$s_{2^3} = 1 + \frac{1}{2} + \left(\frac{1}{3} + \frac{1}{4}\right) + \left(\frac{1}{5} + \cdots + \frac{1}{8}\right)$$

$$> 1 + \frac{1}{2} + 2 \times \frac{1}{4} + 2^2 \times \frac{1}{2^3} = 1 + \frac{3}{2}$$

so that

$$s_{2^k} = 1 + \frac{1}{2} + \left(\frac{1}{3} + \frac{1}{4}\right) + \left(\frac{1}{5} + \cdots + \frac{1}{8}\right) + \cdots + \left(\frac{1}{2^{k-1}+1} + \cdots + \frac{1}{2^k}\right)$$

$$> 1 + \underbrace{\frac{1}{2} + \cdots + \frac{1}{2}}_{k-\text{times}}.$$

Hence $s_{2^k} > 1 + \frac{k}{2}$. It follows that (s_n) is not bounded above. We conclude that the harmonic series $\sum_n \frac{1}{n}$ is not convergent.

Theorem 5.1.10 (Cauchy Criterion). *The series $\sum a_n$ converges iff for each $\varepsilon > 0$ there exists $N \in \mathbb{N}$ such that*

$$n, m \geq N \implies |s_n - s_m| < \varepsilon.$$

Thus, the series $\sum a_n$ converges iff for each $\varepsilon > 0$ there exists $N \in \mathbb{N}$ such that

$$n > m \geq N \implies |a_{m+1} + a_{m+2} + \cdots + a_n| < \varepsilon.$$

This Cauchy criterion is quite useful when we want to show that a series is convergent without bothering to know its sum. See Theorem 5.1.17 for a typical use.

Proof. Let $\sum a_n$ be convergent. Then the sequence (s_n) of its partial sums is convergent. We know that a real sequence is convergent iff it is Cauchy. Hence (s_n) is convergent iff it is Cauchy. The result follows from the very definition of Cauchy sequences. □

Corollary 5.1.11. *If $\sum_n a_n$ converges, then $a_n \to 0$.*

Proof. We need to estimate $|a_n|$. The key observation is $a_n = s_n - s_{n-1}$ and the fact that (s_n) is convergent and hence is Cauchy. (Here (s_n) is as usual the sequence of the partial sums of the series $\sum a_n$.)

Let $\varepsilon > 0$ be given. Since the sum $\sum a_n$ is convergent, the sequence (s_n) of partial sums is convergent and in particular, it is Cauchy. Hence for the given ε there exists $N \in \mathbb{N}$ such that for $n \geq m \geq N$ we have $|s_n - s_m| < \varepsilon$. Now if we take any $n \geq N + 1$, then $a_n = s_n - s_{n-1}$. Note that $n - 1 \geq N$. Hence we obtain $|a_n| = |s_n - s_{n-1}| < \varepsilon$ for $n \geq N + 1$. This proves that $a_n \to 0$. $\square$

The converse of the above proposition is not true. See Example 5.1.9.

Remark 5.1.12. Most often we need the following observation on a convergent series $\sum a_n$. If $\sum_n a_n = s$, then $\sum_{n=N+1}^{\infty} a_n = s - \sum_{k=1}^{N} a_k$.

Now what is the meaning of the symbol $\sum_{n=N+1}^{\infty} a_n$? We define a new sequence (b_k) by setting $b_k := a_{N+k}$. The infinite series associated with the sequence (b_k) is denoted by $\sum_{n=N+1}^{\infty} a_n$ or simply by $\sum_{n \geq N+1} a_n$.

Let s_n denote the partial sums of $\sum a_k$. Let $\sigma_n := \sum_{N+1}^{N+n} a_k = \sum_{k=1}^{n} b_k$. Let $s_N := a_1 + \cdots + a_N$. Then we have $\sigma_n = s_{N+n} - s_N$. Clearly, $\sigma_n \to s - s_N$. The claim follows from this.

An important corollary, which is used most often, is the following.

Corollary 5.1.13. *Given* $\varepsilon > 0$, *there exists* $N \in \mathbb{N}$ *such that the "tail" of the series* $\sum_{n=N+1}^{\infty} a_n < \varepsilon$.

Proof. This is easy. Since $s_n \to s$, for $\varepsilon > 0$ there exists $N \in \mathbb{N}$ such that for $n \geq N$, $s_n \in (s - \varepsilon, s + \varepsilon)$. In particular, $s - \varepsilon < s_N$, that is, $s - s_N < \varepsilon$. By the last remark, $\sum_{n \geq N+1} a_n = s - s_N$. Hence the corollary follows. $\square$

Exercise 5.1.14. Given a sequence (a_n), let us assume the associated infinite series $\sum a_n$ is convergent. Let $N \in \mathbb{N}$ be fixed. Let $b_k \in \mathbb{R}$, $1 \leq k \leq N$ be given. We form a new sequence (c_n) where $c_k = b_k$ for $1 \leq k \leq N$ and $b_k = a_k$ for $k > N$. Let $s = \sum a_n$ and $b := b_1 + \cdots + b_N$. Show that $\sum c_n$ is convergent and that $\sum c_n = s + b - s_N$.

Given two series (whether or not convergent) $\sum a_n$ and $\sum b_n$, we may define their sum as the infinite series associated with the sum $(a_n + b_n)$ of the sequences (a_n) and (b_n). Thus, $\sum a_n + \sum b_n := \sum (a_n + b_n)$. Similarly, given a scalar $\lambda \in \mathbb{R}$, we define the scalar multiple $\lambda \sum a_n$ to be the series $\sum (\lambda a_n)$.

Theorem 5.1.15 (Algebra of Convergent Series). *Let* $\sum a_n$ *and* $\sum b_n$ *be two convergent series with their respective sums* A *and* B, *respectively.*
(i) *Their sum* $\sum (a_n + b_n)$ *is convergent and we have* $\sum (a_n + b_n) = A + B$.
(ii) *The series* $\lambda \sum a_n$ *is convergent and we have* $\lambda \sum a_n = \lambda \cdot A$.
The set of all (real) convergent series is a vector space over $\mathbb{R}$.

Proof. The proofs are straightforward and the reader should go on his own.

Let (s_n), (t_n), and (σ_n) be the partial sums of the series $\sum a_n$, $\sum b_n$, and $\sum(a_n + b_n)$. Observe that using standard algebraic facts about the commutativity and associativity of addition, we obtain

$$\sigma_n = (a_1 + b_1) + \cdots + (a_n + b_n)$$
$$= (a_1 + \cdots + a_n) + (b_1 + \cdots + b_n)$$
$$= s_n + t_n.$$

It follows from the algebra of convergent sequences that $\sigma_n \to A + B$.

(ii) is left to the reader.   □

Remark 5.1.16. The ONLY way to deal with an infinite series is through its partial sums and by using the definition of the sum of an infinite series.

We need to be careful when dealing with infinite series. Mindless algebraic/formal manipulations may lead to absurdities.

Let

$$s = 1 - 1 + 1 - 1 + \cdots + (-1)^{n+1} + \cdots$$

(Note that s has no meaning, if we apply our knowledge of infinite series!) Then

$$-s = -1 + 1 - 1 + 1 + \cdots = 1 + (-1 + 1 + \cdots) - 1 = s - 1.$$

Hence $s = 1/2$. On the other hand

$$s = (1 - 1) + (1 - 1) + \cdots = 0.$$

Hence we arrive at the absurdity $0 = 1/2$.

Proposition 5.1.17. *If $\sum a_n$ is absolutely convergent, then $\sum_n a_n$ is convergent.*

Proof. Let s_n and σ_n denote the partial sums of $\sum a_n$ and $\sum |a_n|$, respectively. It is enough to show that (s_n) is Cauchy. (Why?) We have, for $n > m$,

$$|s_n - s_m| = \left| \sum_{k=m+1}^{n} a_k \right| \leq \sum_{k=m+1}^{n} |a_k| = \sigma_n - \sigma_m,$$

which converges to 0, as (σ_n) is convergent. Hence (s_n) is Cauchy sequence.   □

The converse of the last proposition is not true. See Remark 5.2.8.

Theorem 5.1.18 (Comparison Test). *Let $\sum a_n$ and $\sum_n b_n$ be series of non-negative reals. Assume that $a_n \leq b_n$ for all n. Then:*

(i) *if $\sum b_n$ is convergent, then so is $\sum a_n$,*

(ii) *if $\sum a_n$ is divergent, so is $\sum b_n$,*

(iii) *Let $\sum_n b_n$ be a series of positive reals. Assume that $\sum_n b_n$ is convergent and that there exists $N \in \mathbb{N}$ such that $|a_n| \leq b_n$ for all $n \geq N$. Then $\sum_n a_n$ is absolutely convergent and hence convergent.*

Proof. Let s_n and t_n denote the n-th partial sums of $\sum a_n$ and $\sum b_n$, respectively. Since $a_k \leq b_k$ for $k \in \mathbb{N}$, we see that $s_n \leq t_n$.

Now if $\sum b_n$ is convergent, let $t_n \to t$. We know that $t = \text{lub } \{t_n : n \in \mathbb{N}\}$. Hence $s_n \leq t_n \leq t$ so that t is an upper bound of the set $\{s_n : n \in \mathbb{N}\}$. Thus, the sequence (s_n) of partial sums of the non-negative series $\sum a_n$ is bounded above. It follows that $\sum a_n$ is convergent, say, to s. Since $s = \text{lub } \{s_n\}$, we see that $s \leq t$.

If $\sum a_n$ is divergent, note that its partial sums form an increasing unbounded sequence. Given $M \in \mathbb{R}$, there exists $N \in \mathbb{N}$ such that for $k \geq N$, we have $s_k > M$. Hence $t_k \geq s_k > M$ for such k. We conclude that $\sum b_n$ is divergent.

Note that (iii) is an extension of (i). To prove this, we compare the series $\sum_{n \geq N} |a_n|$ and $\sum_{n \geq N} b_n$ to conclude that $\sum_{n \geq N} a_n$ is absolutely convergent and hence convergent. Since $\sum_{n=1}^{\infty} a_n = \sum_{k=1}^{N-1} a_k + \sum_{k \geq N} a_k$, the result follows. $\square$

Example 5.1.19 (Harmonic p-series). Consider the series $\sum_n \frac{1}{n^p}$. We claim that the harmonic p-series is divergent if $0 < p \leq 1$ and is convergent if $p > 1$.

This is often used in conjunction with the comparison test.

The case when $p = 1$ is already done in Example 5.1.9.

For $0 < p < 1$, observe that each of the terms in the sum of s_n is greater than or equal to n^{-p}. Hence $s_n \geq n \cdot n^{-p} = n^{1-p} \to \infty$. Thus the series $\sum_{n=1}^{\infty} n^{-p}$ is divergent for $0 < p < 1$.

For $p > 1$, observe that

$$\frac{1}{2^p} + \frac{1}{3^p} < \frac{2}{2^p} = \frac{2}{2^p}$$

$$\frac{1}{4^p} + \frac{1}{5^p} + \frac{1}{6^p} + \frac{1}{7^p} < \frac{4}{4^p} = \left(\frac{2}{2^p}\right)^2$$

$$\frac{1}{8^p} + \frac{1}{9^p} + \cdots + \frac{1}{15^p} < \frac{8}{8^p} = \left(\frac{2}{2^p}\right)^3.$$

Now the geometric series $\sum_k 2^k/2^{pk}$ is convergent if $p > 1$. It follows from the comparison test that the p-harmonic series is convergent if $p > 1$.

Exercise Set 5.1.20 (Exercises for comparison test).

(1) Let $b_n > 0$ and $a_n/b_n \to \ell > 0$. Then either both $\sum a_n$ and $\sum b_n$ converge or both diverge.

(2) Let $a_n > 0$ and $b_n > 0$. Assume that $(a_{n+1}/a_n) \leq (b_{n+1}/b_n)$ for all n. Show that
  (i) if $\sum b_n$ converges, then $\sum a_n$ converges, and
  (ii) if $b_n \to 0$ so does a_n.

Remark 5.1.21. The geometric series and the comparison test along with the integral test are the most basic tricks in dealing with absolute convergence of an infinite series.

Theorem 5.1.22 (d'Alembert's Ratio Test). *Let $\sum_n c_n$ be a series of positive reals. Assume that*

$$\lim_n c_{n+1}/c_n = r.$$

Then the series $\sum_n c_n$ is
  (i) *convergent if $0 \le r < 1$,*
  (ii) *divergent if $r > 1$.*
    The test is inconclusive if $r = 1$.

Proof. If $r < 1$, choose an s such that $r < s < 1$. Then there exists $N \in \mathbb{N}$ such that $c_{n+1} \le sc_n$ for all $n \ge N$. Hence $c_{N+k} \le s^k c_N$, for $k \in \mathbb{N}$. The convergence of $\sum c_n$ follows.

  If $r > 1$, then $c_n \ge c_N$ for all $n \ge N$ and hence $\sum c_n$ is divergent as the n-th term does not go to 0.

  Can you think of why the test is inconclusive when $r = 1$? The failure of the test when $r = 1$ follows from looking at the examples $\sum_n 1/n$ and $\sum_n 1/n^2$. □

Theorem 5.1.23 (Cauchy's Root Test). *Let $\sum_n a_n$ be a series of positive reals. Assume that $\lim_n a_n^{1/n} = a$. Then the series $\sum_n a_n$ is convergent if $0 \le a < 1$, divergent if $a > 1$, and the test is inconclusive $a = 1$.*

Proof. If $a < 1$, then choose α such that $a < \alpha < 1$. Then $a_n < \alpha^n$ for $n \ge N$. Hence by comparing with the geometric series $\sum_{n \ge N} \alpha^n$, the convergence of $\sum_n a_n$ follows. If $a > 1$, then $a_n \ge 1$ for all large n and hence, the n-th term does not approach zero.

  Can you think of why the test is inconclusive when $r = 1$?

  The examples $\sum_n 1/n$ and $\sum 1/n^2$ illustrate the failure of the test when $r = 1$. □

Exercise Set 5.1.24.

(1) Show that $\sum_n \frac{2^n n!}{n^n}$ is convergent.

(2) Is $\sum_n \frac{7^{n+1}}{9^n}$ convergent?

(3) Use your knowledge of infinite series to conclude that $\frac{n}{2^n} \to 0$.

(4) Show that the sequence $\left(\frac{n!}{n^n}\right)$ is convergent. Find its limit.

(5) Assume that $\sum a_n$ converges and $\sum a_n = s$. Show that $\sum_n (a_{2k} + a_{2k-1})$ converges and its sum is s.

(6) Let (a_n) be given such that $a_n \to 0$. Show that there exists a subsequence (a_{n_k}) such that the associated series $\sum_k a_{n_k}$ is convergent.

(7) Show that the series $\sum_n \frac{1}{2^n - n}$ is convergent.

(8) Let (a_n) be given. Assume that $a_n > 0$ for all n. Let s_n denote the n-th partial sum of the series $\sum_n a_n$. Show that the series $\sum_n \frac{s_n}{n}$ is divergent. Can you say anything more specific?

(9) Let $\sum a_n$ be absolutely convergent. Assume that $a_n + 1 \neq 0$ for any n. Show that the series $\sum \frac{a_n}{1+a_n}$ is absolutely convergent.

We shall now state and prove the integral test. We shall use some of the results from the theory of integration, which will be established in Chapter 6. (See page 202.)

If $f \colon [a, b] \to \mathbb{R}$ is continuous with $\alpha \le f(x) \le \beta$ for $x \in [a, b]$, then

$$\alpha(b - a) \le \int_a^b f(x)\, dx \le \beta(b - a).$$

We can motivate this inequality geometrically by considering a non-negative function f and using the geometric interpretation of the definite integral.

Theorem 5.1.25 (Integral Test). *Assume that $f \colon [1, \infty] \to [0, \infty)$ is continuous and decreasing. Let $a_n := f(n)$ and $b_n := \int_1^n f(t)\, dt$. Then:*
(i) *$\sum a_n$ converges if (b_n) converges,*
(ii) *$\sum a_n$ diverges if (b_n) diverges.*

Proof. Observe that for $n \ge 2$, we have $a_n \le \int_{n-1}^n f(t)\, dt \le a_{n-1}$ so that

$$\sum_{k=2}^n a_k \le \int_1^n f(t)\, dt \le \sum_{k=1}^{n-1} a_k.$$

If the sequence (b_n) converges, then (b_n) is a bounded increasing sequence. $\sum_{k=2}^n a_k \le b_n$. Hence (s_n) is convergent.

If the integral diverges, then $b_n \to \infty$. Since $b_n \le \sum_{k=1}^{n-1} a_k$, the divergence of the series follows. $\qquad\square$

In the following examples, you will again have to use results such as the fundamental theorem of calculus to compute the integrals.

Exercise Set 5.1.26 (Typical applications of the integral test).

(1) The p-series $\sum_n n^{-p}$ converges if $p > 1$ and diverges if $p \le 1$.

(2) The series $\sum \frac{1}{(n+2)\log(n+2)}$ diverges.

(3) Show that the series $\sum \frac{\log n}{n^p}$ is convergent if $p > 0$.

Theorem 5.1.27 (Cauchy Condensation Test). *If (a_n) is a decreasing sequence of non-negative terms, then $\sum a_n$ and $\sum_k 2^k a_{2^k}$ are either both convergent or both divergent.*

Proof. The argument below is similar to the one in Example 5.1.9. Observe

$$a_3 + a_4 \ge 2a_4$$

$$\vdots$$

$$a_{2^{n-1}+1} + a_{2^{n-1}+2} + \cdots + a_{2^n} \ge 2^{n-1} a_{2^n}.$$

Adding these inequalities, we get

$$\sum_{k=1}^{2^n} a_k > \sum_{k=3}^{2^n} a_k > \sum_{k=1}^{n} 2^{k-1} a_{2^k}. \qquad (5.1)$$

If $\sum_{k=1}^{\infty} 2^k a_{2^k}$ diverges so does $\sum_{k=1}^{\infty} 2^{k-1} a_{2^k}$. Hence $\sum a_k$ diverges.

Note that

$$a_2 + a_3 \leq 2a_2$$

$$\vdots$$

$$a_{2^{n-1}} + a_{2^{n-1}+1} + \cdots + a_{2^n-1} \leq 2^{n-1} a_{2^{n-1}}.$$

Adding these inequalities, we get

$$a_2 + a_3 + \cdots + a_{2^n-1} \leq \sum_{k=1}^{n-1} 2^k a_{2^k}. \qquad (5.2)$$

If $\sum_{k=1}^{\infty} 2^k a_{2^k}$ is convergent, then arguing as above, we conclude that the series $\sum a_k$ is convergent. $\qquad \square$

Exercise 5.1.28 (A typical application of the condensation test)**.** The series $\sum \frac{1}{n^p}$ is convergent if $p > 1$ and divergent if $p \leq 1$.

Theorem 5.1.29 (Abel-Pringsheim)**.** *If $\sum_n a_n$ is a convergent series of non-negative terms with (a_n) decreasing, then $na_n \to 0$.*

Proof. Since $\sum a_n$ is convergent, there exists N such that $|s_n - s_m| < \varepsilon$ for $n, n \geq N$. For $k \geq N$,

$$ka_{2k} \leq a_{k+1} + \cdots + a_{2k} = s_{2k} - s_k.$$

Thus, $\lim(2n)a_{2n} \to 0$. Since $a_{2n+1} \leq a_{2n}$, we have

$$(2n+1)a_{2n+1} \leq (2n+1)a_{2n} \leq (2n)a_{2n} + a_{2n}. \qquad (5.3)$$

Now the series $\sum a_n$ is convergent, the sequence (a_n) converges to 0 and hence the subsequence (a_{2n}) converges to zero. It follows from (5.3) that the sequence $((2n+1)a_{2n+1})$ is convergent. Using Item 13 of Exercise 2.1.28 on page 40, we conclude that $na_n \to 0$. $\qquad \square$

Remark 5.1.30. One may also deduce Abel-Pringsheim from the condensation test. For $\sum 2^k a_{2k}$ is convergent. Note that $2^n a_{2^n} \to 0$. Given k choose n such that $2^n \leq k \leq 2^{n+1}$. Then $ka_k < 2^{n+1} a_{2^n} = 2(2^n a_{2^n}) \to 0$.

What does Abel-Pringsheim say? It says that if (a_n) is a decreasing sequence of non-negative reals and if the associated series $\sum a_n$ is convergent, then $a_n \to 0$ *much faster* than $(1/n)$ going to zero. Compare this also with the convergence/divergence of harmonic p-series.

Example 5.1.31. The series $\sum \frac{1}{an+b}$, $a > 0$, $b \geq 0$ is divergent. Note that (a_n) is decreasing. If the given series is convergent, then $na_n \to 0$ by Abel-Pringsheim. But $na_n = \frac{n}{an+b} \to \frac{1}{a} \neq 0$.

5.2 Abel's Summation by Parts

The tests we have seen so far are for absolute convergence. There are infinite series which are convergent but not absolutely convergent. These are often quite subtle to handle. We give some tests which are useful to deal with such series. The basic tool for these tests is the following Abel's summation formula. If you encounter a series which is not absolutely convergent, then you should resort to summation by parts, if possible.

Theorem 5.2.1 (Abel's Summation by Parts Formula). *Let (a_n) and (b_n) be two sequences of real numbers. Define $A_n := a_1 + \cdots + a_n$. We then have the identity*

$$\sum_{k=1}^{n} a_k b_k = A_n b_{n+1} - \sum_{k=1}^{n} A_k (b_{k+1} - b_k). \tag{5.4}$$

It follows that $\sum_k a_k b_k$ converges if (i) the series $\sum A_k(b_{k+1} - b_k)$ and (ii) the sequence $(A_n b_{n+1})$ converges.

The name should remind us of a similar one which we have seen in the calculus course, namely integration by parts. We claim that (5.4) is similar to integration by parts formula. Given a sequence (s_n) of real numbers, we let $\Delta s_k := s_{k+1} - s_k \equiv \frac{s_{k+1} - s_k}{(k+1) - k}$ (a formula which may be familiar from "Finite differences"). This is a discrete version of a derivative with respect to the variable k. We let (s_n) be the sequence of partial sums of the series $\sum a_n$. Hence the left side of (5.4) assumes the form $\sum_{k=1}^{n} b_k \Delta s_k$. The right side is of the form $s_n b_{n+1} - \sum_{k=1}^{n} s_k \Delta b_k$. This resembles the integration by parts formula

$$\int_1^n b \, ds = sb \, |_1^n - \int_1^n s \, db.$$

Proof. Let $A_0 = 0$. We have

$$\sum_{k=1}^{n} a_k b_k = \sum_{k=1}^{n} (A_k - A_{k-1}) b_k$$

$$= \sum_{k=1}^{n} A_k b_k - \sum_{k=1}^{n} A_k b_{k+1} + A_n b_{n+1}.$$

Equation (5.4) follows from this. The last conclusion is an easy consequence of (5.4). □

Lemma 5.2.2 (Abel's Lemma). *Keep the notation of the last item. Assume further that (i) $m \leq A_n \leq M$ for $n \in \mathbb{N}$ and (ii) (b_n) is a decreasing sequence. Then*

$$mb_1 \leq \sum_{k=1}^{n} a_k b_k \leq Mb_1. \tag{5.5}$$

Proof. The proof of (5.5) follows from the summation formula (5.4), the fact $b_k - b_{k+1} \geq 0$ and telescoping:

$$\sum_{k=1}^{n} a_k b_k \leq M \sum_{k=1}^{n} (b_k - b_{k+1}) + M b_{n+1}$$
$$= M\left[(b_1 - b_2) + (b_2 - b_3) + \cdots + (b_n - b_{n+1}) + b_{n+1}\right]$$
$$= M b_1.$$

The left side inequality in (5.5) is proved in a similar way. □

Theorem 5.2.3 (Dirichlet's Test). *The series $\sum a_k b_k$ is convergent if the sequence (A_n) where $A_n := \sum_{k=1}^{n} a_k$ is bounded and (b_k) is decreasing to zero.*

Proof. Assume that $|A_n| \leq M$ for all n. We have $A_n b_{n+1} \to 0$. In view of the last item, it suffices to prove that $\sum_k A_k(b_{k+1} - b_k)$ is convergent. Since $b_k \searrow 0$, we have

$$|A_k(b_{k+1} - b_k)| \leq M(b_k - b_{k+1}).$$

The series $\sum_k (b_k - b_{k+1})$ is telescoping. We conclude that $\sum_k A_k(b_{k+1} - b_k)$ is absolutely convergent. □

Let us look at an important example.

Example 5.2.4. Let $b_n \searrow 0$. Then $\sum b_n \sin nx$ is convergent for all $x \in \mathbb{R}$ and $\sum b_n \cos nx$ is convergent for all $x \in \mathbb{R} \setminus \{2n\pi : n \in \mathbb{Z}\}$. Instead of using the trigonometric formulas, we use Euler's identity and de Moivre's theorem below, which involve complex numbers. If you do not like this, let us assure you that the trigonometric identities (5.6)–(5.7) can be established using standard trigonometric identities.

Consider the geometric series:

$$\sum_{k=1}^{n} e^{ikx} = e^{ix} \frac{1 - e^{inx}}{1 - e^{ix}} = e^{i\frac{(n+1)x}{2}} \frac{\sin(nx/2)}{\sin(x/2)}.$$

Taking real and imaginary parts, we get

$$\sum_{k=1}^{n} \cos kx = \frac{\sin \frac{nx}{2} \cos(n+1)\frac{x}{2}}{\sin \frac{x}{2}} \tag{5.6}$$

$$\sum_{k=1}^{n} \sin kx = \frac{\sin \frac{nx}{2} \sin(n+1)\frac{x}{2}}{\sin \frac{x}{2}}. \tag{5.7}$$

We thus have the easy estimates

$$\left| \sum_{k=1}^{n} \cos kx \right| \leq \frac{1}{|\sin \frac{x}{2}|} \quad \text{and} \quad \left| \sum_{k=1}^{n} \sin kx \right| \leq \frac{1}{|\sin \frac{x}{2}|} \quad \text{if } \sin \frac{x}{2} \neq 0.$$

Now it is easy to complete the proof. □

Theorem 5.2.5 (Dedekind's Test). *Let* (a_n), (b_n) *be two complex sequences. Let* $A_n := (a_1 + \cdots + a_n)$. *Assume that (i)* (A_n) *is bounded, (ii)* $\sum |b_{n+1} - b_n|$ *is convergent and* $b_n \to 0$. *Then* $\sum a_n b_n$ *is convergent. In fact,* $\sum a_n b_n = \sum A_n (b_{n+1} - b_n)$.

Proof. We use Abel's summation formula (5.4). Argue as in Dirichlet's test. □

Exercise 5.2.6 (Abel's Test). Assume that the series $\sum a_k$ is convergent and the sequence (b_k) is monotone and bounded. Then the series $\sum a_k b_k$ is convergent.
  Proof is quite similar to that of Dirichlet's test 5.2.3.

Theorem 5.2.7 (Leibniz Test or Alternating Series Test). *Let* (t_n) *be a real monotone sequence converging to zero. Then* $\sum (-1)^{n-1} t_n$ *is convergent and we have*

$$t_1 - t_2 \leq \sum (-1)^{n-1} t_n \leq t_1.$$

Proof. The convergence of the series is an immediate corollary of Dirichlet's test.
  We indicate a direct proof which also exhibits the estimates for the sum. Clearly $s_{2n} = (t_1 - t_2) + \cdots (t_{2n-1} - t_{2n})$ is increasing. Also,

$$s_{2n} = t_1 - (t_2 - t_3) - \cdots - (t_{2n-2} - t_{2n-1}) - t_{2n} \leq t_1.$$

Hence the sequence (s_{2n}) is a bounded increasing sequence and hence is convergent, say, to $s \in \mathbb{R}$. We claim that $s_n \to s$. Given $\varepsilon > 0$, find $N \in \mathbb{N}$ such that

$$n \geq N \implies |s_{2n} - s| < \varepsilon/2 \text{ and } |t_{2n+1}| < \varepsilon/2.$$

For $n \geq N$, we have

$$|s_{2n+1} - s| \leq |s_{2n} + t_{2n+1} - s| \leq |s_{2n} - s| + |t_{2n+1}| < \varepsilon.$$

□

Remark 5.2.8. A typical example for the alternating series test is $\sum \frac{(-1)^n}{n}$. In particular, we have an example of a series which is convergent but not absolutely convergent.

5.3 Rearrangements of an Infinite Series

Let an n-tuple $(a_1, \ldots, a_n)$ of real numbers be given. Let $s = a_1 + \cdots + a_n$ be the sum of the finite sequence. Let σ be a permutation of $\{1, \ldots, n\}$. Note that σ is a bijection of the set $\{1, \ldots, n\}$. Consider the new n-tuple $(a_{\sigma(1)}, \ldots, a_{\sigma(n)})$. Let $t := a_{\sigma(1)} + \cdots + a_{\sigma(n)}$. Thanks to the commutativity and associativity of the addition, we know that $s = t$. (Note that we do this often while adding a list of numbers. We do not add them as per the listing. We may find two numbers quite apart but may pair them off as they are easy to add up or they cancel out each other.)

Given an infinite sequence (a_n), assume that the associated infinite series $\sum_n a_n$ is convergent. Fix $N \in \mathbb{N}$. Let now σ be a permutation of $\{1, \ldots, N\}$. Construct a new sequence (b_k) where $b_k = a_{\sigma(k)}$ for $1 \leq k \leq N$ and $b_k = a_k$ for $k > N$. Look at $\sum b_n$, the infinite series associated with the sequence (b_n). Is $\sum b_n$ convergent and if so what is its sum? Let s_n and t_n denote the partial sums of the series $\sum a_n$ and $\sum b_n$, respectively. If $n = N$, the finite sequence (or the N-tuple $(b_1, \ldots, b_N)$ is just a permutation of $(a_1, \ldots, a_N)$. Hence we know that $s_N = t_N$. (Note that it may happen $s_k \neq t_k$ for $k < N$.) Also, $t_n = s_n$ for $n \geq N$. It follows that $t_n \to s := \lim s_n$.

Can we extend this if σ is a permutation of $\mathbb{N}$? What do we mean by this? Let (a_n) be given. Let $\sigma \colon \mathbb{N} \to \mathbb{N}$ be a bijection. (We may think of σ as a permutation of $\mathbb{N}$.) Then we construct a new sequence (b_n) where $b_n := a_{\sigma(n)}$. Our question is if $\sum a_n$ is convergent, is $\sum b_n$ convergent, and if so, what is its sum? Based on our experience with algebra of finite sums, we may be tempted to believe that the answer is $\sum b_n$ converges to $\sum a_n$. But this is not the case. See Remark 5.3.2 below.

Definition 5.3.1. Let $\sum a_n$ and a bijection $\sigma \colon \mathbb{N} \to \mathbb{N}$ be given. Define $b_n := a_{\sigma(n)}$. Then the new series $\sum b_n$ is said to be a *rearrangement* of the series $\sum a_n$.

Remark 5.3.2. Rearranged series $\sum b_n$ may converge to a sum different from that of $\sum a_n$.

Consider the standard alternating series $\sum (-1)^{n+1} n^{-1}$. We know that it is convergent, say, to a sum s. From Theorem 5.2.7 we also know $s \geq t_1 - t_2 = 1/2$. Hence $s \neq 0$. We rearrange the series to get a new series $\sum b_n$ which converges to $s/2$!

Rearrange the given series in such a way that two negative terms follow a positive term:

$$1 - \frac{1}{2} - \frac{1}{4} + \frac{1}{3} - \frac{1}{6} - \frac{1}{8} + \cdots + \frac{1}{2n-1} - \frac{1}{4n-2} - \frac{1}{4n} + \cdots .$$

(Can you write an explicit formula for b_n?) Let s_n be the n-th partial sum of the original series and t_n denote the n-th partial sum of this rearranged series. We have

$$t_{3n} = \left(1 - \frac{1}{2} - \frac{1}{4}\right) + \left(\frac{1}{3} - \frac{1}{6} - \frac{1}{8}\right) + \cdots + \left(\frac{1}{2n-1} - \frac{1}{4n-2} - \frac{1}{4n}\right) + \cdots .$$

In each block of three terms (in the brackets), subtract the second term from the first to get

$$t_{3n} = \left(\frac{1}{2} - \frac{1}{4}\right) + \left(\frac{1}{6} - \frac{1}{8}\right) + \cdots + \left(\frac{1}{4n-2} - \frac{1}{4n}\right) + \cdots = \frac{s_{2n}}{2}.$$

Thus $t_{3n} \to s/2$. Also, $t_{3n+1} = t_{3n} + \frac{1}{2n+1} \to s/2$ etc. Hence we conclude that $t_n \to s/2$. $\qquad\square$

Given a real series $\sum a_n$, we let $a_n^+ := \max\{a_n, 0\}$ and $a_n^- := -\min\{a_n, 0\}$. We call the series $\sum a_n^+$ (respectively, $\sum a_n^-$) as the positive part or the series of positive terms (respectively, the negative part or the series of negative terms) of the given series $\sum a_n$. Note that both these series have non-negative terms.

Proposition 5.3.3. *If $\sum a_n$ is conditionally convergent, then the series of its positive terms and the series of negative terms are both divergent.*

Proof. Let (s_n), (σ_n), α_n^+, and α_n^- denote the sequences of partial sums of the series $\sum a_n$, $\sum |a_n|$, $\sum a_n^+$, and $\sum a_n^-$, respectively. Note that α_n^+ is the sum of the non-negative terms in s_n and $-\alpha_n^-$ is the sum of the negative terms in s_n. Hence we have

$$\sigma_n := \sum_{k=1}^n |a_k| = \alpha_n^+ + \alpha_n^-, \text{ and } s_n = \alpha_n^+ - \alpha_n^-.$$

Let $s_n \to s$. Observe that (α_n^+) and (α_n^-) are increasing. By hypothesis $\sum |a_n|$ is not convergent, which implies that $\sigma_n \to \infty$. Note that

$$\alpha_n^+ = \frac{\sigma_n + s_n}{2} \text{ and } \alpha_n^- = \frac{\sigma_n - s_n}{2}.$$

We shall show that $\alpha_n^+ \to \infty$. Let $R \in \mathbb{R}$ be given. Since $s_n \to s$, the sequence (s_n) is bounded. Let $M > 0$ be such that $-M \leq s_n \leq M$ for $n \in \mathbb{N}$. Since (σ_n) diverges to infinity, there exists N such that for $n \geq N$, we have $\sigma_n > 2R + M$. It follows that $(2R + M) - M \leq \sigma_n + s_n$ for $n \geq N$. In particular,

$$n \geq N \implies \alpha_n^+ = \frac{\sigma_n + s_n}{2} \geq \frac{2R}{2} = R.$$

A similar argument shows that $\alpha_n^- \to \infty$.

We can also argue as follows. Since (α_n^+) is increasing, it is enough to show that it is not bounded. (Why?) If it is bounded, so is $(2\alpha_n^+)$. Since $\sigma_n = 2\alpha_n^+ - s_n$ and since (s_n) is bounded (Why?), we conclude that (σ_n) is bounded. This contradicts the fact that $\sum_n |a_n|$ is divergent. □

Remark 5.3.4. The proof above shows the following. If $\sum a_n$ is a series of real numbers, then $\sum a_n$ converges iff $\sum a_n^+$ and $\sum a_n^-$ converge, in which case we have $s = \alpha^+ - \alpha^-$. (Here, $\sum a_n^+ = \alpha^+$ and $\sum a_n^- = \alpha^-$.)

Using the last result, we can find a rearrangement of the alternating series so that it converges to any pre-assigned real number, say, 2012. We know that the series $\sum a_n^+$ and $\sum a_n^-$ of positive and negative terms of a conditionally convergent series is divergent. So for $n \gg 0$ we have $\alpha_n^+ > 2012$. Let N_1 be the first such that $\sum_{k=1}^{N_1} \frac{1}{2k-1} > 2012$. Hence $\sum_{k=1}^{N_1} \frac{1}{2k-1} - \frac{1}{2} < 2012$. (Can you justify this claim?) Let N_2 be the smallest odd integer so that

$$\left(\sum_{k=1}^{N_1} \frac{1}{2k-1} \right) - \frac{1}{2} + \left(\sum_{k=N_1}^{N_2} \frac{1}{2k-1} \right) > 2012.$$

Necessarily

$$\left(\sum_{k=1}^{N_1} \frac{1}{2k-1}\right) - \frac{1}{2} + \left(\sum_{k=N_1}^{N_2} \frac{1}{2k-1}\right) - \frac{1}{4} < 2012.$$

How much does the sum on the left differ from 2012? What is the next step? Do you see that you can continue this process forever and that at each step you will be closer to 2012?

Riemann's theorem 5.3.5 below says something very dramatic and startling. It should convince us of the danger of manipulating an infinite series without any attention to rigorous analysis.

Theorem 5.3.5 (Riemann's Theorem). *A conditionally convergent series can be made to converge to any arbitrary real number or even made to diverge by a suitable rearrangement of its terms.*

> **Strategy:** Let $\sum a_n$ be conditionally convergent. The crucial fact is that $\sum a_n^+$ and $\sum a_n^-$ diverge to infinity. Let $s \in \mathbb{R}$ be given. Choose the first k_1 such that the sum $\sum_{r=1}^{k_1} a_r^+$ exceeds s. Then we subtract just enough terms from $\{a_n^-\}$ so that it is less than s. And, so on.
>
> We exploit the fact that $a_n \to 0$ to estimate at each step how much the sums differ from s.

Proof. We omit the proof of the theorem, as a rigorous argument will be more on book-keeping than on analysis. The interested reader may read Theorem 3.54 in [7] or Theorem 8.33 in [1].   □

The next theorem is in contrast with Riemann's theorem and it also brings out the reason why we would prefer (given a choice, of course!) to deal with absolutely convergent series. This theorem is very useful and its proof is worth learning well.

Theorem 5.3.6 (Rearrangement of Terms). *If $\sum a_n$ is absolutely convergent and $\sum b_n$ is a rearrangement of $\sum a_n$, then $\sum b_n$ is convergent and we have $\sum a_n = \sum b_n$.*

> **Strategy:** If s_n and t_n denote the partial sums of the series $\sum a_n$ and $\sum b_n$, and if $s_n \to s$, we need to show that $|t_n - s| \to 0$. We know how to estimate $|s_n - s|$. So we employ the curry-leaf trick. We look at an estimate of the form
>
> $$|t_n - s| \le |t_n - s_m| + |s_m - s|.$$
>
> If we fix $m \gg 0$, then $|s_m - s|$ can be made very small. We now choose N large enough so that all the terms a_k, $1 \le k \le m$ appear in the sum t_N. It follows that $t_N - s_m$ is the sum of a_k's where $k > m$. An obvious upper bound for this sum (as well as any sum of the form $t_n - s_m$ for $n \ge N$) via the triangle inequality is $\sum_{k \ge m} |a_k|$, which is a tail of the convergent series $\sum_{k=1}^{\infty} |a_k|$. This tail can be made as small as we please.

Proof. Let t_n denote the n-th partial sum of the series $\sum b_n$. Let $\sum a_n = s$. We claim that $t_n \to s$. Let $\varepsilon > 0$ be given. Choose $n_0 \in \mathbb{N}$ such that

$$(n \geq n_0 \implies |s_n - s| < \varepsilon) \text{ and } \sum_{n_0+1}^{\infty} |a_n| < \varepsilon.$$

Choose $N \in \mathbb{N}$ such that $\{a_1, \ldots, a_{n_0}\} \subset \{b_1, \ldots, b_N\}$. (That is, choose N so that $\{1, \ldots, n_0\} \subseteq \{\sigma(1), \ldots, \sigma(N)\}$ where $\sigma \colon \mathbb{N} \to \mathbb{N}$ is a permutation.) We then have for $n \geq N$,

$$|t_n - s_{n_0}| \leq \sum_{n_0+1}^{\infty} |a_k| < \varepsilon.$$

It follows that for $n \geq N$, we have $|t_n - s| \leq |t_n - s_{n_0}| + |s_{n_0} - s| < 2\varepsilon.$ □

Exercise 5.3.7. An easy corollary is that any rearrangement of series of non-negative terms does not affect the convergence and the sum.

See also Item 23 in Exercise 5.3.9.

Example 5.3.8. Theorem 5.3.6 is quite useful in finding the sum of a series which is absolutely convergent, since it allows us to manipulate the terms of the series whichever way we want so that it becomes tractable.

Consider the series $\sum \frac{(-1)^{n+1}}{n^2}$. It is absolutely convergent. Now see how we find its sum!

$$\sum_{n=1}^{\infty} \frac{(-1)^{n+1}}{n^2} = \sum_{n=1}^{\infty} \frac{1}{n^2} - 2\sum_{k=1}^{\infty} \frac{1}{(2k)^2}$$

$$= \frac{1}{2} \sum_{n=1}^{\infty} \frac{1}{n^2}.$$

You might have learned from the theory of Fourier series that $\sum_{n=1}^{\infty} \frac{1}{n^2} = \frac{\pi^2}{6}$.

Exercise Set 5.3.9. Miscellaneous exercises.

(1) Let $\sum b_n$ be a convergent series of non-negative terms. Let (a_n) be sequence such that $|a_n| \leq M b_n$ for $n \geq N$, for a fixed N and $M > 0$. Show that $\sum a_n$ is convergent.

(2) If (a_n) and (b_n) are sequences of positive terms such that $a_n/b_n \to \ell > 0$. Prove that $\sum a_n$ and $\sum b_n$ either both converge or both diverge.

(3) As an application of the last item, discuss the convergence of
(a) $\sum 1/2n$, (b) $\sum 1/(2n-1)$ and (c) $\sum 2/(n^2+3)$.

(4) Assume that $\sum a_n$ is absolutely convergent and (b_n) is bounded. Show that $\sum a_n b_n$ is convergent.

(5) Show that the sum of two absolutely convergent series and a scalar multiple of an absolutely convergent series are again absolutely convergent. Hence conclude that the set ℓ^1 of all absolutely convergent series is a real vector space.

(6) Let $\sum a_n$ be a convergent series of positive terms. Show that $\sum a_n^2$ is convergent. More generally, show that $\sum a_n^p$ is convergent for $p > 1$.

(7) Let $p > 0$. Show that the series $\sum_n \frac{n^p}{e^n}$ is convergent. Can we take $p = 0$?

(8) Find the values of $x \in [0, 2\pi]$ such that the series $\sum \sin^n(x)$ is convergent.

(9) Let $\sum a_n$ and $\sum b_n$ be convergent series of positive terms. Show that $\sum \sqrt{a_n b_n}$ is convergent.

(10) Give an example of a convergent series $\sum a_n$ such that the series $\sum a_n^2$ is divergent.

(11) Give an example of a divergent series $\sum a_n$ such that the series $\sum a_n^2$ is convergent.

(12) Let (a_n) be a real sequence. Show that $\sum(a_n - a_{n+1})$ is convergent iff (a_n) is convergent. If the series converges, what is its sum?

(13) When does a series of the form $a + (a + b) + (a + 2b) + \cdots$ converge?

(14) Assume that $\left|\frac{a_{n+1}}{a_n}\right| \leq \frac{n^2}{(n+1)^2}$ for $n \in \mathbb{N}$. Show that the series $\sum a_n$ is absolutely convergent.

(15) Prove that if $\sum |a_n|$ is convergent, then $|\sum a_n| \leq \sum |a_n|$.

(16) Prove that if $|x| < 1$,

$$1 + x^2 + x + x^4 + x^6 + x^3 + x^8 + x^{10} + x^5 + \cdots = \frac{1}{1 - x}.$$

(17) Prove that if a convergent series in which only a finite number of terms are negative is absolutely convergent.

(18) If $(n^2 a_n)$ is convergent, then $\sum a_n$ is absolutely convergent.

(19) Assume that (a_n) is a sequence such that $\sum_n a_n^2$ is convergent. Show that $\sum a_n^3$ is absolutely convergent.

(20) If $\sum a_n$ is absolutely convergent, show that $\sum a_n^2$ is convergent. Is the result still true if we assume that only $\sum a_n$ is convergent?

(21) Let $\sum a_n$ be conditionally convergent. Let $p > 1$. Show that $\sum_n n^p a_n$ is divergent.

(22) True or false? If $\sum a_n$ is absolutely convergent, then $\sum \frac{a_n}{n^p}$ is absolutely convergent for all $p \geq 0$.

(23) Let $\sum a_n$ be a convergent series of non-negative terms. Let its sum be s. Let S be the set of all finite sums of (a_n), that is,

$$S := \left\{ \sum_{j \in F} a_j : F \text{ is a finite subset of } \mathbb{N} \right\}.$$

Show that $s = \text{lub } S$.

What can you conclude if $\sum a_n$ is divergent?

Can you use this information to come up with a definition of "sum" of an arbitrary collection $\{a_i : i \in I\}$ of non-negative real numbers.

(24) Is the series $\sum \frac{(k!)^2}{(2k)!}$ convergent?

(25) Let $x > 0$. For what values of x is the series $\sum nx^n$ convergent?

(26) For what values of x is the series $\sum \frac{(n+1)^n x^n}{n!}$ convergent?

(27) Show that the series $\sum \log(1 + \frac{1}{n})$ is divergent.

(28) Assume that the ratio test when applied to $\sum a_n$ establishes the convergence of the series. Show that one can prove that the series converges using the root test. (Loosely speaking, this says that the root test is stronger than the ratio test. Of course, the ratio test is simpler to use.)

(29) Show that the series $\sum a_n$ is absolutely convergent iff $\sum \varepsilon_n a_n$ is convergent for any choice of $\varepsilon_n \in \{\pm 1\}$.

(30) Show that if $\sum a_n^2$ and $\sum b_n^2$ are convergent, then $\sum a_n b_n$ is convergent.

(31) Compute $\sum \frac{1}{n^2(2n-1)}$, assuming that $\sum \frac{1}{n^2} = \frac{\pi^2}{6}$.

(32) Compute $\sum \frac{(-1)^{n-1}}{n^4}$, assuming that $\sum \frac{1}{n^4} = \frac{\pi^4}{90}$.

(33) Let $\sum a_n$ be a convergent series of positive terms. What can you say about the series $\sum \frac{1}{a_n}$?

(34) Show that the sequence $\left(\frac{a^n}{n!}\right)$ is convergent. What is its limit?

(35) Let $a > 0$. Show that the series $\sum \frac{a^n}{(n!)^{1/n}}$ is convergent if $a < 1$.

5.4 Cauchy Product of Two Infinite Series

Given two convergent sequences (a_n) and (b_n), we defined their product as the sequence $(a_n b_n)$ and the product converges to the product $(\lim a_n) \cdot (\lim b_n)$ of the limits.

Given two series $\sum_{n=1}^{\infty} a_n$ and $\sum_{n=1}^{\infty} b_n$, a naive way of defining their product would be $\sum c_n$ where $c_n := a_n b_n$. Why is this not a good definition? If we are given two sums $A := a_1 + \cdots + a_n$ and $B := b_1 + \cdots + b_n$, how do we define their product? Certainly not as the sum $a_1 b_1 + \cdots + a_n b_n$. Can we define it using the distributive law for an infinite sum?

Note that we would like the product of two convergent infinite series $\sum a_n$ and $\sum b_n$ to be another infinite series $\sum c_n$. We may also would like that $\sum c_n$ to be convergent to $(\sum a_n) \cdot (\sum b_n)$.

If we look at two polynomials $P(X) := a_0 + a_1 X + \cdots + a_n X^n$ and $Q(X) := b_0 + b_1 X + \cdots + b_m X^m$, then their product is a polynomials $c_0 + c_1 X + \cdots + c_{n+m} X^{m+n}$ where $c_0 = a_0 b_0$, $c_1 = a_0 b_1 + a_1 b_0$, $c_2 = a_0 b_2 + a_1 b_1 + a_2 b_0$ and more generally,

$$c_k = a_0 b_k + a_1 b_{k-1} + \cdots + a_{k-1} b_1 + a_k b_0 = \sum_{r=0}^{k} a_r b_{k-r}.$$

This suggests the following definition.

Definition 5.4.1. Given two series $\sum_{n=0}^{\infty} a_n$ and $\sum_{n=0}^{\infty} b_n$, their *Cauchy product* is the series $\sum_{n=0}^{\infty} c_n$ where $c_n := \sum_{k=0}^{n} a_k b_{n-k}$.

Remark 5.4.2. In spite of our wishful thinking, in general, the Cauchy product of two convergent series may not be convergent. Consider the series $\sum a_n$ and $\sum b_n$ where $a_n = \frac{(-1)^n}{\sqrt{n+1}}$. Then the c_n, the n-th term of their Cauchy product, is

$$c_n = (-1)^n \sum_{k=0}^{n} \frac{1}{\sqrt{(n-k+1)(k+1)}}.$$

For $k \leq n$, we have

$$(n - k + 1)(k + 1) = \left(\frac{n}{2} + 1\right)^2 - \left(\frac{n}{2} - k\right)^2 \leq \left(\frac{n}{2} + 1\right)^2.$$

Hence $|c_n| \geq \frac{2(n+1)}{n+2} \to 2$. It follows that the Cauchy product is not convergent.

Theorem 5.4.3 (Mertens' Theorem). *Let $\sum_{n=0}^{\infty} a_n$ be absolutely convergent and $\sum_{n=0}^{\infty} b_n$ be convergent. Define $c_n := \sum_{k=0}^{n} a_k b_{n-k}$. If $A := \sum_n a_n$ and $B := \sum_n b_n$, then $\sum_n c_n$ is convergent and we have $\sum_n c_n = AB$.*

Strategy: Let A_n, B_n, and C_n denote the partial sums of the three series. We wish to show that $C_n \to AB$. Obviously we need to find an expression which involves A_n, B_n, A, B, and C. Writing out C_n explicitly and regrouping the terms of the finite sum, we arrive at $C_n = a_0 B_n + \cdots + a_n B_0$. Since $B_n - B \to 0$, we can rewrite this as $C_n = a_0(B_n - B) + \cdots + a_n(B_0 - B) + A_n B = R_n + A_n B$, say. As $A_n B \to A$, we need to show that $R_n \to 0$. In the expression for R_n, if $n \gg 0$, then $B_n - B$ can be made small whereas when we deal with $B_k - B$ for k small, the term a_{n-k} can be made small. So we employ the divide and conquer method. We need to exercise a little more care here.

Proof. Using an obvious notation, we let A_n, B_n, and C_n denote the partial sums of the three series. Let $D_n := B - B_n$.

$$C_n = c_0 + c_1 + \cdots + c_n$$
$$= (a_0 b_0) + (a_0 b_1 + a_1 b_0) + \cdots + (a_0 b_n + \cdots + a_n b_0)$$
$$= a_0(b_0 + b_1 + \cdots + b_n) + a_1(b_0 + b_1 + \cdots + b_{n-1}) + \cdots + a_n b_0.$$

Hence, we have

$$C_n = a_0 B_n + a_1 B_{n-1} + \cdots + a_n B_0 \tag{5.8}$$
$$= a_0(-B + B_n) + a_1(-B + B_{n-1}) + \cdots + a_n(-B + B_0) + B\left(\sum_{k=0}^{n} a_k\right)$$
$$= A_n B - R_n, \tag{5.9}$$

where $R_n := a_0 D_n + a_1 D_{n-1} + \cdots + a_n D_0$ and $D_n = B_n - B$. Let $\alpha := \sum_n |a_n|$. Since $D_n \to 0$, (D_n) is bounded, say by D: $|D_n| \le D$ for all n. Given $\varepsilon > 0$, there exists $N \in \mathbb{N}$ such that $\sum_{n \ge N} |a_n| < \varepsilon$ and $|D_n| \le \varepsilon$. For all $n \ge 2N$, we have

$$|R_n| \le (|a_0| + \cdots + |a_{n-N}|)\varepsilon + (|a_{n-N+1}| + \cdots + |a_n|)D$$
$$\le (\alpha + D)\varepsilon.$$

Hence $R_n \to 0$. Since $A_n B \to AB$, the result follows from (5.9). □

Remark 5.4.4. What happens if we summed up the expression for C_n as

$$C_n = b_0(a_0 + \cdots + a_n) + \cdots + b_n a_0?$$

Do you see the need to change the hypothesis appropriately?

Example 5.4.5. We know that $\sum_{n=0}^{\infty} z^n = 1/(1-z)$ for $|z| < 1$. If we take $a_n = z^n = b_n$ in the theorem, we get $\sum_{n=1}^{\infty} n z^{n-1} = 1/(1-z)^2$ for $|z| < 1$.

Theorem 5.4.6 (Abel). *Let $\sum a_n$ and $\sum b_n$ be convergent, say, with sums A and B, respectively. Assume that their Cauchy product $\sum c_n$ is also convergent to C. Then $C = AB$.*

Proof. Recall the expression (5.8) for C_n:

$$C_n = a_0 B_n + a_1 B_{n-1} + \cdots + a_n B_0.$$

Adding them for $0 \leq k \leq n$, we obtain

$$C_0 + C_1 + \cdots + C_n = A_0 B_n + A_1 B_{n-1} + \cdots + A_n B_0.$$

Since $C_n \to C$, from Corollary 2.5.5 on page 51, we get $\frac{C_0 + \cdots + C_n}{n} \to C$. Since $A_n \to A$ and $B_n \to B$, by the last example, $\frac{A_n B_0 + \cdots + A_0 B_n}{n} \to AB$. The result follows. $\qquad\square$

Existence of Decimal Expansion. We motivate the study of infinite series by means of the example of decimal expansions of a real number. Let $a \in \mathbb{R}$. Let $a_0 := [a]$, the greatest integer less than or equal to a. Write $a = a_0 + x_1$. Then $0 \leq x_1 < 1$. Observe that $a = a_0 + \frac{10x_1}{10}$. Let $a_1 = [10x_1]$. Then $0 \leq a_1 \leq 9$. Also, $10x_1 = a_1 + x_2 = a_1 + \frac{10x_2}{10}$ with $0 \leq x_2 < 1$. Hence

$$a = a_0 + \frac{a_1}{10} + \frac{10x_2}{10^2} \text{ with } 0 \leq 10x_2 < 10.$$

We let $a_2 := [10x_2]$ and so on. Inductively, we obtain

$$0 \leq a - \left(a_0 + \frac{a_1}{10} + \frac{a_2}{10^2} + + \cdots + \frac{a_n}{10^n} \right) = \frac{10x_{n+1}}{10^{n+1}} < \frac{1}{10^n}.$$

Hence $a = \sum_{n=0}^{\infty} \frac{a_n}{10^n}$, which is usually denoted by $a = a_0.a_1 a_2 \cdots a_n \cdots$.

To understand the decimal expansion above, try to find the decimal expansions of $a = 10/9$, $a = 7/4$, and $a = 4/7$ following the algorithm above. We are sure you will get what you already know!

Chapter 6

Riemann Integration

Contents

The notion of integration was developed much earlier than differentiation. The main idea of integration is to assign a real number A, called the "area", to the region R (Figure 6.1) bounded by the curves $x = a$, $x = b$, $y = 0$, and $y = f(x)$, where we assume that f is non-negative. The number, A, the area of the region R, is called the integral of f over $[a, b]$ and denoted by the *symbol $\int_a^b f(x)\,dx$.*

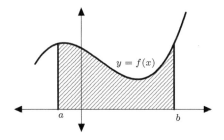

Figure 6.1: Area under the curve.

The most basic geometric region for which the area is known is a rectangle. The area of a rectangle whose sides have lengths ℓ and b is $\ell \cdot b$. This suggests that our definition of an integral should be such that if $f(x) = c$, a constant, then $\int_a^b f(x)\,dx = c(b-a)$.

We use this basic notion of area as the building block to assign an area to the regions under the graphs of bounded functions. To understand the concepts and results of this section, it is suggested that the reader may assume that f is non-negative and draw pictures whenever possible.

Unless specified otherwise, we let $J = [a, b]$ denote a closed and bounded interval and $f, g \colon [a, b] \to \mathbb{R}$ bounded functions.

If $f \colon J \to \mathbb{R}$ is given, and $J_i = [t_i, t_{i+1}]$ is a subinterval of J, we let

$$m_i(f) = \text{glb } \{f(t) : t_i \le t \le t_{i+1}\} \text{ and } M_i(f) = \text{lub } \{f(t) : t_i \le t \le t_{i+1}\}.$$

If f is understood or clear from the context, we simply denote these by m_i and M_i. See Figure 6.2. We also let $m = \text{glb } \{f(x) : x \in [a, b]\}$ and $M = \text{lub } \{f(x) : x \in [a, b]\}$.

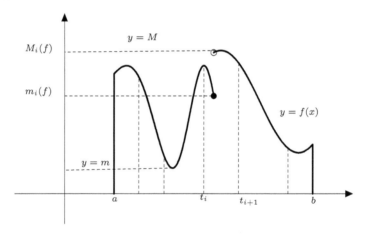

Figure 6.2: $m_i(f)$ and $M_i(f)$.

6.1   Darboux Integrability

Definition 6.1.1. A *partition* or subdivision P of an interval $[a, b]$ is a finite set $\{x_0, x_1, \ldots, x_n\}$ such that $a = x_0 < x_1 < \cdots < x_{n-1} < x_n = b$. The points x_i are called the *nodes* of P.

Example 6.1.2. (i) $P = \{a = x_0, x_1 = b\}$ is the trivial partition of $[a, b]$.
(ii) For any $n \in \mathbb{N}$, let $x_i = a + \frac{i}{n}(b - a)$ for $0 \le i \le n$. Then $\{x_0, \ldots, x_n\}$ is a partition, say, P_n. Note that P_n divides $[a, b]$ in subintervals of equal length. (One often says that P_n divides $[a, b]$ into equal parts.)

Given two partitions P and Q of $[a, b]$, we say that Q is a refinement of P if $P \subset Q$. In the example (ii) above, the partition $P_{2^{k+1}}$ is a refinement of P_{2^k}.

Definition 6.1.3. Given $f : [a, b] \to \mathbb{R}$ and a partition $P = \{x_0, \ldots, x_n\}$ of $[a, b]$, we let

$$L(f, P) := \sum_{i=0}^{n-1} m_i(f)(x_{i+1} - x_i),$$

$$U(f, P) := \sum_{i=0}^{n-1} M_i(f)(x_{i+1} - x_i).$$

Observe that, if $f \geq 0$, $L(f, P)$ (respectively, $U(f, P)$) is the sum of areas of rectangles inscribed inside (respectively, circumscribing) the region bounded by the graph. The numbers $L(f, P)$ and $U(f, P)$ are called the *lower and the upper Darboux sums* of f with respect to the partition P. They approximate the area under the graph from below and from above. See Figure 6.3.

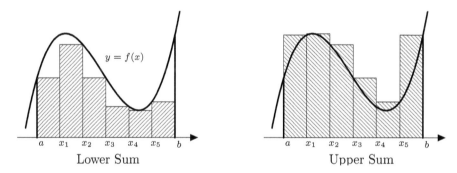

Figure 6.3: Upper and lower Darboux sums.

It is easy to see that $m(b - a) \leq L(f, P) \leq U(f, P) \leq M(b - a)$ for any partition P of $[a, b]$.

Exercise 6.1.4. Let $f : [a, b] \to \mathbb{R}$ be bounded. Let $c \in (a, b)$. Let P_1 (respectively, Show that $P := P_1 \cup P_2$ is a partition of $[a, b]$. Show also that $L(f, P) = L(f, P_1) + L(f, P_2)$ and $U(f, P) = U(f, P_1) + U(f, P_2)$.

A very pedantic formulation is as follows. Let f_1 (respectively, f_2) be the restriction of f to $[a, c]$ (respectively, to $[c, b]$). Prove that $L(f, P) = L(f_1, P_1) + L(f_2, P_2)$ and so on.

Definition 6.1.5. Given a partition $P = \{x_0, \ldots, x_n\}$, we insert a new node, say, t such that $x_i < t < x_{i+1}$ for some i and get a new partition Q.

Then drawing pictures of a non-negative function, it is clear that $L(f, Q) \geq L(f, P)$. Similarly, $U(f, Q) \leq U(f, P)$. Look at Figures 6.4–6.5. (We shall prove this later.) Thus, Q produces a better approximation to the area bound by the graph. This suggests that to get the "real" area we should look at

$$L(f) \equiv \underline{\int_a^b} f(x)\,dx := \text{lub } \{L(f,P) : P \text{ is a partition of } [a,b]\}$$

$$U(f) \equiv \overline{\int_a^b} f(x)\,dx := \text{glb } \{U(f,P) : P \text{ is a partition of } [a,b]\}.$$

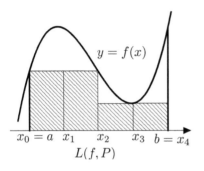

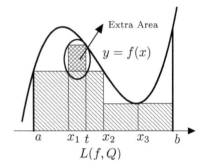

Figure 6.4: $L(f,P) \leq L(f,Q)$.

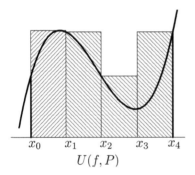

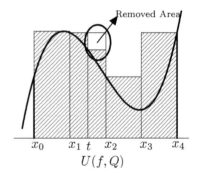

Figure 6.5: $U(f,P) \geq U(f,Q)$.

These numbers exist (why?) and are called the *lower and upper integral* of f on $[a,b]$. The upper integral of f on $[a,b]$ may be understood as the best possible approximation to the area of the region under the graph as approximated from above. How do we understand the lower integral?

Exercise 6.1.6. Show that $L(f) \leq U(f)$ for any bounded function $f : [a,b] \to \mathbb{R}$.

Definition 6.1.7. We say that f is *Darboux integrable (or simply integrable)* on $[a,b]$ if the upper and lower integrals coincide. (This intuitively says that we require that the area should be approximable both from below and from above.) If f is integrable, the common value of the upper and lower integrals is denoted by the symbol $\int_a^b f(x)\,dx$. This is just a notation; we can as well use $I_a^b(f)$ or $\int_a^b f$ etc.

Example 6.1.8. Let $f\colon [a, b] \to \mathbb{R}$ be a constant function with $f(x) = c$ for all x. Let P be any partition of $[a, b]$. Then $m_i = M_i = c$, for all i and $L(f, P) = U(f, P) = c(b-a)$. Thus $L(f) = c(b-a) = U(f)$. That is, f is integrable with its integral $c(b-a)$. This substantiates our idea that the integral of a non-negative function generalizes the area of a rectangle.

Example 6.1.9. Let $f\colon [0, 1] \to \mathbb{R}$ be defined by $f(x) = 1$ if $0 \le x < 1$ and $f(1) = 10^9$. We claim that f is integrable on $[0, 1]$ and $\int_0^1 f = 1$.
    Let $P = \{x_0, \ldots, x_n\}$ be a partition of $[0, 1]$. Note that $m_i(f) = M_i(f) = 1$ for $0 \le i < n - 2$ and $M_{n-1}(f) = 10^9$ and $m_{n-1}(f) = 1$. Hence, we have

$$L(f, P) = 1 \text{ and } U(f, P) = x_{n-1} + 10^9(1 - x_{n-1}).$$

It follows that lub $\{L(f, P) : P\} = 1$ and glb $\{U(f, P) : P\} = 1$. (Why?) This proves that f is integrable with $\int_a^b f(x)\,dx = 1$.

Example 6.1.10. Let us consider the function $f\colon [0, 1] \to \mathbb{R}$ defined by $f(x) = 1$ if $x \in \mathbb{Q}$ and 0 otherwise. This is called Dirichlet's function. Let $P = \{0 = x_1 < x_1 \cdots < x_n = 1\}$ be any partition of $[0, 1]$. By the density of rationals and irrationals, there exist $s_i, t_i \in (x_i, x_{i+1})$ where $s_i \in \mathbb{Q}$ and $t_i \notin \mathbb{Q}$. Hence we conclude that $m_i = 0$ and $M_i = 1$. Also, $L(f, P) = \sum m_i(x_{i+1} - x_i) = 0$ and $U(f, P) = 1$. It follows that $L(f) = 0$ whereas $U(f) = 1$. We therefore conclude that f is not integrable on $[0, 1]$.

Example 6.1.11. Let $f(x) = x$ on $[0, 1]$. Let $P = \{x_i = \frac{i}{4}, i = 0, \ldots 4\}$ and $Q = \{x_j = \frac{j}{8}, j = 0, \ldots, 8\}$. Find $L(f, P)$, $L(f, P)$ and $U(f, Q)$, $U(f, P)$.
    In the subinterval $[x_i, x_{i+1}]$, $m_i(P, f) = x_i$ and $M_i(f, P) = x_{i+1}$.

$$L(f, P) = \sum_{i=0}^{3} \left(\frac{i}{4} \times \frac{1}{4} \right) = \frac{6}{16}$$

$$U(f, P) = \sum_{i=1}^{4} \left(\frac{i}{4} \times \frac{1}{4} \right) = \frac{10}{16}.$$

Similarly, $L(f, Q) = 28/64$ and $U(f, Q) = 36/64$.

    In the above example, notice that Q is a refinement of P. What is the relation between (i) $L(f, P)$ and $L(f, Q)$ and (ii) $U(f, P)$ and $U(f, Q)$? Look at Figures 6.4 and 6.5. They lead us to the next theorem.

Theorem 6.1.12. Let $f\colon [a, b] \to \mathbb{R}$ be a bounded function. Let P and Q be partitions of $[a, b]$. Then we have the following:

(i) *If Q is a refinement of P, then $L(f, P) \le L(f, Q)$ and $U(f, P) \ge U(f, Q)$.*
(ii) *$L(f, P) \le U(f, Q)$ for* any *two partitions P and Q.*
(iii) *The lower integral of f is less than or equal to the upper integral, that is,*
$L(f) \le U(f)$.

Proof. (i) It is enough to prove it when $Q = P \cup \{c\}$, that is, Q contains exactly one extra node. (Draw pictures; you will understand (i) immediately. See also Figures 6.4–6.5.) Let $x_i < c < x_{i+1}$. For $j \ne i$, all the j-th terms, in $L(f, P)$ and $U(f, P)$ will be present in $L(f, Q)$ and $U(f, Q)$. Corresponding to the term $m_i(f)(x_{i+1} - x_i)$, we have two terms in $L(f, Q)$:

$$\text{glb } \{f(x) : x \in [x_i, c]\}(c - x_i) + \text{glb } \{f(x) : x \in [c, x_{i+1}\}(x_{i+1} - c).$$

Using Exercise 3 on page 18, we note that

$$m_i(f) := \text{glb } \{f(x) : x \in [x_i, x_{i+1}]\} \le \text{glb } \{f(x) : x \in [x_i, c]\}(c - x_i)$$
$$m_i(f) := \text{glb } \{f(x) : x \in [x_i, x_{i+1}]\} \le \text{glb } \{f(x) : x \in [c, x_{i+1}]\}(c - x_i).$$

Hence

$$\text{glb } \{f(x) : x \in [x_i, c]\}(c - x_i) + \text{glb } \{f(x) : x \in [c, x_{i+1}]\}(x_{i+1} - c)$$
$$\ge m_i(f)(c - x_i) + m_i(f)(x_{i+1} - c)$$
$$= m_i(f)(x_{i+1} - x_i).$$

It follows that

$$L(f, P) = \left(\sum_{j \ne i} m_j(x_{j+1} - x_j) \right) + m_i(x_{i+1} - x_i) \le L(f, Q).$$

Similarly, we obtain

$$\text{lub } \{f(x) : x \in [x_i, c]\}(c - x_i) + \text{lub } \{f(x) : x \in [c, x_{i+1}]\}(x_{i+1} - c)$$
$$\le M_i(f)(x_{i+1} - x_i)$$

and conclude that $U(f, Q) \le U(f, P)$.

(ii) is easy. Let $P' = P \cup Q$. Note that P' is a refinement of P as well as of Q. Hence, by (i),

$$L(f, P) \le L(f, P') \le U(f, P') \le U(f, Q).$$

(iii) It follows from (ii) that each $U(f, Q)$ is an upper bound for the set $\{L(f, P) : P \text{ a partition of } [a, b]\}$. Hence its lub, namely, the lower integral, will be at most $U(f, Q)$. Thus, the lower integral is a lower bound for the set $\{U(f, Q) : Q \text{ a partition of } [a, b]\}$. Hence its glb, namely, the upper integral is greater than or equal to the lower integral. $\square$

To check integrability of a function directly from the definition can be cumbersome at times. The next result provides a necessary and sufficient condition for integrability of a function.

Theorem 6.1.13 (Criterion for Integrability). *A bounded function $f\colon [a,b] \to \mathbb{R}$ is integrable if and only if for each $\varepsilon > 0$ there exists a partition P such that $U(f,P) - L(f,P) < \varepsilon$.*

Proof. Let the condition be satisfied. We are required to prove that f is integrable. Let I_1 and I_2 be the lower and upper integrals. We need to show $I_1 = I_2$. It is enough to show that for any $\varepsilon > 0$, $|I_1 - I_2| < \varepsilon$.

Since $I_1 \le I_2$ (by Exercise 6.1.6), it is enough to show that $I_2 < I_1 + \varepsilon$ for any $\varepsilon > 0$. Given $\varepsilon > 0$, let P be as in the hypothesis. Observe that

$$I_1 \ge L(f,P) \text{ and } I_2 \le U(f,P).$$

Hence

$$I_2 - I_1 \le U(f,P) - L(f,P) < \varepsilon.$$

Thus we have proved that f is integrable.

To prove the converse, let $\varepsilon > 0$ be given. Let I_1 and I_2 be the lower and upper integrals, then we have $I_1 = I_2$. Since I_1 is the LUB of $L(f,P)$'s, there exists a partition P_1 of $[a,b]$ such that $I_1 - \varepsilon/2 < L(f,P_1)$. That is,

$$I_1 - L(f,P_1) < \varepsilon/2. \tag{6.1}$$

Similarly, there exists a partition P_2 of $[a,b]$ such that

$$U(f,P_2) < I_2 + \varepsilon/2 \text{ so that } U(f,P_2) - I_2 < \varepsilon/2. \tag{6.2}$$

Let $P = P_1 \cup P_2$. Since P is a refinement of P_1 and P_2, we have

$$L(f,P_1) \le L(f,P) \le U(f,P) \le U(f,P_2). \tag{6.3}$$

Now,

$$
\begin{aligned}
U(f,P) &- L(f,P) \\
&= U(f,P) - I_2 + I_2 - L(f,P), &&\text{adding and subtracting } I_2 \\
&= U(f,P) - I_2 + I_1 - L(f,P), &&\text{since } I_1 = I_2 \\
&\le U(f,P_2) - I_2 + I_1 - L(f,P_1), &&\text{by (6.3)} \\
&< \varepsilon/2 + \varepsilon/2, &&\text{by (6.1) and (6.2)} \\
&= \varepsilon.
\end{aligned}
$$

Thus, the condition is necessary. □

Example 6.1.14. Let us look at some examples as applications of the above theorem and check integrability of a given function.

(1) Let $f(x) = x^2$ on $[0,1]$. Let $\varepsilon > 0$ be given. Choose a partition P such that $\max\{x_{i+1} - x_i : 1 \le i \le n-1\} < \varepsilon/2$. Since f is increasing, we obtain

$$m_i(f) = f(x_i) = x_i^2, \text{ and } M_i(f) = f(x_{i+1}) = x_{i+1}^2.$$

It follows that

$$U(f,P) - L(f,P) = \sum_{i=1}^{n} x_i^2(x_i - x_{i-1}) - \sum_{i=1}^{n} x_{i-1}^2(x_i - x_{i-1})$$

$$= \sum_{i=1}^{n} [(x_i - x_{i-1})(x_i + x_{i-1})](x_i - x_{i-1})$$

$$< \sum_{i=1}^{n} \left[\left(\frac{\varepsilon}{2}\right) \times 2\right](x_i - x_{i-1}), \text{ since } 0 \le x_i, x_{i+1} \le 1$$

$$= \varepsilon \sum_{i} (x_i - x_{i-1}) = \varepsilon.$$

Hence, f is integrable by Theorem 6.1.13.

You may try to prove this directly without using Theorem 6.1.13. This may convince you of the significance of the result.

(2) Consider $f \colon [-1,1] \to \mathbb{R}$ defined by

$$f(x) = \begin{cases} a, & -1 \le x < 0 \\ 0, & x = 0 \\ b, & 0 < x \le 1. \end{cases}$$

Look at Figure 6.6.

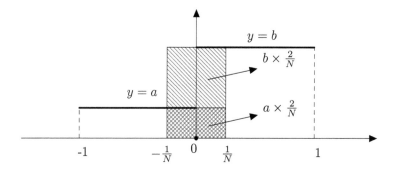

Figure 6.6: Lower and upper sum in the interval $[\frac{-1}{N}, \frac{1}{N}]$.

The trouble is at $x = 0$. We use our divide and conquer trick. We choose a partition in such a way that $x = 0$ is enclosed in a very small subinterval of the partition. Go through the proof with this idea.

Assume that $0 < a < b$. We claim that f is integrable. Given $\varepsilon > 0$, choose $N \in \mathbb{N}$ such that $\frac{2(b-a)}{N} < \varepsilon$. Consider the partition $P = \{-1, \frac{-1}{N}, \frac{1}{N}, 1\}$. Then

$$L(f, P) = a \times \left(1 - \frac{1}{N}\right) + a \times \frac{2}{N} + b \times \left(1 - \frac{1}{N}\right).$$

Similarly, we have

$$U(f, P) = a \times \left(1 - \frac{1}{N}\right) + b \times \frac{2}{N} + b \times \left(1 - \frac{1}{N}\right).$$

Hence, we get

$$U(f, P) - L(f, P) = \frac{2(b-a)}{N} < \varepsilon.$$

(3) The last item can be generalized. Let $P = \{a = x_0, x_1, \dots, x_n\}$ be a partition of $[a, b]$. Let $\sigma \colon [a, b] \to \mathbb{R}$ be defined as

$$\sigma(x) = \begin{cases} c_1, & x \in [a, x_1); \\ c_i, & x \in [x_{i-1}, x_i), \quad 2 \leq i \leq n-1; \\ c_n, & x \in [x_{n-1}, x_n]. \end{cases}$$

Then σ is integrable and we have

$$\int_a^b \sigma(t)\, dt = \sum_{i=1}^n c_i(x_i - x_{i-1}).$$

Functions such as σ are called *step functions*. Observe that next to constant functions, the areas of step functions are easy to write down by looking at their graphs or knowing their definitions!

(4) Recall Thomae's function f on page 76. We claim that f is integrable.

Strategy: Given $\varepsilon > 0$, we must find a partition P such that $U(f, P) - L(f, P) < \varepsilon$. By the density of irrationals, in any subinterval $[x_i, x_{i+1}]$ of a partition P of $[0, 1]$, irrationals exist and hence m_i's are zero and hence $L(f, P) = 0$. Hence we need only show $U(f, P) < \varepsilon$. Let $n \in \mathbb{N}$ be such that $1/n < \varepsilon$. As we argued in Example 3.2.3.9 on page 76, the set $A_n := \{r \in \mathbb{Q} \cap [0, 1] : f(x) > 1/n\}$ is finite. We employ the divide and conquer method. We choose a partition such that each point of A_n is enclosed in a small subinterval the sums of whose lengths is as small as we please. The contribution of the remaining terms in $U(f, P)$ is at most $1/n$.

Let $\varepsilon > 0$ be given. Choose $k \in \mathbb{N}$ such that $\frac{1}{k} < \varepsilon/2$. There exists a finite number, say, N of rational numbers p/q with $q \leq k$. Denote them by $\{r_j : 1 \leq j \leq N\}$. Let $\delta < \varepsilon/(4N)$. Choose a partition $P = \{x_0, \dots, x_n\}$ of $[0, 1]$ such that

$$\max\{|x_{i+1} - x_i| : 0 \leq i \leq n-1\} < \delta.$$

Let $A := \{i : r_j \in [x_i, x_{i+1}],$ for some $j\}$, and $B := \{0, \ldots, n\} \setminus A$. Note that the number of elements in A will be at most $2N$. (Why $2N$? Some r_j could be the left and the right endpoint of adjacent subintervals!) For $i \in A$, we have $M_i \leq 1$. For $j \in B$, we have $M_j < 1/k$. Hence

$$U(f, P) = \sum_{i=0}^{n-1} M_i(x_{i+1} - x_i)$$

$$= \sum_{i \in A} M_i(x_{i+1} - x_i) + \sum_{j \in B} M_j(x_{j+1} - x_j)$$

$$\leq (2N)\delta + \sum_{j \in B} \frac{1}{k}(x_{j+1} - x_j)$$

$$\leq (2N)\delta + \frac{1}{k}$$

$$< \frac{\varepsilon}{2} + \frac{\varepsilon}{2} = \varepsilon.$$

In the above, we used the fact that $\sum_{j \in B}(x_{i+1} - x_i)$ is the sum of the lengths of the disjoint subintervals that lie in B, and hence it is at most 1.

Thus, for any $\varepsilon > 0$, we have found a partition P_ε such that $U(f, P) < \varepsilon$. It follows that glb $\{U(f, P) : P$ is a partition of $[0, 1]\} = 0$. Hence $\int_0^1 f = 0$. Of course, we could use a simpler argument. Since f is integrable, and since each lower sum is zero, it follows that the lower integral is 0. Hence $\int_0^1 f = 0$.

We now give two important classes of integrable functions.

Theorem 6.1.15. *Let* $f : [a, b] \to \mathbb{R}$ *be bounded and monotone. Then* f *is integrable.*

Strategy. We prove the result for an increasing function. If $P = \{a = x_0, x_1, \ldots, x_N = b\}$ is any partition of $[a, b]$, then $m_i = f(x_i)$ and $M_i = f(x_{i+1})$. This means

$$U(f, P) - L(f, P) = \sum_{i=0}^{N-1} f(x_{i+1})(x_{i+1} - x_i) - \sum_{i=0}^{N-1} f(x_i)(x_{i+1} - x_i)$$

$$= \sum_{i=0}^{N-1} [f(x_{i+1}) - f(x_i)](x_{i+1} - x_i).$$

We can choose a partition which divides the interval in N equal parts. In this case, $x_{i+1} - x_i = \frac{b-a}{N}$ for each i. This implies

$$U(f, P) - L(f, P) = \frac{b-a}{N}(f(b) - f(a)).$$

Thus we need to choose N such that $\frac{b-a}{N}(f(b) - f(a)) < \varepsilon$.

Proof. We prove the result for an increasing and bounded function f. Given $\varepsilon > 0$, choose N so that $\frac{(b-a)(f(b)-f(a))}{N} < \varepsilon$ and divide the interval $[a, b]$ in N

equal parts. It is easy to see that in this case, $x_i := a + i\frac{b-a}{N}$, $0 \le i \le N$. Note that $m_i = f(x_i)$ and $M_i = f(x_{i+1})$.

Hence,

$$
\begin{aligned}
U(f,P) &- L(f,P) \\
&= \sum_{i=0}^{N-1} f(x_{i+1})(x_{i+1} - x_i) - \sum_{i=0}^{N-1} f(x_i)(x_{i+1} - x_i) \\
&= \sum_{i=0}^{N-1} [f(x_{i+1}) - f(x_i)](x_{i+1} - x_i) \\
&= \frac{b-a}{N} \left([f(x_1) - f(x_0)] + [f(x_2) - f(x_1)] + \cdots + [f(x_N) - f(x_{N-1})]\right) \\
&= \frac{b-a}{N} (f(b) - f(a)) < \varepsilon.
\end{aligned}
$$

This implies that f is integrable by Theorem 6.1.13.   □

Theorem 6.1.16. *Let $f\colon [a,b] \to \mathbb{R}$ be continuous. Then f is integrable.*

This is the first time where we are seriously using the concept of uniform continuity.

Strategy. For any partition $P = \{a = x_0, x_1, \ldots, x_N = b\}$ of $[a,b]$, we have

$$
U(f,P) - L(f,P) = \sum_{i=0}^{N-1} (M_i - m_i)(x_{i+1} - x_i).
$$

Since f is continuous on $[a,b]$, it is continuous on each subinterval $[x_i, x_{i+1}]$; hence it attains its bounds. In particular, there exist points $t_i, s_i \in [x_i, x_{i+1}]$ such that $m_i = f(t_i)$ and $M_i = f(s_i)$. On the other hand, f is also uniform continuous on $[a,b]$, therefore, for a given $\varepsilon > 0$, there exists $\delta > 0$ such that $|f(y) - f(x)| < \varepsilon$ for $|x - y| < \delta$. This implies

$$
U(f,P) - L(f,P) < \sum_{i=0}^{N-1} \varepsilon(x_{i+1} - x_i) = \varepsilon(b-a).
$$

Now we can make an appropriate choice of δ to complete the proof.

Proof. Let $\varepsilon > 0$ be given. Since f is continuous on $[a,b]$, it is uniformly continuous on $[a,b]$ by Theorem 3.7. Hence there exists $\delta > 0$ such that

$$
|x - y| < \delta \Rightarrow |f(x) - f(y)| < \frac{\varepsilon}{(b-a)}.
$$

Let $N \in \mathbb{N}$ be such that $\frac{1}{N} < \delta$. Let $x_i := a + i\frac{b-a}{N}$, $0 \le i \le N$. Let $P := \{x_i : 0 \le i \le N\}$. Since f is continuous on $[a,b]$, it attains its maximum and minimum on $[x_i, x_{i+1}]$ for each i. Let f attain its maximum and minimum in $[x_i, x_{i+1}]$ at t_i

and s_i, respectively. Since $|t_i - s_i| < 1/N < \delta$, it follows that $M_i - m_i < \varepsilon/(b-a)$ for $0 \le i \le N - 1$.

Therefore,

$$U(f, P) - L(f, P) = \sum_{i=0}^{N-1} (M_i - m_i)(x_{i+1} - x_i)$$

$$< \sum_{i=0}^{N-1} \frac{\varepsilon}{b - a} \frac{b - a}{N}$$

$$= \varepsilon.$$

This implies that f is integrable on $[a, b]$.   □

Exercise 6.1.17. If $f \colon [a, b] \to \mathbb{R}$ is bounded and continuous on $[a, b]$ except at $c \in [a, b]$, show that f is integrable. Can you generalize this result?

Example 6.1.18. Consider $f \colon [0, 1] \to \mathbb{R}$ defined by $f(x) = x^2$. We have already seen that it is integrable in Item 1 in Example 6.1.14. We now compute its integral.

Let $P = \{0 = x_0, x_1, \ldots, x_n = 1\}$ be any partition. Consider $g(x) = x^3/3$. (This is an inspired guess!) Then $g'(x) = f(x)$. By MVT, $g(x_i) - g(x_{i-1}) = f(t_i)(x_i - x_{i-1})$ for some $t_i \in [x_{i-1}, x_i]$. Hence

$$\sum_i f(t_i)(x_i - x_{i-1}) = \sum_i (g(x_i) - g(x_{i-1})) = g(1) - g(0) = 1/3. \qquad (6.4)$$

Since $m_i \le f(t_i) \le M_i$, we see that

$$L(f, P) \le \sum_i f(t_i)(x_i - x_{i-1}) \le U(f, P). \qquad (6.5)$$

It follows from (6.4)–(6.5) that $L(f, P) \le \frac{1}{3} \le U(f, P)$ for all partitions. That is,

$$\text{lub } \{L(f, P) : P\} \le 1/3 \le \text{glb } \{U(f, P) : P\}.$$

Since f is integrable, the first and the third terms are equal. Hence $\int_0^1 f(x)\, dx = 1/3$.   □

Exercise 6.1.19. Show that $f \colon [0, 1] \to \mathbb{R}$ defined by $f(x) = x^n$ is integrable and that $\int_0^1 f = \frac{1}{n+1}$.

6.2   Properties of the Integral

Theorem 6.2.1. Let $f, g \colon [a, b] \to \mathbb{R}$ be integrable functions and $c \in \mathbb{R}$. Then:

(1) cf is integrable and $\int_a^b (cf)(x)\, dx = c \int_a^b f(x)\, dx$.

(2) $f + g$ is integrable and we have $\int_a^b (f + g)(x)\, dx = \int_a^b f(x)\, dx + \int_a^b g(x)\, dx$.

(3) Let $R([a, b])$ denote the set of integrable functions on $[a, b]$. Then $R([a, b])$ is a vector space and the map $f \mapsto \int_a^b f(x)\, dx$ is a linear map.

Proof. First note that if $\emptyset \neq A \subset \mathbb{R}$ is a bounded set and $c \in \mathbb{R}$

$$\text{lub } cA = \begin{cases} c \times \text{lub } A, & \text{if } c \geq 0 \\ c \times \text{glb } A, & \text{if } c < 0. \end{cases} \tag{6.6}$$

Let P be any partition of $[a, b]$; then it is easy to see that (why?)

$$U(cf, P) = \begin{cases} c \times U(f, P), & \text{if } c \geq 0 \\ c \times L(f, P), & \text{if } c < 0. \end{cases} \tag{6.7}$$

and

$$L(cf, P) = \begin{cases} c \times L(f, P), & \text{if } c \geq 0 \\ c \times U(f, P), & \text{if } c < 0. \end{cases} \tag{6.8}$$

Consequently, from (6.6), we have the following

$$U(cf) = \begin{cases} c \times U(f), & \text{if } c \geq 0 \\ c \times L(f), & \text{if } c < 0. \end{cases} \quad \text{and} \quad L(cf) = \begin{cases} c \times L(f), & \text{if } c \geq 0 \\ c \times U(f), & \text{if } c < 0. \end{cases} \tag{6.9}$$

Since f is integrable, we have $U(f) = L(f)$. It follows from (6.9) that $U(cf) = L(cf) = cU(f) = cL(f)$. Thus (i) is proved.

Proof of (ii) We shall find the relations between $U(f + g)$, $U(f)$, and $U(g)$ and, similarly, between $L(f + g, P)$, $L(f, P)$, and $L(g, P)$, respectively. We make use of the following inequalities:

For any nonempty subset $S \subset [a, b]$, we have

$$\text{lub } \{f(x) + g(x) : x \in S\} \leq \text{lub } \{f(x) : x \in S\} + \text{lub } \{g(x) : x \in S\}$$
$$\text{glb } \{f(x) + g(x) : x \in S\} \geq \text{glb } \{f(x) : x \in S\} + \text{glb } \{g(x) : x \in S\}.$$

(Reason: Let F and G denote the GLBs of f and g on S. Then $f(x) + g(x) \geq F + G$ for all $x \in S$ so that $F + G$ is a lower bound for the set $\{f(x) + g(x) : x \in S\}$.)

In particular, $m_i(f+g) \geq m_i(f) + m_i(g)$. Similarly, $M_i(f+g) \leq M_i(f) + M_i(g)$ for each i. It follows that

$$L(f+g, P) \geq L(f, P) + L(g, P) \quad \text{and} \quad U(f+g, P) \leq U(f, P) + U(g, P). \tag{6.10}$$

We recall Example 1.2.12 which says that if $a_i \leq b_i$ for each $i \in I$, then

$$\text{glb } \{a_i : i \in I\} \leq \text{glb } \{b_i : i \in I\} \quad \text{and} \quad \text{lub } \{b_i : i \in I\} \geq \text{lub } \{a_i : i \in I\}.$$

In view of this, we deduce from (6.10) that

$$U(f + g) = \text{glb } \{U(f + g, P) : P\} \leq \text{glb } \{U(f, P) : P\} + \text{glb } \{U(g, P) : P\}$$
$$= U(f) + U(g).$$

Similarly, $L(f + g) \geq L(f) + L(g)$. Thus we have arrived at

$$L(f) + L(g) \leq L(f + g) \leq U(f + g) \leq U(f) + U(g).$$

Since f and g are integrable, $L(f) = U(f)$ and $L(g) = U(g)$. Hence we deduce (1) $L(f+g) = U(f+g)$, that is, $f + g$ is integrable, and (2) $U(f + g) = U(f) + U(g)$ and $L(f + g) = L(f) + L(g)$, that is, $\int_a^b (f + g) = \int_a^b f + \int_a^b g$.

It follows from (i) and (ii) that $R([a, b])$ is a vector space over $\mathbb{R}$ and the map $f \mapsto \int_a^b f(x) \, dx$ is a linear map.   □

Proposition 6.2.2 (Monotonicity of the Integral). *Let $f, g \colon [a, b] \to \mathbb{R}$ be integrable. Assume that $f(x) \leq g(x)$ for $x \in [a, b]$. Then $\int_a^b f(x) \, dx \leq \int_a^b g(x) \, dx$.*

Proof. We make use of Example 1.2.12.

Let P be a partition of $[a, b]$. Let $[t_i, t_{i+1}]$ be a subinterval of the partition. We now apply the first part of Example 1.2.12 to the sets $A := \{f(x) : x \in [t_i, t_{i+1}]\}$ and $B := \{g(x) : x \in [t_i, t_{i+1}]\}$. We conclude that $M_i(f) \leq M_i(g)$. Hence $U(f, P) \leq U(g, P)$. We now apply the second part of the same example to the sets $\{U(f, P) : P\}$ and $\{U(g, P) : P\}$ where P varies over the set of partitions of $[a, b]$. We deduce that

$$U(f) := \mathrm{glb} \, \{U(f, P) : P\} \leq \mathrm{glb} \, \{U(g, P) : P\} = U(g).$$

Since f and g are integrable, we have $U(f) = \int_a^b f(x) \, dx$ and $U(g) = \int_a^b g(x) \, dx$. Hence the result follows.   □

Theorem 6.2.3. *Let f be integrable on $[a, b]$. Assume that $m \leq f(t) \leq M$ for $t \in [a, b]$. Let $g \colon [m, M] \to \mathbb{R}$ be continuous. Then $g \circ f$ is integrable on $[a, b]$.*

Strategy: To prove integrability of $g \circ f$, we need to find a partition P of $[a, b]$ and estimate

$$U(g \circ f, P) - L(g \circ f, P) = \sum (M_i(g \circ f) - m_i(g \circ f))(t_{i+1} - t_i).$$

Note that by the uniform continuity of g, we can estimate the i-th terms for which $M_i(f) - m_i(f)$ is small. Let A denote the set of such i's and B its complement. We resort to the divide and conquer method. We split the sum as a sum over A and another over B. The sum over A is easy to estimate by uniform continuity and can be made less than $\varepsilon(b - a)$ for any preassigned $\varepsilon > 0$. To estimate the j-th term for $j \in B$, we use the crude estimate $M_i(g \circ f) - m_i(g \circ f) < 2C$, where C is a bound for g. So we need to control the sum of the lengths of the j-th subintervals for $j \in B$. This is done in (6.11). To achieve this, we carefully choose the partition P by invoking the integrability of f on $[a, b]$.

Proof. We shall use s, t for elements of $[a, b]$ and x, y for elements of $[m, M]$.

Let $\varepsilon > 0$ be given. Since g is continuous on $[m, M]$, it is uniformly continuous. Hence there exists $\delta > 0$ such that

$$x, y \in [m, M] \quad \text{and} \quad |x - y| < \delta \implies |g(x) - g(y)| < \varepsilon.$$

Let $\delta_1 := \min\{\delta, \varepsilon\}$.

Since f is integrable, by the integrability criterion (Theorem 6.1.13), there exists a partition P of $[a, b]$ such that $U(f, P) - L(f, P) < \eta$, where η is to be specified later.

Let $A := \{i : M_i(f) - m_i(f) < \delta_1\}$ and B its complement.

We claim that if $i \in A$, then $|f(s) - f(t)| < \delta_1$, for $s, t \in [t_i, t_{i+1}]$. We have $f(s) \le M_i(f)$ and $f(t) \ge m_i(f)$. It follows that $f(s) - f(t) \le M_i(f) - m_i(f)$. Interchanging s and t we see that $f(t) - f(s) \le M_i(f) - m_i(f)$. Thus, $|f(s) - f(t)| < M_i(f) - m_i(f)$, Since for $i \in A$, $M_i(f) - m_i(f) < \delta_1$, we conclude that $|f(s) - f(t)| < \delta_1$.

We claim that $M_i(g \circ f) - m_i(g \circ f) \le \varepsilon$ for $i \in A$. If $s, t \in [t_i, t_{i+1}]$, it follows that $|g(f(s)) - g(f(t))| < \varepsilon$ by the definition of δ_1 and uniform continuity of g. In particular, $g(f(s)) - g(f(t)) < \varepsilon$, for $s, t \in [t_i, t_{i+1}]$. For any fixed $t \in [t_i, t_{i+1}]$ and for all $s \in [t_i, t_{i+1}]$, we have,

$$g(f(s)) < g(f(t)) + \varepsilon \implies \text{lub }_{s \in [t_i, t_{i+1}]} g(f(s)) < g(f(t)) + \varepsilon.$$

We arrive at $M_i(g \circ f) < g(f(t)) + \varepsilon$ for any $t \in [t_i, t_{i+1}]$. This leads us to the inequality $M_i(g \circ f) \le m_i(g \circ f) + \varepsilon$. The claim follows now.

If $j \in B$, $M_j(g \circ f) - m_j(g \circ f) \le 2C$ where C is an upper bound for $|g|$ on $[m, M]$. We have

$$\eta > U(f, P) - L(f, P)$$

$$= \sum_{i=0}^{n-1}(M_i - m_i)(t_{i+1} - t_i)$$

$$= \sum_{j \in B}(M_j(f) - m_j(f))(t_{j+1} - t_j) + \sum_{j \notin B}(M_j(f) - m_j(f))(t_{j+1} - t_j)$$

$$\ge \sum_{j \in B}(M_j(f) - m_j(f))(t_{j+1} - t_j)$$

$$\ge \delta_1 \sum_{j \in B}(t_{j+1} - t_j). \tag{6.11}$$

Hence $\sum_{j \in B}(t_{j+1} - t_j) \le \eta/\delta_1$. We therefore obtain

$$U(g \circ f, P) - L(g \circ f, P) = \sum_{i \in A}(M_i(g \circ f) - m_i(g \circ f))(t_{i+1} - t_i)$$

$$+ \sum_{j \in B}(M_j(g \circ f) - m_j(g \circ f))(t_{j+1} - t_j)$$

$$\le \delta_1(b - a) + (2C)(\eta/\delta_1)$$

$$\le \varepsilon(b - a) + (2C)(\eta/\delta_1), \quad \text{since } \delta_1 \le \varepsilon.$$

If we choose $\eta = \delta_1^2$, since $\delta_1 \le \varepsilon$, it follows that,

$$U(g \circ f, P) - L(g \circ f, P) < \varepsilon(b - a + 2C).$$

That is, $g \circ f$ is integrable on $[a, b]$. $\qquad\square$

Have you understood the proof? It is worthwhile for you go through the strategy and the proof once again.

Remark 6.2.4. If we drop the condition that g is continuous, then $g \circ f$ may not be integrable. Consider Thomae's function f. Let $g \colon [0, 1] \to \mathbb{R}$ be defined by $g(x) = 1$ if $x \in (0, 1]$ and $g(0) = 0$. Then g is integrable, but $g \circ f$ is Dirichlet's function in Example 6.1.10 and it is not integrable.

Remark 6.2.5. If g in Theorem 6.2.3 is assumed to be Lipschitz, then the proof is much simpler. Can you see why? Work out the details.

Exercise 6.2.6 (Applications of Theorem 6.2.3). Assume that $f, g \colon [a, b] \to \mathbb{R}$ is integrable. Show that the following are integrable:
(i) $|f|$, (ii) f^2, (iii) fg, (iv) $\max\{f, g\}$ and (v) $\min\{f, g\}$.

Theorem 6.2.7 (Basic Estimate for Integrals). *Let* $f \colon [a, b] \to \mathbb{R}$ *be integrable. Then* $|f|$ *is integrable and we have*

$$\left| \int_a^b f(x)\, dx \right| \le \int_a^b |f(x)|\, dx. \tag{6.12}$$

Proof. We know that $|f|$ is integrable from Theorem 6.2.6. Choose $\varepsilon = \pm 1$ so that $\varepsilon \int_a^b f(t)\, dt = \left| \int_a^b f(t)\, dt \right|$. Then

$$\left| \int_a^b f(t)\, dt \right| = \varepsilon \int_a^b f(t)\, dt$$

$$= \int_a^b \varepsilon f(t)\, dt, \qquad \text{(by linearity of the integral)}$$

$$\le \int_a^b |f(t)|\, dt, \qquad \text{(by monotonicity, since } \pm f \le |f|).$$

$\square$

Exercise 6.2.8. (a) Show that $\int_0^1 \frac{x^4}{\sqrt{1+4x^{90}}} \ge \frac{1}{10\sqrt{2}}$. (b) Prove that $\int_0^3 \frac{x^3(x-4)}{1+x^{10}} \sin(2013x)\, dx \le 81$.

Theorem 6.2.9 (Additivity of the Integral as an Interval Function). *If* f *is integrable on* $[a, b]$, *then* f *is integrable on any subinterval* $[c, d]$ *of* $[a, b]$. *Furthermore,*

$$\int_a^b f(x)\, dx = \int_a^c f(x)\, dx + \int_c^b f(x)\, dx, \tag{6.13}$$

for any $c \in (a, b)$.

Proof. We show that f is integrable on $[a, c]$ for $c \in (a, b)$. The general case is proved in a similar vein. Let $\varepsilon > 0$ be given. Let P be a partition of $[a, b]$ such that $U(f, P) - L(f, P) < \varepsilon$.

Let $Q = P \cup \{c\}$ and $P_1 := Q \cap [a, c]$. Note that P_1 is a partition of $[a, c]$. If we let $P_2 := Q \cap [c, b]$, then $Q = P_1 \cup P_2$. We have $U(f, Q) - L(f, Q) \leq U(f, p) - L(f, P) < \varepsilon$. We observe that

$$U(f, Q) - L(f, Q) = [U(f, P_1) - L(f, P_1)] + [U(f, P_2) - L(f, P_2)]$$
$$\geq U(f, P_1) - L(f, P_1).$$

It follows from the displayed inequalities that

$$U(f, P_1) - L(f, P_1) \leq U(f, Q) - L(f, Q)$$
$$\leq U(f, P) - L(f, P) < \varepsilon.$$

Hence f is integrable on $[a, c]$.

Let P be *any* partition of $[a.b]$. Let $Q_0 = P \cup \{c\}$, $Q_1 = Q \cap [a, c]$ and $Q_2 = Q \cap [c, b]$. We have

$$U(f, P) \geq U(f, Q) = U(f, Q_1) + U(f, Q_2)$$
$$\geq \overline{\int_a^c} f(x)\,dx + \overline{\int_c^b} f(x)\,dx$$
$$= \int_a^c f(x)\,dx + \int_c^b f(x)\,dx.$$

Thus for any partition P, we have proved

$$\int_a^c f(x)\,dx + \int_c^b f(x)\,dx \leq U(f, P).$$

This shows that $\int_a^c f(x)\,dx + \int_c^b f(x)\,dx$ is a lower bound of

$$\{L(f, P); P \text{ is a partition of } [a, b]\}.$$

Hence

$$\int_a^c f(x)\,dx + \int_c^b f(x)\,dx \leq U(f, P)$$
$$\leq \int_a^b f(x)\,dx$$
$$= \text{glb } \{U(f, P)\}. \tag{6.14}$$

Similarly, we can show that

$$L(f, P) \leq L(f, Q) = L(f, Q_1) + L(f, Q_2)$$
$$\leq \underline{\int_a^c} f(x)\,dx + \underline{\int_c^b} f(x)\,dx$$
$$= \int_a^c f(x)\,dx + \int_c^b f(x)\,dx.$$

Hence $\int_a^c f(x)\,dx + \int_c^b f(x)\,dx$ is an upper bound of

$$\{L(f, P); P \text{ is a partition of } [a, b]\}.$$

Hence

$$\int_a^b f(x)dx = \text{lub } \{L(f, P); P \text{ is a partition of } [a, b]\}$$

$$\leq \int_a^c f(x)\,dx + \int_c^b f(x)\,dx. \tag{6.15}$$

Hence by (6.14) and (6.15), we get

$$\int_a^b f(x)\,dx = \int_a^c f(x)\,dx + \int_c^b f(x)\,dx.$$

The rest of the cases can either be reduced to this or proved in a similar way. □

Definition 6.2.10. Let $f\colon [a, b] \to \mathbb{R}$ be integrable. We *define*

$$\int_b^a f(x)\,dx := -\int_a^b f(x)\,dx \text{ and } \int_s^s f(x)\,dx = 0.$$

Using this convention, the following corollary is an easy consequence of Theorem 6.2.9.

Corollary 6.2.11. *For any a, b, c such that f is integrable on the smallest interval containing a, b, c, we have*

$$\int_a^b f(x)\,dx := \int_a^c f(x)\,dx + \int_c^b f(x)\,dx.$$

Proof. If $a < c < b$, the result follows from the last theorem. So let us assume that c lies outside the interval defined by a and b. For definiteness sake, assume $a < b < c$. Then, by the last result,

$$\int_a^c f(x)\,dx = \int_a^b f(x)\,dx + \int_b^c f(x)\,dx.$$

Hence

$$\int_a^b f(x)\,dx = \int_a^c f(x)\,dx - \int_b^c f(x)\,dx$$

$$= \int_a^c f(x)\,dx + \int_c^b f(x)\,dx,$$

by the convention.   □

Exercise Set 6.2.12. Let $\mathcal{R}(I)$ denote the set of integrable functions on the interval I.

(1) Show that $f \in \mathcal{R}(I)$ implies that $f^2 \in \mathcal{R}(I)$.

(2) Let $f^2 \in \mathcal{R}(\mathcal{I})$. Does it imply that $f \in \mathcal{R}(\mathcal{I})$?

(3) If $f, g \in \mathcal{R}(I)$, then show that $fg \in \mathcal{R}(I)$.

(4) Let $f \in C[a, b]$. If $\int_a^b f(x)dx = 0$, then $f(c) = 0$ for at least one $c \in [a, b]$.

(5) Let $f \in \mathcal{R}(I)$, $f \geq 0$, and $\int_I f = 0$. Then $f = 0$ at each point of continuity of f.

(6) If $f \in C[a, b]$, $\int_a^b f(x)g(x) = 0$, for all $g \in C[a, b]$, then $f \equiv 0$.

(7) Let $f > 0$ be continuous on $[a, b]$. Let $M = \max\{f(x) : x \in [a, b]\}$. Then

$$\lim_{n \to \infty} \left(\int_a^b [f(x)]^n \right)^{\frac{1}{n}} = M.$$

Compare this with item 4 of Exercise 2.4.3.

(8) Let $f \colon I := [a, b] \to \mathbb{R}$ be a bounded function. Show that $f \in \mathcal{R}(I)$ iff for all $\varepsilon > 0$, there exist step functions s_1, s_2 on I such that $s_1(x) \leq f(x) \leq s_2(x)$ and $\int_I (s_2 - s_1) < \varepsilon$. (We defined step functions on page 183.)

(9) Let $f \colon [a, b] \to \mathbb{R}$ be continuous. Show that there exists $c \in [a, b]$ such that

$$\frac{1}{b - a} \int_a^b f(x) \, dx = f(c).$$

(10) Suppose $f, g \colon [a, b] \to \mathbb{R}$ are positive and continuous functions. Show that there exists $c \in [a, b]$ such that

$$\int_a^b f(x)g(x) \, dx = f(c) \int_a^b g(x) \, dx.$$

(11) Let $0 < a < b$. Let $f > 0$ be continuous and strictly increasing on $[a, b]$. Prove that

$$\int_a^b f + \int_{f(a)}^{f(b)} f^{-1} = bf(b) - af(a). \tag{6.16}$$

Use this result to evaluate the following: (i) $\int_a^b x^{\frac{1}{3}} dx$, $0 < a < b$, (ii) $\int_0^1 \sin^{-1} x dx$.

(12) Prove the following version of Young's inequality. Let f be a continuous and strictly increasing function for $x \geq 0$ with $f(0) = 0$. Let g be the inverse of f. Then for any $a, b > 0$ we have

$$ab \leq \int_0^a f(x)\,dx + \int_0^b g(y)\,dy. \qquad (6.17)$$

Equality holds if and only if $b = f(a)$.

(13) Take $f(x) := x^\alpha$ in Young's inequality (6.17) to deduce the original Young's inequality: Let $p > 0$, $q > 0$ be such that $\frac{1}{p} + \frac{1}{q} = 1$. Deduce the inequality:

$$(x^p/p) + (y^q/q) \geq xy \text{ for all } x > 0 \text{ and } y > 0. \qquad (6.18)$$

The equality holds if and only if $x^{p-1} = y$ iff $x^{1/q} = y^{1/p}$.

(14) For $x, y \in \mathbb{R}^n$ and for $1 \leq p < \infty$, let $\|x\|_p := \left(\sum_i |x_i|^p\right)^{1/p}$ and for $p = \infty$, let $\|x\|_\infty := \max\{|x_i| : 1 \leq i \leq n\}$. For $p > 1$, let q be such that $(1/p) + (1/q) = 1$. For $p = 1$ take $q = \infty$. Prove Hölder's inequality:

$$\sum_i |a_i|\,|b_i| \leq \|a\|_p\,\|b\|_q, \text{ for all } a, b \in \mathbb{R}^n. \qquad (6.19)$$

When does equality occur?

6.3   Fundamental Theorems of Calculus

We now look at one of the most important results in the theory of integration, namely, the fundamental theorems of calculus. These theorems establish the validity of the computation of integrals via Newtonian calculus, as learned in high school. In some sense, they justify the high-school way of defining integration as finding an anti-derivative.

Theorem 6.3.1 (First Fundamental Theorem of Calculus). *Let $f \colon [a, b] \to \mathbb{R}$ be differentiable. Assume that f' is integrable on $[a, b]$. Then*

$$\int_a^b f'(x)\,dx = f(b) - f(a).$$

Proof. We adapt the proof of Example 6.1.18. Let $P = \{x_0, x_1 \dots, x_n\}$ be any partition of $[a, b]$. By MVT, we obtain

$$f(x_i) - f(x_{i-1}) = f'(t_i)(x_i - x_{i-1}), \text{ for some } t_i \in (x_{i-1}, x_i). \qquad (6.20)$$

It follows that

$$\sum_i f'(t_i)(x_i - x_{i-1}) = \sum_{i=0}^{n-1} (f(x_{i+1}) - f(x_i)) = f(b) - f(a). \qquad (6.21)$$

Arguing as in Example 6.1.18, we arrive at the result. (Do you see where we used the integrability of f'?) □

Remark 6.3.2. Note that this justifies what you learned in school about the integral being anti-derivative. That is, to find $\int_a^b f(x)\,dx$, we find a function g such that $g' = f$ and then in this case we have $\int_a^b f(x)\,dx = g(b) - g(a)$.

Exercise Set 6.3.3.

(1) Let $f : [0, a] \to \mathbb{R}$ be given by $f(x) = x^2$. Find $\int_0^a f(x)dx$.

(2) Show that $\int_0^a f(x)dx = \frac{a^4}{4}$ for $f(x) = x^3$.

(3) Let $0 < a \le 1$. Show that $\int_0^a \sin x = 1 - \cos a$.

Let $f \colon [a, b] \to \mathbb{R}$ be integrable. Then for any $x \in [a, b]$, we know that f is integrable on $[a, x]$. Hence we have a function $F \colon x \mapsto \int_a^x f(t)\,dt$, $x \in [a, b]$. The new function F is called the *indefinite integral* of f. This is the area under the curve $y = f(x)$ between the x-axis, $x = a$, and x. Look at Figure 6.7.

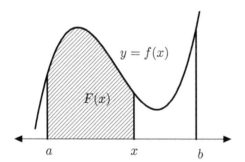

Figure 6.7: Indefinite integral.

Theorem 6.3.4 (Second Fundamental Theorem of Calculus). *Let $f \colon [a, b] \to \mathbb{R}$ be integrable. The indefinite integral F of f is continuous (in fact, Lipschitz) on $[a, b]$ and is differentiable at x if f is continuous at $x \in [a, b]$. In fact, $F'(x) = f(x)$, if f is continuous at x.*

Why is this result plausible? Look at Figure 6.8. It seems that $\int_x^{x+h} f(t)\,dt$ is approximately the area of the rectangle whose base is h and height is $f(x)$. Hence $\frac{1}{h} \int_x^{x+h} f(t)\,dt \approx f(x)$. Observe that

$$\frac{1}{h} \left(\int_a^{x+h} f(t)\,dt - \int_a^x f(t)\,dt \right) = \frac{1}{h} \int_x^{x+h} f(t)\,dt \approx f(x).$$

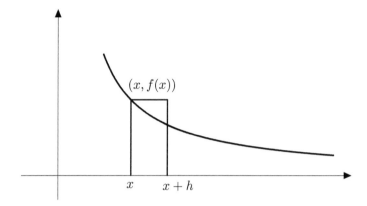

Figure 6.8: $\int_x^{x+h} f(t)dt \approx [(x+h)-x]f(x) = hf(x)$.

Proof. Since F is bounded, there exists M such that $|f(x)| \le M$ for $x \in [a,b]$. Then we have

$$|F(x) - F(y)| = \left| \int_a^x f(t)\,dt - \int_a^y f(t)\,dt \right|$$

$$= \left| \int_x^y f(t)\,dt \right|$$

$$\le M \left| \int_x^y 1\,dt \right|$$

$$= M\,|x-y|\,.$$

Thus F is Lipschitz and hence continuous on $[a,b]$.

Let f be continuous at $c \in [a,b]$. We shall show that F is differentiable at c and $F'(c) = f(c)$. Observe that, for $x > c$,

$$\frac{F(x) - F(c)}{x-c} = \frac{1}{x-c}\int_c^x f(t)\,dt \quad \text{and} \quad f(c) = \frac{1}{x-c}\int_c^x f(c).$$

Hence, we obtain

$$\left| \frac{F(x)-F(c)}{x-c} - f(c) \right| = \left| \frac{1}{x-c}\int_c^x f(t)\,dt - \frac{1}{x-c}\int_c^x f(c)\,dt \right|$$

$$= \left| \frac{1}{x-c}\int_c^x f(t) - f(c)]\,dt \right|$$

$$\le \frac{1}{x-c}\int_c^x |f(t)-f(c)|\,dt. \qquad (6.22)$$

Given $\varepsilon > 0$, by the continuity of f at c, we can find a $\delta > 0$ such that $|f(t) - f(c)| < \varepsilon$ for $|t-c| < \delta$. Hence for $x \in [a,b]$ such that $|x-c| < \delta$, we see that the RHS of (6.22) is estimated above by ε. Similar argument applies when $x < c$.

This shows that F is differentiable at c and $F'(c) = f(c)$. □

Remark 6.3.5. Look at the inequality (6.22). One of the terms in RHS is an integral while the other is a number $f(c)$. We re-wrote this as a sum of two integrals by observing that $f(c)$ is the average $\frac{1}{x-c} \int_c^x f(c)$ and then applied the linearity, continuity, and the standard estimate for the integral. Learn this well as this trick is often used.

Remark 6.3.6. We can deduce a weaker version of the first fundamental Theorem 6.3.1 from the second fundamental theorem of calculus.

Let $f \colon [a, b] \to \mathbb{R}$ be differentiable with f' continuous on $[a, b]$. Then

$$\int_a^b f'(x)\, dx = f(b) - f(a).$$

Proof. Since f' is continuous, it is integrable and its indefinite integral, say, $G(x) = \int_a^x f'(t)\, dt$, exists. By the last item, G is differentiable with derivative $G' = f'$. Hence the derivative of $f - G$ is zero on $[a, b]$ and hence the function $f - G$ is a constant on $[a, b]$. In particular, $f(a) - G(a) = f(b) - G(b)$, that is, $f(a) = f(b) - \int_a^b f'(x)\, dx$. □

Exercise Set 6.3.7. Assume that the domain of functions below are $[-R, R]$, $R > 0$.

(1) Let $f(x) = -1$ if $x < 0$ and $f(x) = 1$ if $x \geq 0$. What is $\int_0^x f(t)\, dt$? What does the second mean value theorem for integrals say in this case?

(2) Let $f(x) = |x|$. What is the indefinite integral of f? Call it f_1. Show that f_1 is C^1, that is, once continuously differentiable. Can you generate functions that are C^n but not C^{n+1}?

(3) Let $f(x) = \begin{cases} 1, & x \neq 1 \\ 0, & x = 1. \end{cases}$ Let g be the indefinite integral of f. Show that g is differentiable at $x = 1$. Note that f is not continuous at $x = 1$.

Theorem 6.3.8 (Integration by Parts). *Let $u, v \colon [a, b] \to \mathbb{R}$ be differentiable. Assume that u', v' are integrable on $[a, b]$. Then*

$$\int_a^b u(x)v'(x)\, dx = u(x)v(x)\,\big|_a^b - \int_a^b u'(x)v(x)\, dx. \qquad (6.23)$$

Strategy: Let $g := uv$. Then g is integrable, and $g' = u'v + uv'$ is integrable. (Why? If we assume that u and v are continuously differentiable, then the integrability of g' etc. are clear.) Apply the first fundamental theorem of calculus to $\int_a^b g'(x)\, dx$ to arrive at the result.

Proof. We assume that u and v are continuously differentiable functions. Then $g = uv$ is continuous and hence integrable. Also $g' = u'v + uv'$. Furthermore, g' is continuous and hence integrable. Applying the first fundamental theorem of calculus, we obtain

$$g(b) - g(a) = u(b)v(b) - u(a)v(a)$$
$$= \int_a^b u(x)v'(x)\, dx + \int_a^b u'(x)v(x)\, dx$$

The term $u(b)v(b) - u(a)v(a)$, we write as $u(x)v(x)\,|_a^b$. Hence we have

$$\int_a^b u(x)v'(x)\, dx = u(x)v(x)\,|_a^b - \int_a^b u'(x)v(x)\, dx.$$

$\square$

One of the most basic tools for computing integration in high school is integration by substitution or the change of variables. The following result justifies this process.

Theorem 6.3.9 (Change of Variables). *Let I, J be closed and bounded intervals. Let $u\colon J \to \mathbb{R}$ be continuously differentiable. Let $u(J) \subset I$ and $f\colon I \to \mathbb{R}$ be continuous. Then $f \circ u$ is continuous on J and we have*

$$\int_a^b f(u(x))u'(x)\, dx = \int_{u(a)}^{u(b)} f(y)\, dy, \quad a, b \in J. \tag{6.24}$$

Proof. Fix $c \in I$. Let $F(y) := \int_c^y f(t)\, dt$. Then by the second fundamental theorem of calculus (Theorem 6.3.4), F is differentiable and $F'(x) = f(x)$. Let $g(x) := (F \circ u)(x)$. Then g is differentiable, and by the chain rule we have

$$g'(x) = F'(u(x))u'(x) = f(u(x))u'(x).$$

We apply the first fundamental theorem of calculus to g':

$$\int_a^b f(u(x))u'(x)\, dx = \int_a^b g'(x)\, dx$$
$$= g(b) - g(a)$$
$$= F(u(b)) - F(u(a))$$
$$= \int_c^{u(b)} f(t)\, dt - \int_c^{u(a)} f(t)\, dt$$
$$= \int_{u(a)}^{u(b)} f(t)\, dt.$$

$\square$

Exercise Set 6.3.10.

(1) Let $f: [a, b] \to \mathbb{R}$ be continuous. Assume that there exist constants α and β such that

$$\forall c \in [a, b], \text{ we have } \alpha \int_a^c f(x)\,dx + \beta \int_c^b f(x)\,dx = 0.$$

Show that $f = 0$.

(2) Let $g : \mathbb{R} \to \mathbb{R}$ be differentiable. Let $F(x) = \int_0^{g(x)} t^2 dt$. Prove that $F'(x) = g^2(x)g'(x)$ for all $x \in \mathbb{R}$. If $G(x) = \int_{h(x)}^{g(x)} t^2 dt$, then what is $G'(x)$?

(3) Let $f: [a, b] \to \mathbb{R}$ be continuous and $g: [c, d] \to [a, b]$ be differentiable. Define $\varphi(x) := \int_a^{g(x)} f(t)\,dt$. Prove that h is differentiable and compute its derivative.

(4) If f'' is continuous on $[a, b]$, show that $\int_a^b xf''(x)dx = [bf'(b) - f(b)] - [af'(a) - f(a)]$.

(5) Let $f > 0$ be continuous on $[1, \infty)$. Let $g(x) := \int_1^x f(t)dt \leq [f(x)]^2$. Prove that $f(x) \geq \frac{1}{2}(x - 1)$.

(6) Let $f: [a, b] \to \mathbb{R}$ be continuously differentiable. Assume that $f' > 0$. Prove

$$\int_a^b f + \int_{f(a)}^{f(b)} f^{-1} = bf(b) - af(a).$$

(7) Let $f(x) = \int_{\sin x}^{\cos x} \sqrt{1 - t^2}\,dt$, for $x \in [0, \pi/2]$. Show that $f(x) = \pi/4 - x$.

6.4 Mean Value Theorems for Integrals

Given an integrable function $f: [a, b] \to \mathbb{R}$, the number $\frac{1}{b-a}\int_a^b f$ is called the *mean* or the *average* of the function on the interval $[a, b]$. Observe that this is based on our intuitive way of thinking that the integral is a "continuous sum" of $f(x)$ as x varies over $[a, b]$, and the average of a finite set of numbers is their sum divided by the number of elements in the set. The next result says that the mean of a continuous function on an interval is taken by the function in the interval.

Theorem 6.4.1 (Mean Value Theorem for Integrals). *Let f be continuous on $[a, b]$. Then there exists $c \in [a, b]$ such that $\frac{1}{b-a}\int_a^b f(x)dx = f(c)$.*

Observe that the result has a geometric interpretation as shown in Figure 6.9. If we interpret the integral as the area of (a continuous non-negative) f, the region bounded by $x = a$, $x = b$ and $y = 0$ and $y = f(x)$, then it is the area of a rectangle whose length is $b - a$ and breadth is $f(c)$ for some $c \in [a, b]$.

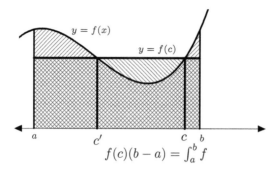

Figure 6.9: Mean value theorem for integrals.

Proof. Let $m = \inf f$, $M = \sup f$ on $[a, b]$. By the extreme values theorem, there exist x_1, x_2 such that $f(x_1) = m$ and $f(x_2) = M$. Then $m \leq f(x) \leq M$ implies

$$\int_a^b m \leq \int_a^b f(x)dx \leq \int_a^b M.$$

This implies

$$m(b - a) \leq \int_a^b f(x)dx \leq M(b - a).$$

Hence

$$f(x_1) = m \leq \frac{1}{b - a}\int_a^b f(x)dx \leq M = f(x_2).$$

Hence, by the intermediate value theorem, there exists c between x_1 and x_2 such that $\frac{1}{b-a}\int_a^b f(x)dx = f(c)$. Since $x_1, x_2 \in [a, b]$, an interval, it follows that $c \in [a, b]$.   □

Theorem 6.4.2 (Weighted/First Mean Value Theorem for Integrals). *Let f, g be continuous on $[a, b]$. Assume that g does not change sign on $[a, b]$. Then for some $c \in [a, b]$ we have $\int_a^b f(x)g(x)dx = f(c)\int_a^b g(x)dx$.*

Proof. Without loss of generality, assume that $g \geq 0$. Let $m = \inf f$, $M = \sup f$ on $[a, b]$. Then $m \leq f(x) \leq M$. Since $g > 0$, this implies

$$mg(x) \leq f(x)g(x) \leq Mg(x).$$

Hence by the monotonicity of the integral,

$$m\int_a^b g(x)dx \leq \int_a^b f(x)g(x)dx \leq M\int_a^b g(x)dx.$$

If $\int_a^b g(x) = 0$, there is nothing to prove. Otherwise, $\int_a^b g(x)dx > 0$. Dividing by $\int_a^b g(x)\,dx$ throughout, we get

$$m \leq \frac{\int_a^b f(x)g(x)\,dx}{\int_a^b g(x)\,dx} \leq M.$$

Arguing as in the last result, by the intermediate value theorem, there exists $c \in [a, b]$ such that

$$\int_a^b f(x)g(x)dx = f(c) \int_a^b g(x)\, dx.$$

If $g \leq 0$, we apply the above argument to $-g$. $\qquad\qquad\qquad\square$

Exercise 6.4.3. Derive Theorem 6.4.1 from Theorem 6.4.2.

Theorem 6.4.4 (Second Mean Value Theorem I). *Let g be continuous on $[a, b]$ and f be continuously differentiable on $[a, b]$. Further assume that f' does not change sign on $[a, b]$. Then there exists $c \in [a, b]$ such that*

$$\int_a^b f(x)g(x)dx = f(a) \int_a^c g(x)dx + f(b) \int_c^b g(x)dx.$$

Proof. Let $G(x) = \int_a^x g(t)dt$. Since g continuous, by Theorem 6.3.4, G is differentiable and $G'(x) = g(x)$. Hence integration by parts gives us:

$$\int_a^b f(x)g(x)dx = \int_a^b f(x)G'(x)dx = f(b)G(b) - \int_a^b f'(x)G(x)dx, \qquad (6.25)$$

since $G(a) = 0$. By the weighted/first mean value theorem (Theorem 6.4.2),

$$\int_a^b f'(x)G(x)dx = G(c) \int_a^b f'(x)dx = G(c)[f(b) - f(a)]. \qquad (6.26)$$

Thus, using Equations (6.25) and (6.26), we get

$$\int_a^b f(x)g(x)dx = f(b)G(b) - G(c)(f(b) - f(a))$$
$$= f(a)G(c) + f(b)[G(b) - G(c)].$$

The desired result follows from this. $\qquad\qquad\qquad\qquad\qquad\square$

We give another version which is very useful in numerical approximation of definite integrals.

Theorem 6.4.5 (Second Mean Value Theorem II). *Let $f : [a, b] \to \mathbb{R}$ be monotone. Then there exists $c \in [a, b]$ such that*

$$\int_a^b f(x)\, dx = f(a)(c - a) + f(b)(b - c).$$

Proof. Assume that f is increasing. We know that $f \in R([a, b])$ by Theorem 6.2.2. We observe that $f(a) \leq f(x) \leq f(b)$ for $x \in [a, b]$. Thus we get,

$$f(a)(b - a) \leq \int_a^b f(x)\, dx \leq f(b)(b - a).$$

Define $h(x) := f(a)(x-a) + f(b)(b-x)$. Then h is continuous. Note that $h(a) = f(b)(b-a)$ and $h(b) = f(a)(b-a)$. Thus, $\int_a^b f(x)\,dx$ is a value lying between the values $h(b)$ and $h(a)$. By the intermediate value theorem applied to h, we conclude that there exists $c \in (a, b)$ such that $h(c) = \int_a^b f(x)\,dx$. This is as required. □

Exercise Set 6.4.6.

(1) Use the first mean value theorem to prove that for $-1 < a < 0$ and $n \in \mathbb{N}$,
$$s_n := \int_a^0 \frac{x^n}{1+x}\,dx \to 0 \text{ as } n \to \infty.$$

(2) Prove that for $0 < a \le 1$ and $n \in \mathbb{N}$, $s_n := \int_0^a \frac{x^n}{1+x}\,dx \to 0$ as $n \to \infty$.

Remark 6.4.7. The various mean value theorems for integrals are quite useful while estimating integrals. It is a commonly held belief among analysts that it is easier to estimate integrals than an infinite series.

Theorem 6.4.8 (Cauchy-Maclaurin Integral Test). *Let $f : [1, \infty) \to \mathbb{R}$ be positive and non-increasing. Let $I_n := \int_1^n f(x)\,dx$ and $s_n := \sum_{k=1}^n f(k)$. Then:*
    (i) *the sequences (I_n) and (s_n) both converge or diverge.*
    (ii) *$\sum_{k=1}^n f(k) - \int_1^n f(x)\,dx \to \ell$ where $0 \le \ell \le f(1)$.*

Proof. Since f is monotone, f is integrable on $[1, n]$ for any $n \in \mathbb{N}$. Observe that $f(k) \le f(x) \le f(k-1)$ for $x \in [k-1, k]$. (See Figure 6.10.) Hence by the monotonicity of the integral, we obtain

$$f(k-1) \ge \int_{k-1}^k f(x)\,dx \ge f(k). \tag{6.27}$$

Adding these inequalities and using the additivity of the integral, we get

$$\sum_{k=1}^{n-1} f(k) \ge \int_1^n f(x)\,dx \ge \sum_{k=2}^n f(k). \tag{6.28}$$

From (6.28), (i) follows from comparison test.

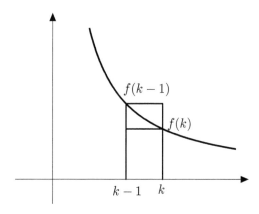

Figure 6.10: $f(k-1)(k-(k-1)) \ge \int_{k-1}^k f(t)\,dt \ge f(k)(k-(k-1))$.

To prove (ii), we define $\varphi(n) := \sum_{k=1}^{n} f(k) - \int_{1}^{n} f(x)\,dx$. Then φ is non-increasing using (6.27). Using (6.28), φ satisfies $0 \le \varphi(n) \le f(1)$. That is, $(\varphi(n))$ is a bounded monotone sequence and therefore is convergent, (by Theorem 2.3.2, on page 43).   □

Example 6.4.9 (Euler's Constant γ). Consider a sequence

$$\gamma_n := 1 + \frac{1}{2} + \frac{1}{3} + \cdots + \frac{1}{n} - \log n.$$

Applying Theorem 6.4.8 with $f(x) = \frac{1}{x}$, we conclude that γ_n is convergent. We denote the limit by γ. Then $0 < \gamma < 1$.

It is not known whether Euler's constant is rational or irrational at the time of this writing.

Exercise Set 6.4.10. Prove that

(1) $\sum_{1}^{\infty} \frac{1}{1+n^2} < \frac{1}{2} + \frac{1}{4\pi}$.

(2) If $-1 < k \le 0$, $1^k + 2^k + \cdots + n^k - \frac{n^{k+1}}{k+1}$ is convergent.

(3) If $-1 < k \le 0$, $\frac{1^k + 2^k + \cdots + n^k}{n^{k+1}} \to \frac{1}{k+1}$.

(4) $p \sum_{1}^{\infty} \frac{1}{n^{1+p}} \to 1$, as $p \to 0+$.

6.5 Integral Form of the Remainder in Taylor's Theorem

We now state and prove a version of Taylor's theorem in which the remainder term is given as an integral. As said earlier in Remark 6.4.7, integrals are easier to estimate. See Section 7.7 for a demonstration of this dictum.

Theorem 6.5.1 (Taylor's Theorem with Integral Form of the Remainder). *Let f be a function on an interval J with $f^{(n)}$ continuous on J. Let $a, b \in J$. Then*

$$f(b) = f(a) + \frac{f'(a)}{1!}(b-a) + \cdots + \frac{f^{(n-1)}(a)}{(n-1)!}(b-a)^{n-1} + R_n \qquad (6.29)$$

where

$$R_n = \int_{a}^{b} \frac{(b-t)^{n-1}}{(n-1)!} f^{(n)}(t)\,dt. \qquad (6.30)$$

Proof. We begin with

$$f(b) = f(a) + \int_a^b f'(t)\, dt.$$

We apply the integration by parts formula $\int_a^b u\, dv = uv\big|_a^b - \int u'\, dv$ to the integral where $u(t) = f'(t)$ and $v = -(b-t)$. (Note the *non-obvious* choice of v!) We get

$$\int_a^b f'(t)\, dt = -f'(t)(b-t)\big|_a^b + \int_a^b f''(t)(b-t)\, dt.$$

Hence we get

$$f(b) = f(a) + f'(a)(b-a) + \int_a^b f''(t)(b-t)\, dt.$$

We again apply integration by parts to the integral where $u(t) = f''(t)$ and $v(t) = -(b-t)^2/2$. We obtain

$$\int_a^b f''(t)(b-t)\, dt = f''(t)\frac{(b-t)^2}{2}\Big|_a^b + \int_a^b f^{(3)}(t)((b-t)^2/2)\, dt.$$

Hence

$$f(b) = f(a) + f'(a)(b-a) + f''(a)\frac{(b-a)^2}{2} + \int_a^b f^{(3)}(t)\frac{(b-t)^2}{2}\, dt.$$

Assume that the formula for R_k is true:

$$R_k = \int_a^b \frac{(b-t)^{k-1}}{(k-1)!} f^{(k)}(t)\, dt.$$

We let

$$u(t) = f^{(k)}(t) \text{ and } v(t) = -(b-t)^{k-1}$$

and apply integration by parts. We get

$$\int_a^b f^{(k)}(t)\frac{(b-t)^{k-1}}{(k-1)!}\, dt = f^{(k)}(t)\frac{(b-a)^k}{k!} + \int_a^b \frac{(b-t)^k}{k!} f^{(k+1)}(t)\, dt.$$

By induction, the formula for R_n is obtained.   □

Example 6.5.2. We use the mean value theorem for the Riemann integral, Theorem 6.4.1, to deduce Cauchy's form (4.38)' of the remainder in the Taylor's theorem.

Applying the mean value theorem for integrals to (6.30), we conclude that there exists $c \in (a, b)$ such that

$$R_n = (b-a)\frac{(b-c)^{n-1}}{(n-1)!} f^{(n)}(c),$$

which is Cauchy's form (4.38)' of the remainder.

6.6   Riemann's Original Definition

In this section, we look at Riemann's original definition of integrability of functions. Apart from historical reasons, this definition is quite intuitively appealing to scientists. For a budding mathematician, it offers insights into the development of the subject. The proof that f is integrable iff it is integrable in the sense of Riemann and that both the integrals are the same is quite instructive.

Definition 6.6.1. Let $P = \{x_0, \dots, x_n\}$ be a partition of $[a, b]$. Let $t_i \in [x_i, x_{i+1}]$, $0 \leq i \leq n - 1$. Then t_i's are called tags. Let $\mathbf{t} = \{t_i : 0 \leq i \leq n - 1\}$ be the set of tags. The pair $(P, \mathbf{t})$ is called a *tagged partition* of $[a, b]$. See Figure 6.11.

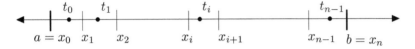

Figure 6.11: Tagged partition.

Definition 6.6.2. Let $f \colon [a, b] \to \mathbb{R}$. The sum defined by

$$S(f, P, \mathbf{t}) := \sum_{i=0}^{n-1} f(t_i)(x_{i+1} - x_i)$$

is called the *Riemann sum* of f for the tagged partition $(P, \mathbf{t})$. See Figure 6.12.

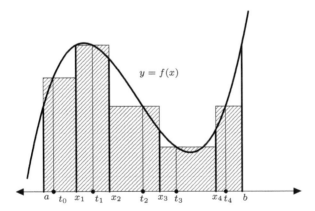

Figure 6.12: Riemann sum for tagged partition.

We say that f is *Riemann integrable* on $[a, b]$ if there exists $A \in \mathbb{R}$ such that for every $\varepsilon > 0$, there exists a partition P such that for any refinement Q of P and for any tag $\mathbf{t}$ of Q, we have

$$|S(f, Q, \mathbf{t}) - A| < \varepsilon.$$

We call A the Riemann integral of f on $[a, b]$. It is easy to see that A is unique. (See Exercise 6.6.3 below.)

Exercise 6.6.3. Show that A, the Riemann integral of f on $[a, b]$ defined above, is unique.

Remark 6.6.4. The condition "for every partition $Q \supset P$" should remind us of a similar condition "for every $n \geq N$" in the definition of convergent sequences. Thus the definition of Riemann integrability may be considered as a limiting process or a question of convergence of the Riemann sums to A. Let us write $Q \geq P$ to mean $Q \supset P$.

Consider an *extremely large* subset $\mathcal{S} := \{S(f, P, \mathbf{t})\} \subset \mathbb{R}$ where P varies over all partitions of $[a, b]$ and $\mathbf{t}$ all tags of P. Then we may rewrite the condition as follows:

> There exists $A \in \mathbb{R}$ such that for any $\varepsilon > 0$ there exists a partition P such that for any $Q \geq P$ and for any set $\mathbf{t}$ of tags in Q, we have $|S(f, Q, \mathbf{t}) - A| < \varepsilon$.

Note that, unlike $\mathbb{N}$, the law of trichotomy does not hold among the set of partitions! We do not offer any further explanation and leave it to you to think over.

Note that in the above definition of Riemann integrability, we do not demand that f is bounded on $[a, b]$. In fact, if f is Riemann integrable on $[a, b]$, then f is bounded on $[a, b]$. See the next proposition.

Proposition 6.6.5. *If $f : [a, b] \to \mathbb{R}$ is Riemann integrable, then it is bounded on $[a, b]$.*

Strategy: The idea is to show that f is bounded on any subinterval of a tagged partition. Since the number of subintervals in a partition is finite, it follows that f is bounded on $[a, b]$.

Proof. Let A be the Riemann integral of f on $[a, b]$. Given $\varepsilon = 1$, let $P = \{a = x_0, x_1, \ldots, x_n = b\}$ be a partition of $[a, b]$ such that, $|S(f, P, \mathbf{t}) - A| < 1$, for any set of tags in P. Fix a set $\mathbf{t}$ of tags $\{t_i\}$ in P. For any set $\{s_i\}$ of tags, we have

$$|S(f, P, \{t_i\}) - S(f, P, \{s_i\})| < 2.$$

Let $x \in [x_0, x_1]$ be an arbitrary point. Let $s_i = t_i$ for $i \geq 1$ and $s_0 = x$. We have

$$|S(f, P, \{t_i\}) - S(f, P, \{s_i\})| = \left| \sum_{i=0}^{n-1} f(t_i)(x_{i+1} - x_i) - \sum_{i=0}^{n-1} f(s_i)(x_{i+1} - x_i) \right|$$
$$= |f(t_0) - f(x)| (x_1 - x_0) < 2.$$

This implies

$$|f(t_0) - f(x)| < \frac{2}{x_1 - x_0}.$$

Hence we have $|f(x)| \leq \frac{2}{(x_1-x_0)} + |f(t_0)|$. That is, f is bounded on $[x_0, x_1]$. Similarly, it is bounded on each of the subintervals of the partition and hence is bounded on $[a, b]$.   □

Proposition 6.6.6. *Let $f: [0, 1] \to \mathbb{R}$ be continuous. Let $c_i \in [\frac{i-1}{n}, \frac{i}{n}]$, $n \in \mathbb{N}$. Then*

$$\lim_{n \to \infty} \frac{1}{n} \sum_{i=1}^{n} f(c_i) = \int_0^1 f(x)dx. \tag{6.31}$$

Strategy: Let $x_i = \frac{i}{n}$. Observe that

$$\frac{1}{n} \sum_{i=1}^{n} f(c_i) = \sum_{i=1}^{n} \int_{x_{i-1}}^{x_i} [(f(c_i) - f(x)) + f(x)] \, dx$$

$$= \int_0^1 f(x) \, dx + \sum_{i=1}^{n} \int_{x_{i-1}}^{x_i} (f(c_i) - f(x)) \, dx. \tag{6.32}$$

(Did you notice that we used the trick mentioned in Remark 6.3.5?)
If $n \gg 1$ (n is sufficiently large), the terms $|f(x) - f(c_i)|$ can be estimated using uniform continuity.

Proof. Given $\varepsilon > 0$, we need to find N such that for $n \geq N$, we have

$$\left| \frac{1}{n} \sum_{i=1}^{n} f(c_i) - \int_0^1 f(x)dx \right| < \varepsilon.$$

Since f is continuous on $[0, 1]$, it is uniformly continuous on $[0, 1]$. Hence there exists $\delta > 0$ such that $|f(x) - f(y)| < \varepsilon$ for $x, y \in [0, 1]$ with $|x - y| < \delta$. By Archimedean property, there exists $N \in \mathbb{N}$ such that $\frac{1}{N} < \delta$. Let $n \geq N$. We subdivide $[0, 1]$ into n parts as in the statement. Note that for any $x, y \in [x_i, x_{i+1}]$, we have $|x - y| \leq 1/N < \delta$ so that $|f(x) - f(y)| < \varepsilon$.

Let $c_i \in [x_i, x_{i+1}]$, $0 \leq i \leq n - 1$. From (6.32), we deduce

$$\left| \frac{1}{n} \sum_{i=1}^{n} f(c_i) - \int_0^1 f(x)dx \right| = \left| \sum_{i=1}^{n} \int_{x_{i-1}}^{x_i} (f(c_i) - f(x)) \, dx \right|$$

$$\leq \sum_{i=1}^{n} \int_{x_{i-1}}^{x_i} |f(c_i) - f(x)| \, dx, \qquad \text{by (6.12)}$$

$$< \sum_{i=1}^{n} \left(\frac{1}{n} \times \varepsilon \right)$$

$$= \varepsilon.$$

This completes the proof.   □

Remark 6.6.7. Integrals are used to compute areas and the mass of an object once we know its density, etc. Most often, physicists and engineers take sample densities at various points of an object and from these data, they try to estimate or approximate the mass. Now, (6.31) gives some kind of justification to their method.

The next result generalizes the last one.

Theorem 6.6.8. *Let $f \colon [a,b] \to \mathbb{R}$ be a bounded function. Then f is integrable iff it is Riemann integrable, in which case we have $\int_a^b f(x)\,dx$ as the Riemann integral.*

> **Strategy:** Let f be integrable and $\varepsilon > 0$. To prove Riemann integrability we need a partition P such that for any $Q \supset P$ and any set $\mathbf{t}$ of tags, we must show that the Riemann sum lies in $(I - \varepsilon, I + \varepsilon)$.
> Using the integrability of f and ε, we get P. The key observation here is that for any choice of $t_i \in [x_i, x_{i+1}]$, we have
>
> $$m_i(f) \leq f(t_i) \leq M_i(f).$$
>
> It is easy to show that the Riemann sum lies between $I \pm \varepsilon$.
> Now for the converse part. Let f be Riemann integrable. For a given ε we need to find a partition P such that $U(f,P) - L(f,P) < \varepsilon$. For this $\varepsilon > 0$, Riemann integrability comes up with a partition. To estimate $U(f,P) - L(f,P)$, we need to estimate $M_i - m_i$. Using the LUB and GLB nature of M_i and m_i, we can find points s_i and t_i such that $f(s_i)$ and $f(t_i)$ are close to m_i and M_i by a fraction of ε. This will lead us to the integrability of f.
> To show that $I = A$, we need to estimate $|A - I|$. Given $\varepsilon > 0$, we find partitions P_1 and P_2 that correspond to the integrability. We let $Q = P_1 \cup P_2$. Then we estimate
>
> $$|A - I| \leq |A - S(f,Q,\mathbf{t})| + |S(f,Q,\mathbf{t}) - L(f,Q)| + |L(f,Q) - I|.$$

Proof. Assume that f is integrable on $[a,b]$. Let $I = \int_a^b f(x)\,dx$. Let $\varepsilon > 0$ be given. Since f is integrable, there exists a partition P of $[a,b]$ (by Theorem 6.1.13) such that

$$L(f,P) > I - \varepsilon \quad \text{and} \quad U(f,P) < I + \varepsilon. \tag{6.33}$$

Let $Q = \{x_0, \ldots, x_n\}$ be any refinement of P. Then (6.33) holds true when P is replaced by Q. We claim that for any set of tags in Q, we have

$$\left| \sum_{i=0}^{n-1} f(t_i)(x_{i+1} - x_i) - I \right| < \varepsilon. \tag{6.34}$$

This will prove that f is Riemann integrable.

The key observation to prove (6.34) is that for any choice of $t_i \in [x_i, x_{i+1}]$, we have

$$m_i(f) \leq f(t_i) \leq M_i(f).$$

It follows that

$I - \varepsilon$

$$< L(f, Q), \qquad \text{using (6.33) and the fact } L(f, P) \leq L(f, Q) \tag{6.35}$$

$$\leq \sum_{i=0}^{n-1} f(t_i)(x_{i+1} - x_i) \tag{6.36}$$

$$\leq U(f, Q) \tag{6.37}$$

$$< I + \varepsilon, \qquad \text{using (6.33) and the fact } U(f, Q) \leq U(f, P). \tag{6.38}$$

That is, for any tagged partition $(Q, \mathbf{t})$ with $Q \supset P$, we have

$$I - \varepsilon < \sum_{i=0}^{n-1} f(t_i)(x_{i+1} - x_i) < I + \varepsilon.$$

We have thus established (6.34) and hence f is Riemann integrable on $[a, b]$ with Riemann integral I.

Let f be Riemann integrable with Riemann integral A. Given $\varepsilon > 0$, there exists a partition $P = \{x_i : 0 \leq i \leq n\}$ such that

$$\left| \sum_{i=0}^{n-1} f(t_i)(x_{i+1} - x_i) - A \right| < \varepsilon/3,$$

for *any* set of tags $\mathbf{t} = \{t_i : 0 \leq i < n\}$. There exist $s_i, u_i \in [x_{i-1}, x_i]$ such that

$$f(s_i) < m_i(f) + \frac{\varepsilon}{6(b-a)} \quad \text{and} \quad f(u_i) > M_i(f) - \frac{\varepsilon}{6(b-a)}$$

so that

$$M_i(f) - m_i(f) \leq [f(u_i) - f(s_i)] + \frac{\varepsilon}{3(b-a)}.$$

We now obtain

$$U(f, P) - L(f, P)$$

$$= \sum_{i=0}^{n-1} [M_i(f) - m_i(f)](x_{i+1} - x_i)$$

$$< \sum_{i=0}^{n-1} (f(u_i) - f(s_i))(x_{i+1} - x_i) + \frac{\varepsilon}{3(b-a)} \sum_{i=0}^{n-1} (x_{i+1} - x_i)$$

$$\leq \left| \sum_{i=0}^{n-1} f(u_i)(x_{i+1} - x_i) - A \right| + \left| A - \sum_{i=0}^{n-1} f(s_i)(x_{i+1} - x_i) \right| + \frac{\varepsilon}{3}$$

$$< \frac{\varepsilon}{3} + \frac{\varepsilon}{3} + \frac{\varepsilon}{3} = \varepsilon.$$

Hence f is integrable.

Let $I = \int_a^b f(x)\,dx$ and A be the Riemann integral of f on $[a, b]$. We need to show that $A = I$. Let $\varepsilon > 0$ be given. Since f is Riemann integrable, there exists a partition P_1 such that for *any* refinement Q of P_1 we have

$$|S(f, P, \mathbf{t}) - A| < \varepsilon/3, \text{ for any set of tags } \mathbf{t}. \tag{6.39}$$

Since f is integrable, there exists a partition P_2 such that

$$U(f, P_2) - L(f, P_2) < \varepsilon/3 \text{ so that } U(f, P_2) < L(f, P_2) + \varepsilon/3. \tag{6.40}$$

Let $Q = P_1 \cup P_2$.

We observe that (using (6.40)),

$$L(f, Q) \leq S(f, Q, \mathbf{t}) \leq U(f, Q) \leq L(f, Q) + \frac{\varepsilon}{3}. \tag{6.41}$$

Again, using (6.40), we have

$$L(f, Q) \leq I \leq U(f, Q) \leq L(f, Q) + \frac{\varepsilon}{3}. \tag{6.42}$$

Using (6.39)–(6.42), we obtain

$$|A - I| \leq |A - S(f, Q, \mathbf{t})| + |S(f, Q, \mathbf{t}) - L(f, Q)| + |L(f, Q) - I|$$
$$< \frac{\varepsilon}{3} + \frac{\varepsilon}{3} + \frac{\varepsilon}{3}.$$

Since $\varepsilon > 0$ is arbitrary, this shows that $A = I$. $\qquad \square$

6.7 Sum of an Infinite Series as a Riemann Integral

The Riemann definition of integration is useful in finding the sum of an infinite series. The main tool here is Proposition 6.6.6. Keep the notation of the proposition. Take any $c_i \in [\frac{i-1}{n}, \frac{i}{n}]$. Then

$$\lim_{n \to \infty} \frac{1}{n} \sum_{i=1}^{n} f(c_i) = \int_0^1 f(x)\,dx. \tag{6.43}$$

We may take $c_i = \frac{i}{n}$ in above the integral. In particular,

$$\lim_{n \to \infty} \frac{1}{n} \sum_{i=1}^{n} f\left(\frac{i}{n}\right) = \int_0^1 f(x)\,dx. \tag{6.44}$$

Example 6.7.1. Find $\lim_{n\to\infty} \sum_{k=1}^{n} \frac{n}{k^2+n^2}$.

$$\lim_{n\to\infty} \sum_{k=1}^{n} \frac{n}{k^2+n^2} = \lim_{n\to\infty} \frac{1}{n} \sum_{k=1}^{n} \frac{n^2}{k^2+n^2}$$

$$= \lim_{n\to infty} \frac{1}{n} \sum_{k=1}^{n} \frac{1}{(\frac{k}{n})^2+1}$$

$$= \lim_{n\to\infty} \frac{1}{n} \sum_{k=1}^{n} f\left(\frac{k}{n}\right)$$

where $f(x) = \frac{1}{x^2+1}$. By (6.44), we have

$$\lim_{n\to\infty} \sum_{k=1}^{n} \frac{n}{k^2+n^2} = \int_0^1 \frac{1}{x^2+1}\, dx$$

$$= \tan^{-1} x \, |_0^1 = \frac{\pi}{4}.$$

Exercise Set 6.7.2. Applications to sum of an infinite series. Show that

(1) $s_n = \sum_{r=1}^{n} \frac{r}{r^2+n^2} \to \log\sqrt{2}$ as $n \to \infty$.

(2) for $a > -1$, $s_n = \frac{1^a+2^a+\cdots+n^a}{n^{1+a}} \to \frac{1}{1+a}$.

(3) $s_n = \frac{1}{2n+1} + \frac{1}{2n+2} + \cdots + \frac{1}{3n} \to \log(3/2)$.

(4) $\lim_{n\to\infty} \frac{\sqrt{1}+\sqrt{2}+\cdots+\sqrt{n}}{\sqrt{n^3}} = 2/3$.

(5) $\sum_{k=1}^{n} \frac{1}{(n^2+k^2)^{\frac{1}{2}}} \to \log(1+\sqrt{2})$.

Exercise Set 6.7.3. Obtain the limits of the sequences whose n-th term is

(1) $\frac{1}{n+1} + \frac{1}{n+2} + \cdots + \frac{1}{2n}$.

(2) $\frac{1}{n+1} - \frac{1}{n+2} + \cdots + \frac{(-1)^{n-1}}{2n}$.

Exercise Set 6.7.4. Miscellaneous Exercises on Integration.

(1) Let $\int_a^b f(x)\, dx$ exist and be positive. Show that there exists an interval $J \subset [a, b]$ and a positive constant m such that for $x \in J$, we have $f(x) \geq m$.

(2) Let f, g be integrable on $[a, b]$. Show that $\max\{f, g\}$, $\min\{f, g\}$ are integrable on $[a, b]$.

(3) Can you think of a sandwich lemma for integrable functions?

(4) Assume that $f\colon [a, b] \to \mathbb{R}$ is integrable and that $f(x) \geq m > 0$ for $x \in [a, b]$. Show that $g := 1/f$ is integrable on $[a, b]$.

(5) Let $f\colon [0, 1] \to \mathbb{R}$ be bounded. Assume that f is integrable on $[\delta, 1]$ for each $0 < \delta < 1$. True or false: f is integrable on $[0, 1]$.

(6) Let $f\colon [a, b] \to \mathbb{R}$ be continuous and non-negative. If $\int_a^b f(x)\, dx = 0$, then $f = 0$.

Is this true if f is not assumed to be continuous?

(7) Let $f\colon [0, 1] \to \mathbb{R}$ be continuous. Assume that $\int_0^c f(t)\, dt = 0$ for each $c \in [0, 1]$. Show that $f = 0$. Note that we have not assumed that $f \geq 0$.

(8) Let $S := \{1/n : n \in \mathbb{N}\}$. Let $f\colon [0, 1] \to \mathbb{R}$ be defined by $f(x) = 1$ if $x \in S$ and 0 otherwise. Show that f is integrable.

Is f still integrable if we assume $f(x) = x$ for $x \in S$ and 0 otherwise?

(9) Assume that f is integrable on $[a, b]$. Fix $c \in [a, b]$. Assume that $g\colon [a, b] \to \mathbb{R}$ is such that $g(x) = f(x)$ and $g(c) \neq f(c)$. Show that g is integrable on $[a, b]$. What is $\int_a^b g(x)\, dx$?

(10) Let $f\colon [a, b] \to \mathbb{R}$ be bounded. Assume that there exists a sequence (P_n) of partitions of $[a, b]$ such that $U(f, P_n) - L(f, P_n) \to 0$. Show that f is integrable on $[a, b]$. What is $\int_a^b f(t)\, dt$?

(11) Let $f\colon \mathbb{R} \to \mathbb{R}$ be continuous. Let $g(x) := \int_{x-1}^{x+1} f(t)\, dt$ for $x \in \mathbb{R}$. Is g differentiable? If so, what is $g'(x)$?

(12) Let $I_n := \int_0^1 \frac{x^n}{1+x}$. Show that $I_n \to 0$. Compare this with Exercise 6.4.6.

(13) Let $f\colon [0, 1] \to \mathbb{R}$ be continuous. Show that $\int_0^1 x^n f(x) \to 0$.

(14) Let $f\colon [a, b] \to \mathbb{R}$ be continuously differentiable. Let $a_n := \int_a^b f(t) \sin(nt)\, dt$. Show that $a_n \to 0$.

(15) Let $f\colon [0, 1] \to \mathbb{R}$ be non-negative. Let $a_n := n \int_0^1 (f(t))^n\, dt$.
(a) Assume that $f(t) < 1$ for $t \in [0, 1]$. Show that $a_n \to 0$.
(b) Assume that $f(t) > 1$ for $t \in [0, 1]$. Show that $a_n \to \infty$.

6.8 Logarithmic and Exponential Functions

We develop the logarithmic and exponential functions using many of the results proved so far. This will serve two purposes: (1) to put these two functions on a rigorous footing and (2) to put to use many of the results proved in the course. We shall do this as a series of graded exercises, each being either an observation or requiring an easy and short proof.

Exercise Set 6.8.1 (Logarithmic Function).

(1) Define $L(x) := \int_1^x \frac{1}{t}\, dt$ for $x > 0$. Then L is differentiable with $L'(x) = \frac{1}{x}$.

(2) $L(1) = 0$.

(3) L is a (strictly) increasing function and hence is one-one.

(4) $L(xy) = L(x) + L(y)$ for any $x > 0$ and $y > 0$. Hint: $\int_1^{xy} = \int_1^x + \int_x^{xy}$. In the second integral, use a suitable change of variable.

(5) $L(1/x) = -L(x)$ for $x > 0$.

(6) The sequence $(L(n))$ diverges to ∞, that is, $L(n) \to \infty$. Hint: The proof of the integral test may give you some idea.

(7) L maps $(0, \infty)$ bijectively onto $\mathbb{R}$. Hint: L is continuous. Use the last item and the intermediate value theorem.

We call L, the logarithmic function and write $\log x$ for $L(x)$.

Exercise Set 6.8.2 (Exponential Function).

(1) We know that $L \colon (0, \infty) \to \mathbb{R}$ is a bijection. We define $E = L^{-1}$, the inverse of the function L. Then E is a continuous bijection of $\mathbb{R}$ onto $(0, \infty)$. Hint: Review the section on monotone functions!

(2) Note that $E(0) = 1$ and E is strictly increasing and takes only positive values.

(3) E is differentiable and we have $E'(x) = E(x)$. Hint: Inverse function theorem.

We claim $E(1) = e$, the Euler number. We sketch a proof. Observe that

$$\lim_{h \to 0} \frac{L(x+h) - L(x)}{h} = \frac{1}{x}.$$

That is,

$$\lim_{h \to 0} \frac{1}{h} L\left(\frac{x+h}{h}\right) = \lim_{h \to 0} \frac{1}{h} L\left(1 + \frac{x}{h}\right) = \frac{1}{x}.$$

Since E is a continuous function, we obtain

$$\lim_{h \to 0} E\left(\frac{1}{h} L\left(1 + \frac{x}{h}\right)\right) = E(1/x).$$

What result did you use here? Substitute $x = 1$ in the last displayed equation to obtain $\lim_{h \to 0}(1 + h)^{1/h} = E(1)$. Let h vary through the sequence $1/n$ to arrive at the claim. Which definition of the Euler number was used here?

Let $x > 0$ and $\alpha \in \mathbb{R}$. We define $x^\alpha := E(\alpha L(x))$. In particular, $e^x = E(xL(e)) = E(x \times 1) = E(x)$. We have thus arrived at the exponential function, as we know!

6.9   Improper Riemann Integrals

We extend the concept of Riemann integral to functions defined on unbounded intervals or to unbounded functions. First of all, let us look at the following observation.

Proposition 6.9.1. *Let $f\colon [a,b] \to \mathbb{R}$ be integrable. Then*

$$\int_a^b f(t)\,dt = \lim_{c \to a+} \left(\lim_{d \to b-} \int_c^d f(t)\,dt \right).$$

Proof. Let $g(x) := \int_a^x f(t)\,dt$. Then g is continuous on $[a,b]$. Therefore,

$$\int_a^b f(t)\,dt = g(b) - g(a) = \lim_{c \to a+} \left(\lim_{d \to b-} [g(d) - g(c)] \right)$$

$$= \lim_{d \to b-} \left(\lim_{c \to a+} [g(d) - g(c)] \right).$$

$\qquad\qquad\qquad\qquad\qquad\qquad\qquad\qquad\qquad\qquad\qquad\qquad\qquad\square$

Proposition 6.9.1 motivates the following definition.

Definition 6.9.2. Let (a,b) be a nonempty, open, possibly unbounded interval and $f\colon (a,b) \to \mathbb{R}$.

(i) We say that f is *locally integrable* on (a,b) if f is integrable on each closed subinterval $[c,d]$ of (a,b).

(ii) We say that the *improper Riemann integral* of f exists or is convergent on (a,b) if

$$\lim_{c \to a+} \left(\lim_{d \to b-} \int_c^d f(t)\,dt \right)$$

exists. The limit is denoted by $\int_a^b f(t)\,dt$.

Proposition 6.9.3. *The order in which limits are taken in the last definition does not matter.*

Proof. Let $t_0 \in (a,b)$ be fixed. We observe

$$\lim_{c \to a+} \left(\lim_{d \to b-} \int_c^d f(t)\,dt \right) = \lim_{c \to a+} \left(\int_c^{t_0} f(t)\,dt + \lim_{d \to b-} \int_{t_0}^d f(t)\,dt \right)$$

$$= \lim_{c \to a+} \int_c^{t_0} f(t)\,dt + \lim_{d \to b-} \int_{t_0}^d f(t)\,dt$$

$$= \lim_{d \to b-} \left(\lim_{c \to a+} \int_c^d f(t)\,dt \right).$$

$\qquad\qquad\qquad\qquad\qquad\qquad\qquad\qquad\qquad\qquad\qquad\qquad\qquad\square$

Remark 6.9.4. In view of Proposition 6.9.3, we use the notation

$$\lim_{\substack{c \to a+ \\ d \to b-}} \int_c^d f(t)\, dt \quad \text{to stand for} \quad \lim_{c \to a+} \left(\lim_{d \to b-} \int_c^d f(t)\, dt \right).$$

If we deal with intervals of the form $(a, b]$, we may simplify the notation

$$\int_a^b f(t)\, dt = \lim_{c \to a+} \int_c^b f(t)\, dt.$$

The integral of f on an interval of the form $[a, b)$ can be defined analogously.

Example 6.9.5. Following are some examples of improper integrals:

(1) The function $f(x) := x^{-1/2}$ has an improper integral on $(0, 1]$.

(2) The function $f(x) := e^{-x^2}$ has an improper integral on $[0, \infty)$.

(3) The function $f(x) := \frac{1}{x(x-1))}$ has an improper integral on $(0, 1)$.

(4) The function $f(x) := \frac{1}{1+x^2}$ has an improper integral on $(-\infty, \infty)$.

Example 6.9.6. Let us evaluate $\int_0^\infty e^{-x}\, dx$ if it exists. Let $a > 0$. Then

$$\int_0^a e^{-x}\, dx = -e^{-x} \big|_0^a = -e^{-a} + 1.$$

We have

$$\lim_{a \to \infty} \int_0^a e^{-x}\, dx = \lim_{a \to \infty} [1 - e^{-a}] = 1.$$

Hence $\int_0^\infty e^{-x}\, dx$ exists and is 1.

Example 6.9.7. Discuss the convergence of the integral $\int_1^\infty \frac{1}{x^p}\, dx$ for various values of p.

Let $t > 1$. Then

$$\int_1^t \frac{1}{x^p}\, dx = \frac{x^{1-p}}{1 - p} \quad \text{if } p \neq 1.$$

We shall discuss the integral when $p = 1$ separately. If $p > 1$, then $1 - p < 0$, hence $x^{1-p} \to 0$ as $x \to \infty$. Thus for $p > 1$,

$$\int_1^\infty \frac{1}{x^p}\, dx = \lim_{t \to \infty} \int_1^t \frac{1}{x^p}\, dx = \lim_{t \to \infty} \frac{x^{1-p}}{1 - p} = \frac{1}{p - 1}.$$

If $p < 1$, then $x^{1-p} \to \infty$ as $x \to \infty$. That is, the improper integral $\int_1^\infty \frac{1}{x^p}\, dx$ does not exists if $p < 1$.

When $p = 1$, then $\int_1^\infty \frac{1}{x}\, dx = \lim_{t \to \infty} \int_1^t \frac{1}{x} = \lim_{t \to \infty} [\log t]$, which does not exist. (Why?)

Now that we have defined improper integrals, we can now reformulate the integral test as follows.

Theorem 6.9.8 (Integral Test). *Assume that $f\colon [1, \infty] \to [0, \infty)$ is continuous and decreasing. Let $a_n := f(n)$ and $b_n := \int_1^n f(t)\,dt$. Then:*
(i) $\sum a_n$ *converges if the improper integral $\int_1^\infty f(t)\,dt$ exists.*
(ii) $\sum a_n$ *diverges if he improper integral $\int_1^\infty f(t)\,dt$ does not exist.*

Proof. We leave the proof as an exercise to the reader. ☐

Theorem 6.9.9. *If the improper integrals of f, g exist on (a, b) and $\alpha, \beta \in \mathbb{R}$, then the improper integral of $\alpha f + \beta g$ exists on (a, b) and we have*

$$\int_a^b (\alpha f(t) + \beta g(t))\,dt = \alpha \int_a^b f(t)\,dt + \beta \int_a^b g(t)\,dt.$$

Proof of this theorem is easy and follows from the fact

$$\int_c^d \alpha f + \beta g = \alpha \int_c^d f + \beta \int_c^d \quad \text{for any } [c, d] \subset (a, b).$$

Theorem 6.9.10 (Comparison theorem). *Let f, g be locally integrable on (a, b). Assume that $0 \le f(t) \le g(t)$ for $t \in (a, b)$ and that the improper integral of g exists on (a, b). Then the improper integral of f exists on (a, b) and we have*

$$\int_a^b f(t)\,dt \le \int_a^b g(t)\,dt.$$

Proof. Fix $c \in (a, b)$. Let $F(d) := \int_c^d f(t)\,dt$ and $G(d) := \int_c^d g(t)\,dt$ for $d \in [c, b)$. We have $F(d) \le G(d)$. Since $f \ge 0$, the function F is increasing on $[c, b]$. Hence $F(b-)$ exists. Thus, the improper integral of f exists on (c, b) and we get

$$\int_c^d f(t)\,dt = F(b-) \le G(b-) = \int_c^b g(t)\,dt.$$

A similar argument works for the case $c \to a+$ ☐

Example 6.9.11. We now look at some typical examples of improper integrals.

(1) The function $f(x) := \left| x^{-3/2} \sin x \right|$ has an improper R-integral on $(0, 1]$. *Hint:* Observe that $0 \le f(x) \le x^{-3/2} |x| = x^{-1/2}$ on $(0, 1]$. Note that $\int_0^1 x^{-1/2}\,dx$ exists. Hence $\int_0^1 x^{-3/2} \sin x\,dx$ also exists.

(2) The function $f(x) := x^{-5/2} \log x$ has an improper R-integral on $[1, \infty)$. *Hint:* Note that $0 \le f(x) \le x^{-5/2} x^{1/2}$ for all $x > M$ for some $M > 0$.

(3) The function $f(x) := e^{-x^2}$ has an improper integral on $[1, \infty)$. *Hint:* Note that $0 \le e^{-x^2} \le e^{-x}$ for all $x \ge 1$. Since $\int_1^\infty e^{-x}\,dx$ exists, $\int_1^\infty e^{-x^2}\,dx$ also exists.

Exercise 6.9.12. If f is bounded and locally integrable on (a, b), and g has an improper R-integral on (a, b), then $|fg|$ has an improper R-integral on (a, b).

Definition 6.9.13. Let $f\colon (a, b) \to \mathbb{R}$. We say that f is *absolutely integrable* on (a, b) if $|f|$ has an improper integral on (a, b). The function f is said to be *conditionally integrable* on (a, b) if it is integrable on (a, b), but not absolutely integrable.

Theorem 6.9.14. *If f is absolutely integrable on (a, b), then the improper integral of f on (a, b) exists and we have*

$$\left| \int_a^b f(t)\, dt \right| \le \int_a^b |f(t)|\, dt.$$

Proof. Note that $f(x) \le |f(x)|$ for all x. This implies that f is locally integrable on (a, b). Note that $0 \le |f(x)| + f(x) \le 2|f(x)|$. Hence the improper integral of $|f| + f$ exists on (a, b) (by comparison theorem). By Theorem 6.9.9, the improper integral of $f = (|f| + f) - |f|$ also exists on (a, b). Also,

$$\left| \int_c^d f(t)\, dt \right| \le \int_c^d |f(t)|\, dt \text{ for all } a < c < d < b.$$

Now finish the proof by taking limits as $c \to a+$ and $d \to b-$. $\square$

Theorem 6.9.15. *Let f be bounded and locally integrable on (a, b). Assume that g is absolutely integrable on (a, b). Then fg is absolutely integrable on (a, b).* $\square$

Example 6.9.16 (An important example). We show that $f(x) = \dfrac{\sin x}{x}$ is conditionally integrable on $[1, \infty)$.

Let $R > 1$. Then integration by parts yields

$$\int_1^R \frac{\sin x}{x}\, dx = -\frac{\cos x}{x} \Big]_1^R - \int_1^R \frac{\cos x}{x^2}\, dx$$

$$= \cos 1 - \frac{\cos R}{R} - \int_1^R \frac{\cos x}{x^2}\, dx. \tag{6.45}$$

Since x^{-2} is absolutely integrable on $[1, \infty)$ and since $|\cos x| \le 1$, it follows that $\frac{\cos x}{x^2}$ is absolutely integrable on $[1, \infty)$ by the last item. We obtain from (6.45)

$$\int_1^\infty \frac{\sin x}{x}\, dx = \cos 1 - \int_1^\infty \frac{\cos x}{x^2}\, dx.$$

We now show that $\dfrac{\sin x}{x}$ is not absolutely integrable on $[1, \infty)$. We observe that

$$\int_1^{n\pi} \frac{|\sin x|}{x}\, dx \geq \sum_{k=2}^{n} \int_{(k-1)\pi}^{k\pi} \frac{|\sin x|}{x}\, dx$$

$$\geq \sum_{k=2}^{n} \frac{1}{k\pi} \int_{(k-1)\pi}^{k\pi} |\sin x|\, dx$$

$$= \sum_{k=2}^{n} \frac{1}{k\pi} \int_{0}^{\pi} \sin x\, dx$$

$$= \sum_{k=2}^{n} \frac{2}{k\pi} \to \infty.$$

Remark 6.9.17. Go through the solution of Exercise 14 on page 212 and the last example. You may notice that we used the integration by parts to gain better control of the integral. This is a standard technique in analysis. To use integration by parts to estimate an integral, analysts often impose a further condition of continuous differentiability on the integrand and then resort to some limiting/density arguments to derive the general case.

Now we deal with another example (gamma function), as an improper integral. It is ubiquitous in the sense that it appears not only in mathematics but also in physics and engineering.

Example 6.9.18 (Gamma Function). We shall show that the improper integral of the function $f(t) := t^{x-1}e^{-t}$ exists on $(0, \infty)$.
Observe that $t^{x-1}e^{-t} \leq t^{x-1}$ for $t > 0$. Hence the improper integral of f exists on $(0, 1)$ for $x > 0$ (why?).
Also, since $t^{x+1}e^{-t} \to 0$ as $t \to \infty$, it follows that

$$t^{x-1}e^{-t} \leq Ct^{-2} \text{ for } t \geq 1 \text{ for some } C > 0.$$

Hence the improper integral of f exists on $[1, \infty)$.
The *gamma function* is defined on $(0, \infty)$ by the formula

$$\Gamma(x) := \int_0^{\infty} t^{x-1}e^{-t}\, dt.$$

It may be considered as a generalization of the factorial $n!$ for $x \in \mathbb{R}$. See Item 3 below.

Exercise Set 6.9.19. Prove the following:

(1) $\Gamma(1) = 1$.

(2) $\Gamma(\alpha + 1) = \alpha\Gamma(\alpha)$.

(3) If m is a positive integer, then $\Gamma(m + 1) = m!$

(4) We know that $\int_{-\infty}^{\infty} e^{-u^2} \, du = \sqrt{\pi}$. Using this, it follows that $\Gamma\left(\frac{1}{2}\right) = \sqrt{\pi}$.

Exercise Set 6.9.20.

(1) For each of the following, find the values of $p \in \mathbb{R}$ for which the improper integral exists on the specified interval I.

    (a) $f(x) = x^{-p}$, $I = (1, \infty)$.

    (b) $f(x) = x^{-p}$, $I = (0, 1)$.

    (c) $f(x) = 1/(1 + x^p)$, $I = (0, \infty)$.

    (d) $f(x) = 1/(x \log^p x)$, $I = (e, \infty)$.

(2) Decide whether the improper integral of $f(x) := (2 + x^8)^{-1/4}$ exists on $(1, \infty)$.

(3) Decide whether the improper integral of $f(x) := (\pi + x^3)^{-1/4}$ exists on $(0, \infty)$.

(4) Decide whether the improper integral of $f(x) := \frac{e^x}{1 + e^{2x}}$ exists on $\mathbb{R}$.

(5) Show that the improper integral of $f(x) := |x|^{-1/2}$ exists on $[-1, 1]$ and its improper integral is 4.

(6) Decide which of the following functions have an improper integral on I:

    (a) $f(x) = \sin x$, $I = (0, \infty)$.

    (b) $f(x) = x^{-2}$, $I = [-1, 1]$.

    (c) $f(x) = x^{-1} \sin(x^{-1})$, $I = (1, \infty)$.

    (d) $f(x) = \log x$, $I = (0, 1)$.

(7) Assume that the improper integral of f exists on $[1, \infty)$ and that $\lim_{x \to \infty} f(x) = L$ exists. Prove that $L = 0$.

(8) True or false: $\int_0^\pi \sec^2 x \, dx = 0$?

Chapter 7

Sequences and Series of Functions

Contents

In this chapter, we shall deal with convergence of sequence of functions. Let X be a nonempty set, not necessarily a subset of $\mathbb{R}$. Let $f_n \colon X \to \mathbb{R}$ be a function, $n \in \mathbb{N}$. We then say (f_n) is a sequence of functions on X.

Example 7.0.1. Let $X = [0,1]$. Define $f_n(x) := x/n$, $x \in [0,1]$. Then (f_n) is a sequence of functions on $[0,1]$.

Example 7.0.2. Let $X = \{1,2,3\}$. Let $f_n(k) := n(\mathrm{mod}\, k)$, $k = 1,2,3$ where $n(\mathrm{mod}\, k)$ is the remainder when n is divided by k. For instance, $f_2(2) = 0$, $f_3(2) = 1$, $f_{12}(2) = 0$ and so on.

7.1 Pointwise Convergence of Sequence of Functions

Convention: In the sequel X will denote a nonempty set.

Let $f_n, f \colon X \to \mathbb{R}$ be functions on X, $n \in \mathbb{N}$. What do we mean by saying that the sequence (f_n) converges to f? So far, we know what is meant by a sequence of real numbers (a_n) converging to a real number a. The standard trick in mathematics is to reduce the new problem to an old problem which we know how to deal with. Fix $a \in X$. We then have a real number $f_n(a)$ for $n \in \mathbb{N}$. Thus we obtain a sequence $(f_n(a))$.

For example, in Example 7.0.1, if we fix $a = 1/2$, the sequence $(f_n(a))$ is $(1/2n)$. What is the sequence if $a = 0$?

Let $f(x) = x$ for $x \in [0.1]$. Then we may ask whether $f_n(a) \to f(a)$. If $a = 1/2$, then $f(a) = 1/2$ whereas $f_n(a) = 1/2n \to 0$. Thus $f_n(a)$ does not converge to $f(a)$. But if we let $a = 0$, then $(f_n(a))$ is the constant sequence (0) and it converges to $f(0) = 0$.

Consider Example 7.0.2. Let $a = 1$. Then $f_n(1) = 0$ for $n \in \mathbb{N}$ and hence $f_n(1) \to 0$. On the other hand, $(f_n(2)) = (1, 0, 1, 0, \ldots)$ and hence the sequence is not convergent.

These examples lead us the following definition.

Definition 7.1.1. Let $f_n \colon X \to \mathbb{R}$ be a sequence of functions from a set X to $\mathbb{R}$. We say that f_n converges to f *pointwise* on X if for each $x \in X$, the sequence $(f_n(x))$ of real numbers converges to the real number $f(x)$ in $\mathbb{R}$.

Like we say a sequence (x_n) is convergent, we may also define a sequence of functions (f_n) is pointwise convergent on X. This means that there exists a function $f \colon X \to \mathbb{R}$ such that (f_n) is pointwise convergent to f. There is an extra problem for us in this situation. We need to find the limit function f and then show that $f_n \to f$ pointwise. Note that this means that we fix $a \in X$ first and form the sequence $(f_n(a))$ of real numbers. For any given $\varepsilon > 0$, we have to find an $n_0 \in \mathbb{N}$ such that for $n \geq n_0$ we have $|f_n(a) - f(a)| < \varepsilon$. Thus n_0 may depend not only on ε but also on a.

Example 7.1.2. We now look at a few examples and examine their pointwise convergence. Pay attention to the graphs of these functions to get an idea of what is going on. As far as possible, we shall investigate whether the given sequence is pointwise convergent and if so, we shall determine the limit function.

(1) Let $f_n(x) = \frac{x}{n}$, $x \in \mathbb{R}$. See Figure 7.1.

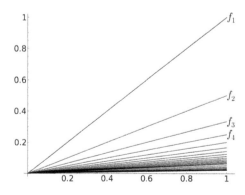

Figure 7.1: Graph of $f_n(x) = \frac{x}{n}$.

Let us play with a few points. Let $a = 0$. Then the sequence $(f_n(a))$ is the constant sequence (0). Hence $f_n(a) \to 0$. If $a = 2012$, then $f_n(a) = 2012/n$. Clearly, $f_n(a) \to 0$. More generally, if $a \in \mathbb{R}$, we get (a/n) as the pointwise sequence. It is convergent to 0. Thus we may take the limit function to be $f = 0$ to conclude that $f_n \to f$ pointwise on $\mathbb{R}$.

Fix $a \in \mathbb{R}$. Let $\varepsilon > 0$ be given. Then we need to make sure $|a/n| < \varepsilon$ for $n \geq N_a$. As the sequence depends on a, we decided to use N_a rather than N. Now $|a|/n < \varepsilon$ is assured if $n > |a|/\varepsilon$. Hence we may take $N_a > |a|/\varepsilon$. See Figure 7.2.

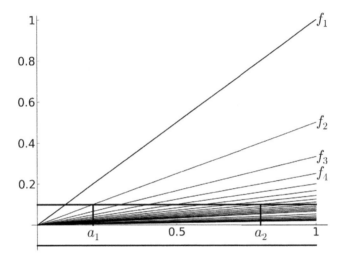

Figure 7.2: Graph of $f_n(x) = \frac{x}{n}$.

We draw an ε-band around the (graph of the) limit function. When does the point $(a, f_n(a))$ lie in ε-band? You may observe that the larger the value of $|a|$, the larger the n such that $(a, f_n(a))$ lies in the band.

(2) Let us look at the sequence in Example 7.0.2. For $a = 1$, the sequence $(f_n(1))$ is the constant sequence 0 and hence is convergent. If $a = 2$, the $(f_n(a))$ is $(1, 0, 1, 0 \ldots)$ and is not convergent. Hence the sequence (f_n) is not pointwise convergent on X.

(3) Let $f_n(x) = \begin{cases} 0, & -\infty < x \leq 0 \\ nx, & 0 \leq x \leq \frac{1}{n} \\ 1, & x \geq \frac{1}{n}. \end{cases}$

Draw the graphs of these functions. See Figure 7.3. Let $a \leq 0$. Then, for $n \in \mathbb{N}$, $f_n(a) = 0$. Hence the sequence $f_n(a) \to 0$ for $a \leq 0$.

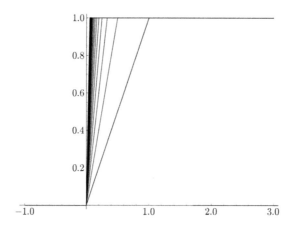

Figure 7.3: Graph of $f_n(x)$ in item 3 of Example 7.1.2.

Let $a \geq 1$. Then $a \geq 1/n$ so that $f_n(a) = 1$ for $n \in \mathbb{N}$. Hence for $a \geq 1$, $f_n(a) \to 1$.

What happens when $0 < a < 1$? Let us take $a = 1/k$ for some $k \in \mathbb{N}$. What is $f_n(a)$? The first few terms of the sequence are $1/k, 2/k, 3/k, \ldots$. But the moment $n \geq k$, we start getting $f_n(1/k) = 1$ as $x = 1/k \geq 1/n$ for $n \geq k$. Hence the sequence $f_n(1/k)$ is eventually a constant sequence and hence $f_n(1/k) \to 1$.

If $0 < a < 1$, by the Archimedean property, we can find a positive integer N_a such that $1/N_a < a$. Hence for $n \geq N_a$, we have $f_n(a) = 1$. Thus we conclude for $0 < a < 1$, we have $f_n(a) \to 1$. If we define $f(x) = 0$ for $x \leq 0$ and $f(x) = 1$ for $x > 0$, we then conclude that $f_n \to f$ pointwise.

(4) Consider for $n \geq 2$.

$$f_n(x) = \begin{cases} nx, & \text{if } 0 \leq x \leq \frac{1}{n} \\ n(\frac{2}{n} - x), & \text{if } \frac{1}{n} \leq x \leq \frac{2}{n} \\ 0, & \text{if } x \geq \frac{2}{n} \text{ and } x < 0. \end{cases}$$

Clearly, if $a \leq 0$ or if $a \geq 1$, $f_n(a) = 0$ and hence $f_n(a) \to 0$ for such a.

Let $0 < a < 1$. Look at Figure 7.4. The picture shows that if n is such that $2/n < a$, then $f_n(x) = 0$. By the Archimedean property, there exists $N_a \in \mathbb{N}$ such that $N_a > 2/a$. For any $n \geq N_a$, we have $2/n < a$ so that $f_n(a) = 0$. Hence the sequence $(f_n(a))$ is eventually zero sequence and hence $f_n(a) \to 0$. Therefore if we define $f = 0$, then $f_n \to f$ pointwise on $\mathbb{R}$.

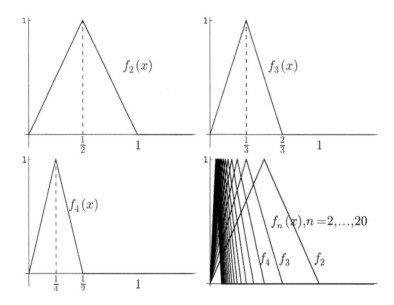

Figure 7.4: Graph of $f_n(x)$ item 4 of Example 7.1.2.

(5) $f_n(x) = \begin{cases} 1, & -n \le x \le n \\ 0, & |x| > n. \end{cases}$

Draw the graph of $f_n(x)$. Almost similar reasoning as in the last example shows that $f_n(a)$ is 1 if $n > |a|$. Hence we may take the limit function $f = 1$.

(6) See Figure 7.5. It is the graph of a function $f_n : \mathbb{R} \to \mathbb{R}$. Can you write down the expression for the function explicitly?

Define

$$f_n(x) = \begin{cases} 0, & \text{if } |x| > 1/n \\ n(1 + nx), & \text{if } -1/n \le x \le 0 \\ n(1 - nx), & \text{if } 0 \le x \le 1/n. \end{cases}$$

By looking at the graph, can you guess the limits of $(f_n(a))$?

If $|x| \ge 1$, clearly, $f_n(a) = 0$ for $n \in \mathbb{N}$. Hence we may define $f(a) = 0$ for such a's.

If $0 < |a| < 1$, by the Archimedean property, we can find $N_a > 1/|a|$ so that for $n \ge N_a$, $f_n(a) = 0$. Again, we may define $f(a) = 0$ for $0 < |a| < 1$.

If $a = 0$, we find that $f_n(0) = n$ and hence the sequence $(f_n(0))$ is not bounded and hence not convergent. Thus we conclude that the sequence (f_n) does not converge pointwise on $\mathbb{R}$.

Let $\mathbb{R}^* = \mathbb{R} \setminus \{0\}$. Let g_n be the restriction of f_n to $\mathbb{R}^*$. The sequence (g_n) converges pointwise to $g = 0$ on $\mathbb{R} \setminus \{0\}$.

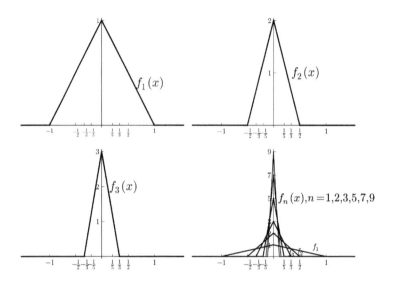

Figure 7.5: Graph of $f_n(x)$ in item 6 of Example 7.1.2.

(7) See Figure 7.6. It is the graph of a function $f_n \colon \mathbb{R} \to \mathbb{R}$. Can you write down the function explicitly? Does it converge pointwise to any function?

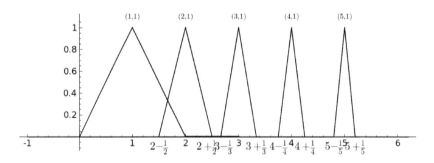

Figure 7.6: Graph of $f_n(x)$ in item 7 of Example 7.1.2.

The function f_n is defined as follows.

$$f_n(x) = \begin{cases} 0, & x < n - \frac{1}{n} \text{ or } x > n + \frac{1}{n} \\ n\left(x - \left(n - \frac{1}{n}\right)\right), & n - \frac{1}{n} \leq x \leq n \\ \frac{1}{n}\left(n + \frac{1}{n} - x\right), & n \leq x \leq n + \frac{1}{n}. \end{cases}$$

(8) Let $f_n \colon \mathbb{R} \to \mathbb{R}$ be defined by $f_n(x) = x^n$. Since (a^n) is not bounded if $|a| > 1$, it follows that the sequence (f_n) is not convergent on $\mathbb{R}$. See Figure 7.7.

For $a = 1$, the sequence is the constant sequence 1 and hence is convergent whereas for $a = -1$, it is $(-1, 1, -1, 1, \ldots)$.

We have seen (on page 48) that if $|a| < 1$, $a^n \to 0$. Thus, for any a with $|a| < 1$, the sequence $(f_n(a)) \to 0$. Thus the sequence (f_n) converges on $(-1, 1]$ to the function f defined by $f(x) = 0$ for $x \in (-1, 1)$ and $f(1) = 1$.

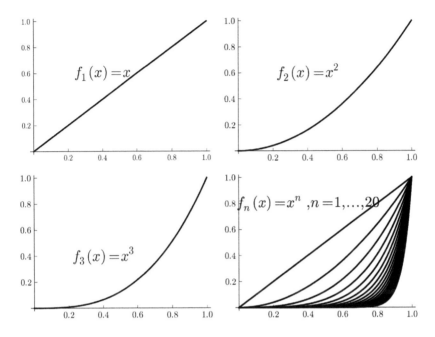

Figure 7.7: Graph of $f_n(x)$ in item 8 of Example 7.1.2.

What can you say if we modify f_n as $f_n(x) := nx^n$ for $x \in (-1, 1]$ or for $x \in (-1, 1)$?

(9) Consider $f_n \colon (0, \infty) \to (0, \infty)$ defined as $f_n(x) = x^{1/n}$. For $a > 0$ we know that $a^{1/n} \to 1$ (see page 48). Hence the sequence f_n converges pointwise on $(0, \infty)$ to the constant function 1.

(10) Let $f_n \colon \mathbb{R} \to \mathbb{R}$ be defined by $f_n(x) = \frac{\sin nx}{n}$. Since $|\sin t| \leq 1$ for any $t \in \mathbb{R}$, we have the obvious estimate for $|f_n(a)| \leq \frac{1}{n}$ for any $a \in \mathbb{R}$. Hence $f_n(a) \to 0$. We conclude that f_n converges pointwise on $\mathbb{R}$ to the zero function. See Figure 7.8.

In each of the cases, we identified the limit function as the limit of the sequence $(f_n(a))$, **if** the sequence is convergent. When we wanted to prove the pointwise convergence, for a given $\varepsilon > 0$, we wanted to find an N_a such that for $n \geq N_a$, $|f_n(a) - f(a)| < \varepsilon$. This shows that in general N_a depends on a, as the sequence itself depends on a. The definition of pointwise convergence when cast in terms of quantifiers exhibits this.

We say that $f_n \to f$ pointwise on X if

$$\forall a \in X \, (\forall \varepsilon > 0 \, (\exists N \in \mathbb{N} \, (\forall n \geq N \, (|f_n(a) - f(a)| < \varepsilon)))). \tag{7.1}$$

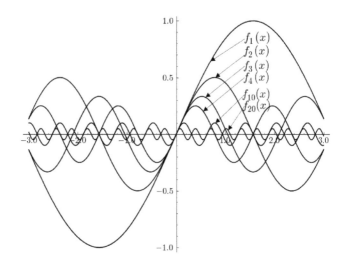

Figure 7.8: Graph of $f_n(x)$ in item 10 of Example 7.1.2.

7.2   Uniform Convergence of Sequence of Functions

The question arises whether we can choose N which will work for all $a \in X$. (Compare this with the notion of uniform continuity of a function.) Note that just because the way we got N_a does not yield a common N, we cannot conclude that there exists no such common N. For instance, in the first example, we noticed that our N_a depended upon a. But if the domain of f_n's is restricted to $[-R, R]$, we have a uniform estimate for $|f_n(a)|$, which is independent of a:

$$|f_n(a) - 0| \leq |a|/n \leq R/n.$$

Hence if we take $N > R/\varepsilon$, then for any $n \geq N$, $|f_n(a)| < \varepsilon$.

These considerations lead us to the following definition.

Definition 7.2.1. A sequence $f_n : X \to \mathbb{R}$ is said to *converge uniformly* on X to f if given $\varepsilon > 0$ there exists an $n_0 \in \mathbb{N}$ such that for all $x \in X$ and $n \geq n_0$, we have $|f_n(x) - f(x)| < \varepsilon$. If $f_n \to f$ uniformly on X we denote it by $f_n \rightrightarrows f$ on X.

In terms of quantifiers, $f_n \to f$ uniformly on X iff

$$\forall \varepsilon > 0 \left(\exists n_0 = n_0(\varepsilon) \left(\forall n \geq n_0 \left(\forall x \in X \left(|f_n(x) - f(x)| < \varepsilon \right) \right) \right) \right). \tag{7.2}$$

Compare (7.1) and (7.2) and observe the position of $\forall x \in X$ in this formulation.

It is clear that (i) uniform convergence implies pointwise convergence and (ii) uniform convergence on X implies the uniform convergence on Y where $Y \subseteq X$.

We interpret the uniform convergence in a geometric way. Let $X \subset \mathbb{R}$, say, an interval. Draw the graphs of f_n and f. Put a band of width ε around the graph of f. The uniform convergence $f_n \rightrightarrows f$ is equivalent to asserting the existence of N such that the graphs of f_n over X will lie inside this band, for $n \geq N$. See Figure 7.9.

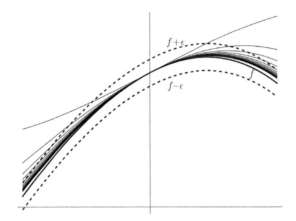

Figure 7.9: Geometric interpretation of uniform convergence.

Example 7.2.2. Let X be a finite set, say $\{a_1, \ldots, a_k\}$. Let (f_n) be a sequence of real-valued functions on X converging pointwise on X to $f \colon X \to \mathbb{R}$. Then the convergence is uniform.

For a given $\varepsilon > 0$, by the convergence of $f_n(a_j) \to f(a_j)$, we pick an N_j, $1 \leq j \leq k$ such that

$$k \geq N_j \implies |f_k(a_j) - f(a_j)| < \varepsilon.$$

Then, $N = \max\{N_j\}$ does the job:

$$k \geq N \text{ and } 1 \leq j \leq k \implies |f_k(a_j) - f(a_j)| < \varepsilon.$$

Hence $f_n \rightrightarrows f$ on X.

Example 7.2.3. We claim that the sequence (f_n) in item 10 of Example 7.1.2 is uniformly convergent to the constant function 0 on $\mathbb{R}$. This means that we need to find an estimate $|f_n(x) - f(x)|$ which may depend on n but must be independent of x. Here it is easy.

$$\left| \frac{\sin nx}{n} - 0 \right| = \left| \frac{\sin nx}{n} \right| \leq \frac{1}{n},$$

where we used the fact that $|\sin t| \leq 1$ for any $t \in \mathbb{R}$.

Let $\varepsilon > 0$ be given. Choose $N \in \mathbb{N}$ such that $\frac{1}{N} < \varepsilon$. Then for any $k \geq N$, we have for any $x \in \mathbb{R}$,

$$\left| \frac{\sin nx}{k} \right| \leq \frac{1}{k} \leq \frac{1}{N} < \varepsilon.$$

Example 7.2.4. We prove that none of the sequences in Example 7.1.2 (except the last one) are uniformly convergent.

Let us consider the sequence in item 1 of Example 7.1.2. From Figure 7.1, it is clear that the convergence is not uniform. Assume that $f_n \to 0$ uniformly on $\mathbb{R}$. Fix $\varepsilon > 0$, say, $\varepsilon = 1$. Let $N \in \mathbb{N}$ be such that

$$\forall n \geq N \text{ and } \forall x \in \mathbb{R} \implies \left| \frac{x}{n} \right| < 1.$$

The picture suggests that for $a \gg 0$ (or $a \ll 0$), the graph of f_n does not lie in the ε-band around the x-axis, the graph of the limit function f. So we look for $a \gg 0$. For example, if we take $a = 2N$ and $n = N$, we end up with $|f_N(a)| = 2$. This contradiction shows that $f_n \to 0$ on $\mathbb{R}$ but the convergence is not uniform.

Item 2 of Example 7.1.2: The sequence does not converge pointwise on X. Hence the question of its uniform convergence does not arise.

Item 3 of Example 7.1.2: Look at Figure 7.3. It is clear that if $a > 0$ is very near to zero, we need $n \gg 0$ so that $f_n(a)$ comes closer to $y = 1$. We exploit this observation to show that the convergence is not uniform. Assume the contrary. Let $\varepsilon = 1/2$. Let $N \in \mathbb{N}$ correspond to ε as in the definition of uniform convergence. Hence for $n \geq N$ and for all $a > 0$ we should expect $|1 - f_n(a)| < 1/2$. Let $a = \frac{1}{2N}$. Then $0 < a < 1/N$ and hence $f_N(a) = Na = 1/2$ so that $1 - f_N(a) = 1/2$. This contradiction shows that the convergence is not uniform.

Item 4 of Example 7.1.2: Look at Figure 7.4. For points $a > 0$ near zero, we need a large n so that $f_n(a)$ comes closer the x-axis. Assume that f_n is uniformly convergent on $\mathbb{R}$. Let $\varepsilon = 1/2$ and $N \in \mathbb{N}$ correspond to ε. If we take $a = 1/N$, then $f_N(a) = 1$ so that $|f_N(a) - f(a)| = 1$, not less than $1/2$, as stated.

Item 5 of Example 7.1.2: As can be deduced from the graphs of f_n's, if $|a| \gg 0$, it requires $n \gg 0$ so that $f_n(a)$ is close to 1. Let it be uniformly convergent on $\mathbb{R}$. For $\varepsilon = 1/2$ and a corresponding N, take $a = 2N$.

Item 6 of Example 7.1.2: The sequence is not convergent on $\mathbb{R}$ and hence the question of its uniform convergence is meaningless.

What can we say about the convergence of (f_n) if we take the domain of f_n's to be $\mathbb{R}^*$? Figure 7.5 shows that if $|a|$ is near to zero, it takes a large n to bring $f_n(a)$ close to zero. So we expect the convergence to be non-uniform. Can you write down a textbook proof now?

Item 7 of Example 7.1.2: Figure 7.6 clearly shows that the "ripples" in the form of "tents" persist forever in the positive side of the x-axis. We expect it to be non-uniformly convergent. We shall prove this by contradiction. Let $\varepsilon = 1$ and let N correspond to ε. Let $a = 2N$. Then $f_{2N}(2N) = 1$ and $|f_{2N}(2N) - f(2N)| = f_{2N}(2N) = 1 \not< 1$.

Item 8 of Example 7.1.2: Let $f_n(x) = x^n$, $x \in [0, 1]$ and

$$f(x) = \begin{cases} 0, & \text{if } 0 \leq x < 1 \\ 1, & \text{if } x = 1. \end{cases}$$

Then f_n converges to f pointwise but not uniformly on $(0,1)$ and hence not on $[0,1]$.

Again, Figure 7.7 shows that the convergence is not uniform. The trouble is located at 1.

If N does the job for $\varepsilon = \frac{1}{2}$, we then have

$$|f_N(x) - f(x)| = x^N < \frac{1}{2} \text{ for } 0 \leq x < 1.$$

Let $x \to 1_-$. Since f_N is continuous, we see that $\lim_{x \to 1_-} f_N(x) = 1$ and hence for x near to 1, the stated inequality cannot be true. We conclude that (f_n) does not converge uniformly on $(0, 1)$ and hence certainly not on $[0,1]$.

In case, you did not like the argument above, let us choose a such that $1 > a > 1/2^{1/N}$, then $a^N > 1/2$, a contradiction. (How do we know such an a exists? If $1/2^{1/N} < 1$, certainly such an a exists, for example, a could be their midpoint. If $1/2^{1/N} \geq 1$, raising both sides to their N-th power, we get $1/2 \geq 1$.)

Item 9 of Example 7.1.2 Let us work this out without recourse to pictures! As we are taking n-th roots and in spite of how large $a \gg 0$, we know $a^{1/n} \to 1$. This should suggest that we take a very large and an n-th power, say, $a = 2^n$. Now we can visualize the graphs of f_n. Though they pointwise approach the line $y = 1$, their graphs can be faraway from it when the $x \gg 0$. Hence we should expect the convergence to be non-uniform. Let us prove this.

Let $\varepsilon = 1$ and let N correspond to ε. Then $|f_N(a) - f(a)| < 1$ for all $a > 0$. If we take $a = 2^N$, then $|f_N(a) - f(a)| = 2 - 1 = 1 \nless 1$.

Even near 0, the problem arises. Assume that the series is uniformly convergent, on, say, (0,1). Let N correspond to $\varepsilon = 1/2$. If $0 < a < 2^{-N}$, then we should have

$$|1 - f_N(a)| = 1 - f_N(a) < 1/2 \iff 1/2 < a^{1/N} \iff \frac{1}{2^N} < a,$$

a contradiction. $\qquad\qquad\qquad\qquad\qquad\qquad\qquad\qquad\qquad\qquad \Box$

7.3   Consequences of Uniform Convergence

The next theorem is easy and an often used tool to establish the non-uniform convergence of a sequence of (necessarily continuous) functions. See Example 7.3.2.

Theorem 7.3.1. *Let $f_n\colon J \subseteq \mathbb{R} \to \mathbb{R}$ converge uniformly on J to f. Assume that f_n are continuous at $a \in X$. Then f is continuous at a.*

Strategy. Fix $a \in X$. To establish continuity of f at a, we need to estimate $|f(x) - f(a)|$. We know how to estimate $|f_n(x) - f(x)|$ and $|f_n(a) - f(a)|$. This suggests that we consider

$$|f(x) - f(a)| = |f(x) - f_n(x) + f_n(x) - f_n(a) + f_n(a) - f(a)|$$
$$\leq |f(x) - f_n(x)| + |f_n(x) - f_n(a)| + |f_n(a) - f(a)|.$$

We now use the curry-leaf trick. We fix n and use the continuity of f_n at a to estimate the middle term. To estimate the first and the third term, we use the uniform convergence of f_n.

Proof. To prove continuity of f at a, let $\varepsilon > 0$ be given. For this ε, using the uniform convergence of f_n to f, there exists N such that

$$\forall n \geq N \text{ and } x \in X \implies |f_n(x) - f(x)| < \varepsilon/3. \tag{7.3}$$

For N as above using the continuity of f_N at a, we can choose a $\delta > 0$ such that

$$x \in J \text{ and } |x - a| < \delta \implies |f_N(x) - f_N(a)| < \varepsilon/3. \tag{7.4}$$

Observe that for $x \in (a - \delta, a + \delta) \cap J$, we have

$$|f(x) - f(a)| \leq |f(x) - f_N(x)| + |f_N(x) - f_N(a)| + |f_N(a) - f(a)|$$
$$< \varepsilon/3 + \varepsilon/3 + \varepsilon/3,$$

where we used (7.3) to estimate the first and the third term and (7.4) to estimate the middle term. $\qquad \square$

Example 7.3.2. Theorem 7.3.1 gives an alternate proof of the fact that the convergence of (f_n) of item 8 of Example 7.1.2 is not uniform on $[0,1]$. However, it cannot be used to prove that the convergence of (f_n) to f is not uniform on $(0,1)$. Why?

Remark 7.3.3. If $f_n \to f$ pointwise and if f_n's and f are continuous, we cannot conclude that the convergence is uniform. Can you find examples to illustrate this in Example 7.1.2?

Exercise 7.3.4. If f_n's are assumed to be uniformly continuous in Theorem 7.3.1, and if $f_n \rightrightarrows f$ on J, can we conclude that f is uniformly continuous?

Example 7.3.5.

(1) Consider $f_n \colon \mathbb{R} \to \mathbb{R}$ given by $f_n(x) = 0$ if $|x| \leq n$ and $f_n(x) = n$ if $|x| > n$. Then $f_n \to 0$ pointwise but not uniformly. Let $a \in \mathbb{R}$. By the Archimedean property, we can choose $N \in \mathbb{N}$ such that $N > |a|$. Hence for $k \geq N$, $f_k(a) = 0$. Thus, the sequence $(f_k(a))$ is eventually the constant sequence 0. Hence

$f_n(a) \to 0$. We may therefore take $f(a) = 0$. This proves the pointwise convergence of f_n to the constant function 0.

Is the convergence uniform? Look at Figure 7.10. The picture shows that it is not.

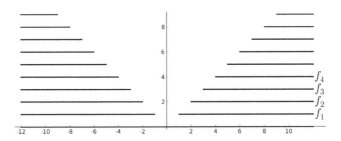

Figure 7.10: Graph of $f_n(x)$ in item 1 of Example 7.3.5.

Let us prove this rigorously. Let us assume that the convergence is uniform. Let $\varepsilon = 1/2$. Then there exists $N \in \mathbb{N}$ such that for all $x \in \mathbb{R}$, we have $|f_k(x)| < 1$ if $k \geq N$. In particular, if we take $k = N$ and $x = N + 1$, then $f_N(N+1) = N > 1/2$, a contradiction. Hence the convergence is not uniform.

(2) Consider $f_n(x) = \frac{nx}{1+n^2x^2}$ and $f(x) = 0$ for $x \in \mathbb{R}$. For $x \neq 0$, we can rewrite this $f_n(x) = \frac{1}{(nx)+\frac{1}{nx}}$. This reminds us of $t + \frac{1}{t} \geq 2$ for $t > 0$ and equality iff $t = 1$. Hence we chose $x_n = 1/n$ so that $f_n(1/n) = 1/2$.

What do these observations lead us to? If we enclose the graph of $f = 0$ in an $\varepsilon > 0$ band, and if we take $0 < \varepsilon \leq 1/2$, then each $n \in \mathbb{N}$, the graph of f_n does not lie within this band. So we expect that the convergence is not uniform. Can you write down a textbook proof now?

(3) Consider $f_n(x) = \frac{x^n}{n+x^n}$, $x \geq 0$. See Figure 7.11. Let us discuss the convergence of this sequence of functions. If $x \in [0, 1]$, we have $|f_n(x)| \leq 1/n$ and hence we conclude that f_n converges to f uniformly on $[0, 1]$.

If $x > 1$, we recast $f_n(x) = \frac{x^n}{x^n} \frac{1}{1+\frac{n}{x^n}}$. From Exercise 2.5.2, we know that $\frac{n}{x^n} \to 0$. Hence $f_n(x) \to 1$ for any $x > 1$. Thus the sequence (f_n) converges pointwise on $[0, \infty)$ to the function

$$f(x) = \begin{cases} 0, & \text{if } 0 \leq x \leq 1 \\ 1, & \text{if } x > 1. \end{cases}$$

Since the limit function is not continuous, we conclude that the convergence is not uniform on $[0, \infty)$.

How about on $(1, \infty)$? Our discussion so far indicates the source of the problem is at $a = 1$. Thus, even after excluding 1 from the domain, points near

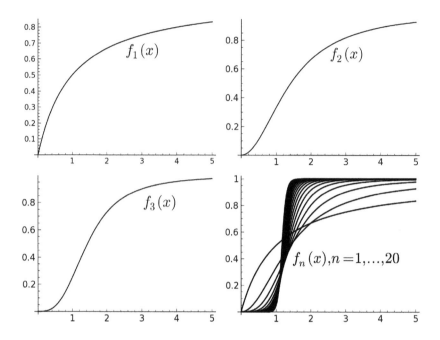

Figure 7.11: Graph of $f_n(x) = \frac{x^n}{n+x^n}$.

to 1 may exhibit bad behavior. Furthermore, since f_n is continuous and since $f_n(1) \to 0$, we expect that for each n, there will be points a_n near to 1 such that $f_n(a_n) = 1/2$. This will lead us to conclude that for each n, the graph of f_n may not lie within the 1/2-band around $y = 1$.

Note that

$$\frac{x^n}{n+x^n} = \frac{1}{\frac{n}{x^n}+1} = \frac{1}{2} \text{ if } x^n = n.$$

Hence at points $a_n = n^{1/n}$, we have $f_n(a_n) = 1/2$ so that $|f_n(a_n) - 1| = 1/2$ for any $n \in \mathbb{N}$. So, if $\varepsilon < 1/2$, we shall not be able to find any $N \in \mathbb{N}$ satisfying the definition of uniform convergence.

Did you observe that $a_n \to 1$? Hence our intuition about the bad behavior of f_n at points near to 1 is vindicated.

Recall that given a real sequence (x_n), we have seen that $x_n \to x$ iff $|x_n - x| \to 0$. Now, if (f_n) is a sequence of functions on a set X, then $f_n \to f$ pointwise iff for every $x \in X$, we must have $|f_n(x) - f(x)| \to 0$. What should be an analogue if we want to characterize uniform convergence of f_n to f?

Proposition 7.3.6. *Let $f_n, f \colon X \to \mathbb{R}$ be functions. Let*

$$M_n := \text{lub } \{|f_n(x) - f(x)| : x \in X\},$$

if it exists. Then $f_n \rightrightarrows f$ iff $M_n \to 0$.

Proof. Let M_n exist and converge to 0. We need to prove that $f_n \rightrightarrows f$. We need to find a uniform estimate for $|f_n(x) - f(x)|$ for $x \in X$. The definition of M_n's suggest a way.

Let $\varepsilon > 0$ be given. Then there exists $N \in \mathbb{N}$ such that for $k \geq N$, we have $|M_k| < \varepsilon$, that is, $M_k < \varepsilon$. We therefore have

$$k \geq N \implies M_k = \text{lub } \{|f_k(x) - f(x)| : x \in X\} < \varepsilon.$$

It follows that for each $x \in X$ and $k \geq N$, we obtain $|f_k(x) - f(x)| < \varepsilon$. Thus $f_n \rightrightarrows f$ on X.

Conversely, let $f_n \rightrightarrows f$ on X. Assume that M_n exists for $n \in \mathbb{N}$. We need to prove that $M_n \to 0$. Let $\varepsilon > 0$ be given. Since $f_n \rightrightarrows f$ on X, for $\varepsilon/2$, there exists $N \in \mathbb{N}$ such that for $x \in X$ and $k \geq N$, we have $|f_k(x) - f(x)| < \varepsilon/2$. It follows that for $k \geq N$, lub $\{|f_k(x) - f(x)| : x \in X\} \leq \varepsilon/2$. Hence for $k \geq N$, we obtain $M_k \leq \varepsilon/2 < \varepsilon$. That is, $M_k \to 0$. $\qquad\square$

Note that the second part of the proof shows that M_k's exist for $k \gg 0$. The result gives a quite useful algorithm whenever applicable.

Example 7.3.7. Consider $f_n(x) = x^2 e^{-nx}$ on $[0, \infty)$. Note that $f_n(x) > 0$ for $x > 0$, $f_n(0) = 0$ and $f_n(x) \to 0$ as $x \to \infty$. Hence 0 must be the global minimum for f. We expect f to attain a positive maximum in $(0, \infty)$. We find the maximum value of f_n using calculus. We have

$$f_n'(x) = xe^{-nx}(2 - nx).$$

Hence the critical points are at 0 and $x = 2/n$. Our analysis shows that $x = 2/n$ must be a point of maximum. Let us verify this by computing the second derivative.

$$\begin{aligned} f_n''(x) &= 2e^{-nx} - 2nxe^{-nx} - 2nxe^{-nx} + n^2x^2e^{-nx} \\ &= e^{-nx}\left(2 - 4nx + n^2x^2\right) \\ &= (nx - 2)^2 - 2. \end{aligned}$$

Hence $f_n''(2/n) = -2 < 0$. It follows that $M_n := f_n(2/n) = \left(\frac{2}{e}\right)^2 \frac{1}{n^2}$. As $M_n \to 0$, it follows the convergence $f_n \rightrightarrows 0$ on $[0, \infty)$, by Proposition 7.3.6.

Example 7.3.8. In this example we look at three sequences together, as they offer us some useful insights into analysis. Consider $f_n, g_n, h_n \colon [0, 1] \to \mathbb{R}$ defined by

$$f_n(x) := x^n(1 - x), \quad g_n(x) = x^n(1 - x^n) \text{ and } h_n(x) := x^n e^{-nx}.$$

The first observation is that the first two are modifications of the functions x^n by multiplying functions which vanish at $x = 1$. The factors $(1 - x)$ and $(1 - x^n)$ are chosen to "kill" the bad behavior x^n at $x = 1$, while e^{-nx} may do so at infinity. Study Figures 7.12–7.14.

Since $1 - x \to 0$ as fast as $x \to 1$ and for x near to 1, the rate at which x^n goes to 0 is slower, we may hope to show that $f_n \rightrightarrows 0$ on $[0, 1]$.

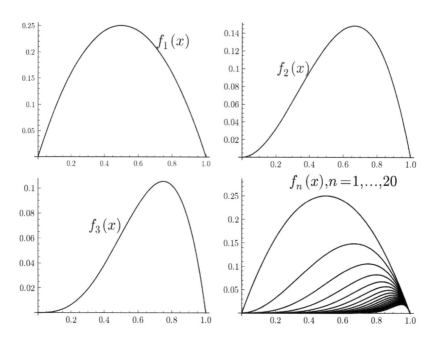

Figure 7.12: Killing the bad behavior at 1 of x^n by $(1-x)$.

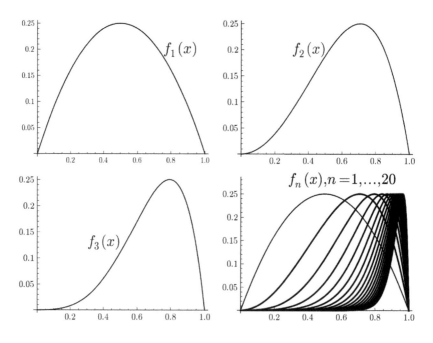

Figure 7.13: Killing the bad behavior at 1 of x^n by $(1-x^n)$.

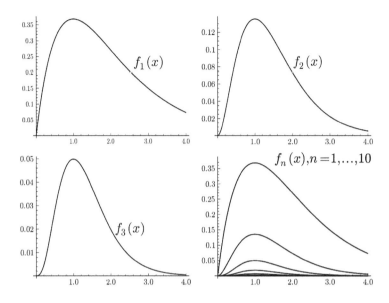

Figure 7.14: Killing the bad behavior at 1 of x^n by e^{-nx}.

In the case of the second factor $1 - x^n$, observe that $1 - x^n \to 0$ at a slower rate than that of $1 - x$. For example, if we take $x_1 = .9$, $x_2 = .99$ and so on $x_1^2 = .81$, $x_2^2 = .729$. Hence $1 - x_n^n \to 0$ much slower than $1 - x \to 0$. So we may not be able to conclude the uniform convergence of g_n.

In the third case, our previous exposure to indeterminate forms which deal with comparing the rates of decay to 0 of functions will lead us to believe that $e^{-nx} \to 0$ much faster than $x^n - 1 \to 0$. In this case, we may be able to conclude the uniform convergence $h_n \rightrightarrows 0$.

After these speculations, we need rigorous proofs. Fortunately, all the functions are non-negative, (infinitely) differentiable, and tend to vanish at the end points 0 and 1. Their minimum is already attained at 0. Hence we may expect to find their global maximum as a local maximum. Proposition 7.3.6 will help us resolve the issues.

We find $M_n(f_n) = \frac{n}{n+1}\frac{1}{n+1} < \frac{1}{n+1}$, $M_n(g_n) = 1/4$ and $M_n(h_n) = e^{-n}$. This proves our guesses are correct.

By the way, could you have guessed that $M_n(g_n) = 1/4$ without a serious computation? What is the maximum of $t(1 - t)$ on $[0, 1]$?

Exercise 7.3.9. What can you say about the convergence of the sequence of functions $f_n(x) := x^n(1-x)^2$ on $[0, 1]$? If you have understood our line of thinking, the answer should come in a flash! How about $x^n(1 - x)^n$?

Exercise Set 7.3.10. If nothing is specified in the following exercise, you are required to check for uniform convergence:

(1) Let $f_n(x) := nx^n$, $x \in [0,1)$. Show that the sequence (f_n) converges pointwise but not uniformly on $[0,1)$.

(2) Let $f_n(x) := nx^n(1-x)$, $x \in [0,1]$. Show that $f_n \to 0$ pointwise but not uniformly on $[0,1]$.

(3) $f_n(x) := (x/n)^n$ for $x \in [0,\infty)$.

(4) $f_n(x) = \frac{x}{1+nx}$ and $f(x) = 0$ for $x \geq 0$.

(5) $f_n(x) = \frac{nx}{1+n^2x^2}$ and $f(x) = 0$ for $x \in \mathbb{R}$.

(6) $f_n(x) = \frac{x}{1+nx^2}$ and $f(x) = 0$ for $x \in \mathbb{R}$.

(7) $f_n(x) = \frac{x^n}{n+x^n}$ and $f(x) = 0$ if $0 \leq x \leq 1$, and $f(x) = 1$ if $x > 1$ for $x \in [0,\infty)$. Show that f_n converges to f uniformly on $[0,1]$ but not on $[0,\infty)$.

(8) $f_n(x) := \frac{nx}{n+x}$ for $x \geq 0$.

(9) $f_n(x) = n^2x^n(1-x)$ and $f(x) = 0$ for $x \in [0,1]$.

(10) Let $f_n(x) = \frac{1}{1+x^n}$, $x \in [0,\infty)$. Show that $f_n \to f$ pointwise but not uniformly

on the domain where $f(x) = \begin{cases} 1, & 0 \leq x < 1 \\ 1/2, & x = 1 \\ 0, & x > 0. \end{cases}$

(11) Discuss the pointwise and uniform convergence of the sequence (f_n) where $f_n(x) = \frac{x^n}{1+x^n}$, $x \in [0,\infty)$.

(12) Complete the sentence: The sequence of functions h_n on $[0,A]$ defined by $h_n(x) = \frac{nx^3}{1+nx}$ converges to .... Is the convergence uniform?

(13) $f_n(x) := \frac{nx}{2n+x}$ for $x \in [0,\infty)$. Show that f_n converge uniformly on $[0,R]$ for any $R > 0$ but not on $[0,\infty)$.

(14) Let $f_n(x) := \frac{x^n}{1+x^{2n}}$ for $x \in [0,1)$. Show that f_n is pointwise convergent on $[0,1)$ but is not uniformly convergent on $[0,1)$.

(15) Show that $f_n : \mathbb{R} \to \mathbb{R}$ defined by $f_n(x) = \frac{x^n}{1+x^{2n}}$ converges uniformly on $[a,b]$ iff $[a,b]$ does not contain either of ± 1.

(16) Let $f_n : [0,1] \to \mathbb{R}$ be defined by $f_n(x) = \frac{nx}{1+n^2x^p}$, for $p > 0$. Find the values of p for which the sequence f_n converges uniformly to the limit.

(17) $f_n(x) = \begin{cases} nx, & 0 \leq x \leq \frac{1}{n} \\ 2 - nx, & \frac{1}{n} \leq x \leq \frac{2}{n} \\ 0, & \frac{2}{n} < x, \end{cases}$ for all $x \geq 0$.

(18) $f_n(x) = xe^{-nx}$ on $[0,\infty)$.

(19) On $[0,1]$, define $g_n(x) = nxe^{-nx^2}$. Discuss its convergence on $[0,1]$.

(20) Show that $f_n(x) = \sin \frac{x}{n}$ is convergent to 0 on $[0,1]$. Is the convergence uniform?

(21) Let $g_n(x) = \frac{\sin nx}{nx}$ on $(0,1)$. Is the sequence convergent? If so, what is the limit? Is the sequence uniformly convergent?

(22) Let $f_n(x) := \frac{1+2\cos^2 nx}{\sqrt{n}}$, $x \in \mathbb{R}$. Show that f_n converges uniformly on $\mathbb{R}$.

(23) $f_n(x) = \begin{cases} n^2, & 0 \le x \le 1/n \\ n^2 - n^3(x - 1/n), & 1/n \le x \le 2/n \\ 0, & \text{otherwise.} \end{cases}$

(24) Consider $f_n(x) = x$ and $g_n(x) = 1/n$ for $x \in \mathbb{R}$. Show that f_n and g_n are uniformly convergent on $\mathbb{R}$ but their product is not.

(25) Let $f_n(x) = x\left(1 - \frac{1}{n}\right)$ and $g_n(x) = 1/x^2$ for $x \in (0,1)$. Show that (f_n) and (g_n) are uniformly convergent on $(0,1)$ but their product is not.

(26) Let $\{r_n\}$ be a sequence consisting of all rational numbers. Define

$$f_n(x) = \begin{cases} 1, & \text{if } x = r_n \\ 0, & \text{otherwise.} \end{cases}$$

Show that f_n converges pointwise to $f = 0$, but not uniformly on *any* interval of $\mathbb{R}$.

(27) Let $f_n(x) = \int_0^x \cos nt\, dt$ for $x \in \mathbb{R}$. The sequence $f_n \rightrightarrows 0$ on $\mathbb{R}$. What can you say if cosine is replaced by sine?

We know that any sequence of real numbers is convergent iff it is Cauchy. Is there an analogue in the context of the sequence of functions? What does it mean to say that (f_n) is pointwise Cauchy and uniformly Cauchy on X?

Definition 7.3.11. Let (f_n) be a sequence of real-valued functions on a set X. We say that (f_n) is Cauchy if for each $x \in X$, the sequence $(f_n(x))$ of real numbers is Cauchy in $\mathbb{R}$. This can be written in terms of quantifiers as follows.

$$\forall x \in X \, (\forall \varepsilon > 0 \, (\exists N_x \in \mathbb{N} \, (\forall m, n \ge N_x \, (|f_n(x) - f_m(x)| < \varepsilon)))). \qquad (7.5)$$

Thus the integers N_x depend on x (and of course on ε too). If we can choose N independent of x (so that it depends only on ε and not on x), we say that (f_n) is uniformly Cauchy on X. We formulate it as follows.

A sequence (f_n) of functions on a set X is said to be *uniformly Cauchy* on X if for a given $\varepsilon > 0$ there exists $N = N_\varepsilon \in \mathbb{N}$ such that for all $m, n \ge N$ and for all $x \in X$ we have $|f_n(x) - f_m(x)| < \varepsilon$.

In terms of quantifiers we have:

$$\forall \varepsilon > 0 \, (\exists N_\varepsilon \in \mathbb{N} \, (\forall m, n \geq N_\varepsilon \, (\forall x \in X \, (|f_n(x) - f_m(x)| < \varepsilon)))). \qquad (7.6)$$

Compare the two formulations in (7.5) and (7.6) and pay attention to the placement of $\forall x \in X$.

Theorem 7.3.12 (Cauchy Criterion for Uniform Convergence). *Let $f_n \colon X \to \mathbb{R}$ be a sequence of functions from a set X to $\mathbb{R}$. Then f_n is uniformly convergent iff the sequence (f_n) is uniformly Cauchy on X.*

Strategy: Observe that for $x \in X$, $(f_n(x))$ is Cauchy in $\mathbb{R}$ and hence converges to a real number $f(x)$. To estimate $|f_n(x) - f(x)|$ we use the curry-leaves trick. Fix $x \in X$.

$$|f_n(x) - f(x)| \leq |f_n(x) - f_m(x)| + |f_m(x) - f(x)|, \text{ for any } m \in \mathbb{N}.$$

where m is the curry leaf. The first term can be made small if $m, n \gg 0$ and the second one if $m \gg 0$ is chosen using the pointwise convergence of $f_n(x) \to f(x)$. So m should satisfy both the requirements.

Proof. Let $f_n \rightrightarrows f$ on X. Let $\varepsilon > 0$ be given. Choose N such that

$$n \geq N \, \& \, \forall x \in X \implies |f_n(x) - f(x)| < \varepsilon/2.$$

Then for $n, m \geq N$, and for all $x \in X$, we get

$$|f_n(x) - f_m(x)| \leq |f_n(x) - f(x)| + |f(x) - f_m(x)| < 2 \times \varepsilon/2 = \varepsilon.$$

Let (f_n) be uniformly Cauchy on X. We need to prove that (f_n) is uniformly convergent on X. This uses our curry-leaves trick. It is clear that for each $x \in X$, the sequence $(f_n(x))$ of real numbers is Cauchy and hence by Cauchy completeness (of $\mathbb{R}$) there exists a unique $r_x \in \mathbb{R}$ such that $f_n(x) \to r_x$. We define $f \colon X \to \mathbb{R}$ by setting $f(x) = r_x$. (Note that we need the uniqueness of the limit of the sequence $(f_n(x))$ to show that f is a well-defined function!) We claim that $f_n \rightrightarrows f$ on X.

Let $\varepsilon > 0$ be given. Since $f_n \to f$ pointwise on X, for a given $x \in X$, there exists $N_x = N_x(\varepsilon)$ such that

$$n \geq N_x \implies |f_n(x) - f(x)| < \varepsilon/2.$$

Also, since (f_n) is uniformly Cauchy on X, we can find N such that

$$m, n \geq N \, \& \, \forall x \in X \implies |f_n(x) - f_m(x)| < \varepsilon/2.$$

Fix $x \in X$. We observe, for $n \geq N$

$$
\begin{aligned}
|f_n(x) - f(x)| &\leq |f_n(x) - f_m(x)| + |f_m(x) - f(x)| \text{ for any } m \\
&\leq |f_n(x) - f_m(x)| + |f_m(x) - f(x)| \text{ for } m > \max\{N, N_x\} \\
&< \varepsilon/2 + \varepsilon/2 = \varepsilon.
\end{aligned}
$$

Thus for any $x \in X$ and $n \geq N$, we have shown $|f_n(x) - f(x)| < \varepsilon$, that is, f is uniformly convergent to f on X. $\qquad \square$

The following is an application of the Cauchy criterion for uniform convergence (Theorem 7.3.12).

Theorem 7.3.13. *Let $f_n \colon J := (a,b) \to \mathbb{R}$ be differentiable. Assume that $f'_n \rightrightarrows g$ uniformly. Further assume that there exists $c \in J$ such that the real sequence $(f_n(c))$ converges. Then the sequence (f_n) converges uniformly to a continuous function $f \colon J \to \mathbb{R}$.*

> **Strategy:** How do we arrive at f? If we could show that (f_n) is uniformly Cauchy, then the (uniform) limit function will be our choice of f. Since f_n's are differentiable and hence continuous, f will be continuous. To show that (f_n) is uniformly Cauchy, we need to find a uniform estimate for $|f_n(x) - f_m(x)|$. If we wish to estimate f in terms of its derivatives, the mean value theorem is our tool. Since $f'_n \rightrightarrows g$, we know how to estimate $|f'_n - g|$ and hence $|f'_n - f'_m|$. This suggests the use of the mean value theorem to estimate $|f_n(x) - f_m(x) - (f_n(c) - f_m(c))|$.

Proof. Fix $x \in J$. We claim that (f_n) is uniformly Cauchy. Applying the mean value theorem to the function $f_n - f_m$ at the points x and c, we have

$$(f_n - f_m)(x) - (f_n - f_m)(c) = (f_n - f_m)'(t)(x - c).$$

That is,

$$f_n(x) - f_m(x) - (f_n(c) - f_m(c)) = [f'_n(t) - f'_m(t)](x - c), \qquad (7.7)$$

for some t between x and c. Note that the point t depends on m, n, x, c.

Given $\varepsilon > 0$, there exists $n_1 \in \mathbb{N}$ such that

$$m, n \geq n_1 \implies |f_n(c) - f_m(c)| < \varepsilon/2. \qquad (7.8)$$

Also, since $f'_n \rightrightarrows g$, the sequence (f'_n) is uniformly Cauchy and hence there exists $n_2 \in \mathbb{N}$ such that

$$n \geq n_2, s \in J \implies |f'_n(s) - f'_m(s)| < \frac{\varepsilon}{2(b-a)}. \qquad (7.9)$$

If $N := \max\{n_1, n_2\}$, we use (7.9) in (7.7) to obtain

$$|(f_n(x) - f_m(x)) - (f_n(c) - f_m(c))| < \frac{\varepsilon}{2(b-a)} |x - c| \leq \frac{\varepsilon}{2(b-a)} |b - a| = \frac{\varepsilon}{2},$$

for all $n \geq N$ and $x \in J$. Now we use the triangle inequality and (7.8) to arrive at

$$n \geq N \implies |f_n(x) - f_m(x)| < \varepsilon, \text{ for all } x \in J.$$

That is, (f_n) is uniformly Cauchy on J and hence is uniformly convergent to a function $f \colon J \to \mathbb{R}$. Since f_n are continuous, so is f by Theorem 7.3.1. □

Remark 7.3.14. Note that if we simply assumed that $f'_n \rightrightarrows g$, we cannot conclude that (f_n) is uniformly Cauchy. For example, consider $f_n(x) = n$ on $J = [a, b]$. Then $f'_n = 0$ and obviously $f'_n \rightrightarrows 0$. But (f_n) is not Cauchy, let alone uniformly Cauchy. So the second condition that there exists c such that $(f_n(c))$ is convergent is essential.

Let $f_n\colon J \to \mathbb{R}$ be differentiable. Assume that $f_n \rightrightarrows f$ on J. Since f_n's are continuous, f is continuous. However, f need not be differentiable. For example, we can exhibit $x \mapsto |x|$ as the uniform limit of a sequences of differentiable functions f_n on $[-1,1]$. Note that $f(x) = |x|$ is uniformly continuous on $[-1,1]$ (in fact, on $\mathbb{R}$). We rewrite $f(x) = (x^2)^{1/2}$. Consider

$$f_n(x) := f\left(x + \frac{1}{n}\right) = \left[\left(x + \frac{1}{n}\right)^2\right]^{1/2}.$$

Then f_n are differentiable and $f_n \rightrightarrows f$ on $[-1,1]$. (See Item 6 in Exercise 7.3.21.) The next result gives us a sufficient condition under which we can conclude the differentiability of the uniform limit f of differentiable f_n's.

Theorem 7.3.15. *Let $f_n\colon (a,b) \to \mathbb{R}$ be differentiable. Assume that there exist $f, g\colon (a,b) \to \mathbb{R}$ such that $f_n \rightrightarrows f$ and $f_n' \rightrightarrows g$ on (a,b). Then f is differentiable and $f' = g$ on (a,b).*

Strategy: Fix $c \in (a,b)$. We show that the auxiliary functions f_{n1} of Theorem 4.1.3 on page 112 characterizing the differentiability at c are uniformly Cauchy. (How? Of course, by the mean value theorem!) We introduce an auxiliary function φ for f with $\varphi(c) = g(c)$. It is then more or less clear that f_{n1} converge uniformly to φ. This will establish the differentiability of f at c and that $f'(c) = g(c)$.

Proof. Fix $c \in (a,b)$. Consider $g_n := f_{n1}$ in the notation of Theorem 4.1.3, that is,

$$g_n(x) = \begin{cases} \frac{f_n(x) - f_n(c)}{x - c}, & \text{for } x \neq c \\ f_n'(c), & \text{for } x = c. \end{cases}$$

Then g_n are continuous and they converge pointwise to $\varphi(x) = \frac{f(x) - f(c)}{x - c}$ for $x \neq c$ and $\varphi(c) = g(c)$.

We claim that g_n are uniformly Cauchy on (a,b) and hence uniformly convergent to a continuous function $\psi\colon (a,b) \to \mathbb{R}$. For, by the mean value theorem, we have, for some ξ between $x \neq c$ and c,

$$g_n(x) - g_m(x) = \frac{(f_n - f_m)'(\xi)(x - c)}{x - c} = f_n'(\xi) - f_m'(\xi).$$

Since (f_n') converge uniformly on $J = (a,b)$, it is uniformly Cauchy and hence (g_n) is uniformly Cauchy on J. It follows from Theorem 7.3.12 that g_n converge uniformly to a function, say, ψ. The function ψ is continuous by Theorem 7.3.1. By the uniqueness of the pointwise limits, we see that $\varphi(x) = \psi(x)$ for $x \in J$. Hence φ is continuous or it is the "f_1" (in the notation of Theorem 4.1.3 on page 112) for the function f at c. Thus, f is differentiable at c with $f'(c) = \varphi(c) = g(c)$. $\square$

Corollary 7.3.16. *Let $f_n\colon [a,b] \to \mathbb{R}$. Assume that there is some $x_0 \in [a,b]$ such that $(f_n(x_0))$ converges and that f_n' exists and converges uniformly to g on $[a,b]$. Then f_n converges uniformly to some f on $[a,b]$ such that $f' = g$ on $[a,b]$.*

Proof. This is an immediate consequence of the last two theorems. □

The last couple of results can be summarized as follows: Under suitable conditions, we have

$$f' \equiv (\lim f_n)' = \lim f_n' \quad \text{OR} \quad \frac{d}{dx}(\lim f_n) = \lim\left(\frac{d}{dx}f_n\right).$$

That is, we are able to interchange the limit process and the differentiation.

The next three results deal with uniform convergence of a sequence (f_n) of integrable functions and the process of integration. In essence, under suitable conditions, we want to say

$$\lim \int_a^b f_n = \int_a^b \lim f_n.$$

Proposition 7.3.17. *Let (f_n) be a sequence of continuous functions on $[a,b]$. Assume that $f_n \to f$ uniformly on $[a,b]$. Then f is continuous and hence integrable on $[a,b]$. Furthermore $\int_a^b f(x)\,dx = \lim \int_a^b f_n(x)\,dx$. That is,*

$$\lim_n \int_a^b f_n(t)\,dt = \int_a^b \lim_n f_n(t)\,dt.$$

Proof. We use the linearity of the integral (Theorem 6.2.1) and the basic estimate (6.12) for the integrals. Observe that

$$\left| \int_a^b f(x)\,dx - \int_a^b f_n(x)\,dx \right| \tag{7.10}$$

$$= \left| \int_a^b (f(x) - f_n(x))\,dx \right|$$

$$\leq \int_a^b |f(x) - f_n(x)|\,dx$$

$$\leq (b-a) \times \text{lub } \{|f(x) - f_n(x)| : x \in [a,b]\}. \tag{7.11}$$

Since $f_n \rightrightarrows f$ on $[a,b]$, given $\varepsilon > 0$, there exists $N \in \mathbb{N}$ such that for $n \geq N$ and $x \in [a,,b]$, we have $|f_n(x) - f(x)| < \frac{\varepsilon}{b-a}$. Hence we conclude that lub $\{|f(x) - f_n(x)| : x \in [a,b]\} \leq \frac{\varepsilon}{b-a}$. Plugging this in (7.11) yields the desired result. □

Exercise 7.3.18. Let $f_n \colon [a,b] \to \mathbb{R}$ be a sequence of continuously differentiable functions converging uniformly to f on $[a,b]$. Assume that f_n' converge uniformly on $[a,b]$ to g. Show that f is differentiable and $g = f'$ on $[a,b]$.

Compare this with Corollary 7.3.16. Note that the hypothesis is more stringent than that in Corollary 7.3.16. This is a very useful result in higher aspects of analysis.

Theorem 7.3.19. *Let (f_n) be a sequence of integrable functions on $[a, b]$. Assume that $f_n \to f$ uniformly on $[a, b]$. Then f is integrable on $[a, b]$ and $\int_a^b f(x)\, dx = \lim \int_a^b f_n(x)\, dx$. That is, $\lim \int_a^b f_n(x)\, dx = \int_a^b \lim f_n(x)\, dx$.*

Strategy: The crucial step is to establish the integrability of f. We invoke the criterion for integrability. Given $\varepsilon > 0$, we need to find a partition P such that $U(f, P) - L(f, P) < \varepsilon$. Our hope is if $N \gg 0$, by the criterion applied to the integrability of f_N will give rise a partition P, so that $U(f_N, P) - L(f_N, P) < \varepsilon$. Since f_N is uniformly close to f, we hope to have control over $U(f, P) - L(f, P)$.

Proof. Given $\varepsilon > 0$ there is $N \in \mathbb{N}$ such that

$$\text{lub } \{|f_n(x) - f(x)| : x \in [a, b]\} < \frac{\varepsilon}{4(b - a)} \text{ for } n \geq N.$$

Since f_N is integrable, there is a partition $P := \{x_0, x_1, \ldots, x_n\}$ of $[a, b]$ such that $U(f_N, P) - L(f_N, P) < \frac{\varepsilon}{2}$. Note that $M_j(f) \leq M_j(f_N) + \varepsilon/4(b - a)$, where $M_j(f) := \sup \{f(x) : x \in [x_{j-1}, x_j]\}$. Hence $U(f, P) \leq U(f_N, P) + \varepsilon/4$ and $L(f, P) + \varepsilon/4 \geq L(f_N, P)$. Hence $U(f, P) - L(f, P) < \varepsilon$. Hence f is integrable in $[a, b]$. The rest of the proof follows that of Proposition 7.3.17. □

Exercise Set 7.3.20.

(1) Let $\{r_n\}$ be an enumeration of all the rationals in $[0, 1]$. Define $f_n \colon [0, 1] \to \mathbb{R}$ as follows:
$$f_n(x) := \begin{cases} 1, & \text{if } x = r_1, \ldots, r_n, \\ 0, & \text{otherwise.} \end{cases}$$
Show that f_n is integrable for $n \in \mathbb{N}$ and that $f_n \to f$ pointwise, where f is Dirichlet's function. Conclude that the pointwise limit of a sequences (f_n) of integrable functions need not be integrable.

(2) Let f_n, f be as in Item 4 of Example 7.1.2. Compute $\lim_n \int_0^1 f_n(t)\, dt$ and $\int_0^1 \lim f_n(t)\, dt$. Here the convergence is not uniform but still the limit of the integrals is the same as the integral of the limit.

(3) Let $f_n \colon [0, 1] \to \mathbb{R}$ be given by $f_n(x) = nxe^{-nx^2}$. Show that $f_n \to 0$ pointwise. Compute $\lim_n \int_0^1 f_n(t)\, dt$ and $\int_0^1 \lim f_n(t)\, dt$.

(4) Let $f_n(x) = nx(1 - x^2)^n$, $x \in [0, 1]$. Find the pointwise limit f of f_n. Does $\int_0^1 f_n(x)\, dx \to \int_0^1 f(x)\, dx$?

(5) Let $f_n(x) = \frac{n^2 x^2}{1 + n^3 x^3}$ on $[0, 1]$. Show that f_n does not satisfy the conditions of Corollary 7.3.16, but that the derivative of the limit function exists on $[0, 1]$ and is equal to the limit of the derivatives.

(6) If $f_n(x) = \frac{x}{1+n^2x^2}$ on $[-1, 1]$, show that f_n is uniformly convergent, and that the limit function is differentiable, but $f' \neq \lim f_n'$ on $[-1, 1]$.

(7) If $f_n(x) = (1+x^n)^{\frac{1}{n}}$ on $[0, 2]$, then show that f_n is differentiable on $[0, 2]$ and converges uniformly to a function which is not differentiable at 1.

(8) Let $f(x) = |x|$ on $\mathbb{R}$. We replace part of the graph of f on the interval $[-1/n, 1/n]$ by a part of the parabola that has correct values and the correct derivatives (so that the tangents match) at the end point $\pm(1/n)$. Let

$$f_n(x) := \begin{cases} \frac{nx^2}{2} + \frac{1}{2n}, & -1/n \leq x \leq 1/n \\ |x|, & |x| > 1/n. \end{cases}$$

Show that $f_n \to f$ pointwise. Is the convergence uniform? Note that f_n are differentiable, but f is not.

(9) Let $f_n(x) := x^n(x - 2)$, $x \in [0, 1]$. Show that $f_n \to g$, where $g(x) = 0$ for $0 \leq x < 1$ and $g(1) = -1$. Can g be the derivative of any function?

Exercise Set 7.3.21. Theoretical Exercises.

(1) Let $f_n: J \subseteq \mathbb{R} \to \mathbb{R}$ converge uniformly on J to f. Assume that f_n are uniformly continuous at $a \in X$. Then show that f is uniformly continuous at a.

(2) (Dini) Let $f_n: [a, b] \to \mathbb{R}$ be continuous for each n. Assume that the sequence (f_n) converges pointwise to a *continuous* function $f: [a, b] \to \mathbb{R}$. Assume that the sequence is *monotone*, that is, for each $x \in [a, b]$, we have $(f_n(x))$ is a monotone sequence of real numbers. Then $f_n \to f$ uniformly on $[a, b]$.

(3) Let $f(x) = \sqrt{x}$ on $[0, 1]$. Let $f_0 = 0$ and $f_{n+1}(x) := f_n(x) + [x - (f_n(x))^2]/2$ for $n \geq 0$. Show that i) f_n is a polynomial, ii) $0 \leq f_n \leq f$, iii) $f_n \to f$ pointwise, and iv) $f_n \to f$ uniformly on $[0, 1]$.

(4) Let $f_n: \mathbb{R} \to \mathbb{R}$ be defined by

$$f_n(x) = \begin{cases} 0, & x \leq n \\ x - n, & n \leq x \leq n + 1 \\ 1, & x \geq n + 1. \end{cases}$$

Show that $f_n \geq f_{n+1}$, $f_n \to 0$ pointwise but not uniformly. Compare this with Dini's theorem (Item 2 above).

(5) Let $f: [0, 1] \to \mathbb{R}$ be continuous. Consider the partition $\{0, 1/n, \ldots, \frac{n-1}{n}, 1\}$ of $[0, 1]$. Define

$$f_n(t) := \begin{cases} f(k/n), & (k - 1)/n \leq t \leq k/n \\ f(1/n), & t = 0 \end{cases}$$

for $1 \leq k \leq n$. Then f_n is a step function taking the value $f(k/n)$ on $(k - 1/n, k/n]$. Show that $f_n \to f$ uniformly.

(6) Let $f: \mathbb{R} \to \mathbb{R}$ be uniformly continuous. Let $f_n(x) = f(x + \frac{1}{n})$. Show that $f_n \to f$ uniformly on $\mathbb{R}$.

(7) Let (f_n) be a sequence of real-valued functions converging uniformly on X. Let $|f_n(x)| \leq M$ for all $n \in \mathbb{N}$ and $x \in X$. Assume that $g: [-M, M] \to \mathbb{R}$ is continuous. Show that $(g \circ f_n)$ is uniformly convergent on X.

(8) Let $\phi : [0, 1] \to \mathbb{R}$ be continuous. Let $f_n : [0, 1] \to \mathbb{R}$ be defined by $f_n(x) = x^n \phi(x)$. Prove that f_n converges uniformly on $[0, 1]$ iff $\phi(1) = 0$.

(9) Let $f_n : X \to \mathbb{R}$. We say that the sequence (f_n) is uniformly bounded on X if there exists $M > 0$ such that

$$|f_n(x)| \leq M \text{ for all } x \in X \text{ and } n \in \mathbb{N}.$$

Assume that f_n's are bounded and that $f_n \rightrightarrows f$ on X. Show that the sequence (f_n) is uniformly bounded and that f is bounded.

(10) Let g be a continuously differentiable function on $\mathbb{R}$. Let

$$f_n(x) := n \left(g(x + \frac{1}{n}) - g(x) \right).$$

Then $f_n \to g'$ uniformly on $[-R, R]$ for each $R > 0$.

(11) Let $f_n \rightrightarrows f$ and $g_n \rightrightarrows g$ on X. Let M be such that $|g_n(x)| \leq M$ and $|f(x)| \leq M$. Show that $f_n g_n \rightrightarrows fg$ on X.

(12) Let $f_n : [a, b] \to \mathbb{R}$ be a sequence of continuous functions converging to f uniformly on $[a, b]$. Let (x_n) be a sequence in $[a, b]$ such that $x_n \to x$. Show that $f_n(x_n) \to f(x)$.

7.4   Series of Functions

Given a sequence (a_n) of real numbers we associated an infinite series, denoted by $\sum_n a_n$. Recall that it stands for the limit of the sequence (s_n) of partial sums where $s_n := \sum_{k=1}^{n} a_k$. Now, given a sequence (f_n) of functions on a set X, how do we define an infinite series $\sum_n f_n$?

Definition 7.4.1. Let $f_n : X \to \mathbb{R}$ be a sequence of functions from a set X to $\mathbb{R}$. Then the associated series, denoted by $\sum_n f_n$, is the sequence (s_n) of partial sums where $s_n := \sum_{k=1}^{n} f_k$. We say the series $\sum f_n$ is *uniformly convergent* (respectively, pointwise convergent) on X if the sequence (s_n) of partial sums $s_n := \sum_{k=1}^{n} f_k$ is uniformly convergent (respectively, pointwise convergent) on X. If f is the uniform limit of (s_n) we write $\sum f_n = f$ uniformly on X.

We say the series $\sum f_n$ is absolutely convergent on X if the sequence $(\sigma_n(x))$ of partial sums $\sigma_n(x) := \sum_{k=1}^{n} |f_k(x)|$ is convergent for each $x \in X$.

Question. If $\sum_n f_n$ is pointwise (respectively, uniformly) convergent on X and if $Y \subset X$, what can you say about the convergence of $\sum_n f_n$ on Y?

Example 7.4.2. Let $X = [0,1]$ and $f_n(x) := x^n$ on X, $n \in \mathbb{Z}_+$. Then the partial sums are

$$s_n(x) = \begin{cases} \frac{1-x^{n+1}}{1-x}, & 0 \le x < 1 \\ n+1, & x = 1. \end{cases}$$

We conclude that the infinite series of functions $\sum_{n=0}^{\infty} x^n$ is pointwise convergent to $1/(1-x)$ for $x \in [0,1)$ and not convergent when $x = 1$.

The infinite series is not uniformly convergent on $[0,1)$. Assume the contrary. Then given $\varepsilon > 0$, there exists $N \in \mathbb{N}$ such that

$$k \ge N \implies \forall x \in [0,1), \left| \frac{1}{1-x} - \frac{1-x^{k+1}}{1-x} \right| < \varepsilon.$$

This implies that $\left| \frac{x^{k+1}}{1-x} \right| < \varepsilon$ for $k \ge N$ and $x \in [0,1)$. Since $\left| x^{k+1} \right| \le \left| \frac{x^{k+1}}{1-x} \right|$, this leads us to conclude that the sequence (x^n) is uniformly convergent to 0 on $[0,1)$. We know that this is false.

Example 7.4.3. Let $p > 0$ be fixed. Let $f_n(x) = \frac{\sin nx}{n^p}$, $x \in \mathbb{R}$. We want to discuss the convergence properties of the series $\sum f_n$ on $\mathbb{R}$. Let $a \in \mathbb{R}$ be fixed. We look at $\sum_n \frac{\sin na}{n^p}$. As stated earlier, our first impulse would be to check for absolute convergence. Note that $|f_n(a)| = \left| \frac{\sin na}{n^p} \right| \le \frac{1}{n^p}$. We now use the comparison test. If $p > 1$, we know that the series $\sum n^{-p}$ is convergent. Hence we conclude that the series $\sum f_n(a)$ is absolutely convergent and hence convergent. Since a is arbitrary, we conclude that the series $\sum f_n$ of functions is (absolutely) convergent on $\mathbb{R}$.

Now we investigate uniform convergence of the series. For this, it is enough if we show that the sequence (s_n) of partial sums of the series $\sum f_n$ is uniformly Cauchy. Thus we need to find a uniform estimate for $|s_n - s_m|$. Observe that, for $n > m$,

$$\begin{aligned} |s_n(x) - s_m(x)| &= |f_{m+1}(x) + \cdots + f_n(x)| \\ &\le |f_{m+1}(x)| + \cdots + |f_n(x)| \end{aligned} \tag{7.12}$$

$$\le \sum_{k=m+1}^{n} \frac{1}{k^p}. \tag{7.13}$$

Since the harmonic p-series is convergent for $p > 1$, given $\varepsilon > 0$, we can find $N \in \mathbb{N}$ such that $\sum_{m+1}^{n} \frac{1}{k^p} < \varepsilon$. (Note that this finite sum is the difference between the n-th and m-th partial sums of $\sum n^{-p}$.) Hence we deduce from (7.13) for $x \in \mathbb{R}$ we obtain

$$n > m \ge N \text{ and } x \in \mathbb{R} \implies |s_n(x) - s_m(x)| < \varepsilon.$$

Hence the sequence (s_n) is uniformly Cauchy on $\mathbb{R}$ and hence the series $\sum f_n$ is uniformly convergent on $\mathbb{R}$.

Note that it follows from (7.12) that $\sum |f_n|$ is also convergent on $\mathbb{R}$. Let σ_n denote the n-th partial sum of the infinite series $\sum |f_k|$. Then the right-hand side of (7.12) is $\sigma_n - \sigma_m$ for $n > m$. Now (7.13) shows that (σ_n) is uniformly Cauchy. Therefore, the series $\sum f_n$ is absolutely and uniformly convergent on $\mathbb{R}$.

Remark 7.4.4. Go through the last example. Can you identify the key ideas of the proof? What is the heart of the proof? It is just that there exists M_n such that (1) $|f_n(x)| \leq M_n$ for all $x \in \mathbb{R}$ and (2) $\sum M_n$ is convergent. We shall return to this later.

Can you formulate the analogue of the Cauchy criterion for the uniform convergence of an infinite series of real-valued functions?

Proposition 7.4.5 (Cauchy Criterion for the Uniform Convergence of a Series). *Let $\sum_n f_n$ be an infinite series of functions on a set X. Then it is uniformly convergent on X iff for any given $\varepsilon > 0$ there exists $N \in \mathbb{N}$ such that*

$$\forall m, n \geq N, (m \leq n), \ and \ \forall x \in X \implies \left| \sum_{k=m+1}^{n} f_k(x) \right| < \varepsilon.$$

Proof. Apply Theorem 7.3.12 to the sequence (s_n) of partial sums. It is an easy exercise for the reader. □

If you have answered the questions in Remark 7.4.4, you would have arrived at the following result.

Theorem 7.4.6 (Weierstrass M-test). *Let f_n be a sequence of real-valued functions on a set X. Assume that there exist $M_n \geq 0$ such that $|f_n(x)| \leq M_n$ for all $n \in \mathbb{N}$ and $x \in X$ and that $\sum_n M_n$ is convergent. Then the series $\sum_n f_n$ is absolutely and uniformly convergent on X.*

Proof. We show that the series $\sum_n |f_n|$ satisfies the Cauchy criterion. Let $\varepsilon > 0$ be given. Since $\sum M_n$ is convergent, we can find $N \in \mathbb{N}$ such that for $n \geq m \geq N$, we have $\sum_{k=m+1}^{n} M_k < \varepsilon$. Then for all m, n we have

$$\sum_{k=m+1}^{n} |f_k(x)| \leq \sum_{k=m+1}^{n} M_k < \varepsilon.$$

Hence the series $\sum_n |f_n|$ is uniformly convergent on X. □

Example 7.4.7. Let us look at some typical applications.

(1) Fix $0 < r < 1$. The series $\sum_{n=1}^{\infty} r^n \cos nt$ and $\sum_{n=1}^{\infty} r^n \sin nt$ are uniformly convergent on $\mathbb{R}$.

For, $|r^n \cos nt| \leq r^n$. So, we take $M_n = r^n$. Then $\sum_n M_n$ is convergent and the result follows from the M-test.

(2) The series $\sum_{n=1}^{\infty} \frac{x}{n(1+nx^2)}$ is uniformly convergent on any interval $[a, b]$.

To start with, we shall assume $a > 0$. The key observation is the presence of n^2-term in the denominator. We work toward exploiting this. Observe that for $x \in [a, b]$,

$$n^2 x^2 \geq n^2 a^2 \implies \frac{x}{n + n^2 x^2} \leq \frac{b}{n + n^2 a^2} \leq \frac{b}{a^2} \frac{1}{n^2}.$$

So we may take $M_n := \frac{b}{a^2} \frac{1}{n^2}$, appeal to the M-test, and arrive at the uniform convergence (on $[a, b]$) of the given series.

What happens if $x \leq 0$?

Exercise Set 7.4.8.

(1) Show that if $w_n(x) = (-1)^n x^n (1 - x)$ on $(0,1)$, then $\sum w_n$ is uniformly convergent.

(2) Let $u_n(x) = x^n(1 - x)$ on $[0, 1]$. Does the series $\sum u_n$ converge? Is the convergence uniform?

(3) If $v_n(x) = x^n(1 - x)^2$ on $[0, 1]$, does $\sum v_n$ converge? Is the convergence uniform?

(4) Show that $\sum x^n(1 - x^n)$ converges pointwise but not uniformly on $[0, 1]$. What is the sum?

(5) Prove that $\sum_{n=1}^{\infty} \frac{\sin nx}{n^2}$ is continuous on $\mathbb{R}$.

(6) Prove that $\sum_{n=1}^{\infty} \frac{1}{1+x^n}$ is continuous for $x > 1$.

(7) Prove that $\sum_{n=1}^{\infty} e^{-nx} \sin nx$ is continuous for $x > 0$.

The next two exercises are analogues of Theorem 7.3.15 and Theorem 7.3.19 for series.

(8) Let $f_n \colon (a, b) \to \mathbb{R}$ be differentiable. Let $x_0 \in (a, b)$ be such that $\sum f_n(x_0)$ converges. Assume further that there is $g \colon (a, b) \to \mathbb{R}$ such that $\sum f_n' = g$ uniformly on (a, b). Then

   (a) There is an $f \colon (a, b) \to \mathbb{R}$ such that $\sum f_n = f$ uniformly on (a, b).

   (b) $f'(x)$ exists for all $x \in (a, b)$ and we have $\sum f_n' = f'$ uniformly on (a, b).

(9) Let $f_n, f \colon [a, b] \to \mathbb{R}$ be such that $\sum f_n = f$ uniformly on $[a, b]$. Assume that each f_n is integrable. Then

   (a) f is integrable.

   (b) $\int_a^b f(t)\, dt = \sum \int_a^b f_n(t)\, dt$.

(10) Justify:

$$\frac{d}{dx} \sum_{n=1}^{\infty} \frac{\sin nx}{n^3} = \sum_{n=1}^{\infty} \frac{\cos nx}{n^2}.$$

We have analogues of Dirichlet's (Theorem 5.2.3) and Abel's tests (Theorem 5.2.6) for uniform convergence too. The proofs follow along the same lines as earlier and from the observation that the estimates obtained earlier are uniform. We encourage the reader to review the above quoted results and prove the following results on their own.

Theorem 7.4.9 (Dirichlet's Test). *Let (f_n) and (g_n) be two sequences of real-valued functions on a set X. Let $F_n(x) := \sum_{k=1}^{n} f_k(x)$. Assume that*
(i) (F_n) *is uniformly bounded on X, that is, there exists $M > 0$ such that* $|F_n(x)| \le M$ *for all $x \in X$ and $n \in \mathbb{N}$.*
(ii) (g_n) *is monotone decreasing to 0 uniformly on X, that is, for each $x \in X$, the sequence $(g_n(x))$ is decreasing to 0 and that the convergence is uniform. Then the series $\sum_{n=1}^{\infty} f_n g_n$ is uniformly convergent on X.*

Strategy: Let $s_n(x) := \sum_{k=1}^{n} f_k(x)g_k(x)$ be the partial sums of the series. We show that the sequence (s_n) is uniformly Cauchy using Abel's partial summation formula.

Proof. Let $s_n(x) := \sum_{k=1}^{n} f_k(x)g_k(x)$. By the partial summation formula (5.5) on page 163, we obtain

$$s_n(x) = \sum_{k=1}^{n} F_k(x)\left(g_k(x) - g_{k+1}(x)\right) + g_{n+1}(x)F_n(x).$$

For $n > m$, we get

$$s_n(x) - s_m(x)$$
$$= \sum_{k=m+1}^{n} F_k(x)\left(g_k(x) - g_{k+1}(x)\right) + g_{n+1}(x)F_n(x) - g_{m+1}(x)F_m(x).$$

We now show that (s_n) is uniformly Cauchy:

$$|s_n(x) - s_m(x)| \le M \sum_{k=m+1}^{n} (g_k(x) - g_{k+1}(x)) + Mg_{n+1}(x) + Mg_{m+1}(x)$$
$$= M\left(g_{m+1}(x) - g_{n+1}(x)\right) + Mg_{n+1}(x) + Mg_{m+1}(x)$$
$$= 2Mg_{m+1}(x).$$

We have used the fact that $(g_n(x))$ decreases to replace $|g_k(x) - g_{k+1}(x)|$ by $g_k(x) - g_{k+1}(x)$. Since $g \searrow 0$ uniformly, given $\varepsilon > 0$, there exists $N \in \mathbb{N}$ such that for $m \ge N$ and $x \in X$, we have $|Mg_{m+1}(x)| < \varepsilon$. This proves that (s_n) is uniformly Cauchy and hence the series $\sum_n f_n g_n$ is uniformly convergent on X. $\qquad\square$

Theorem 7.4.10 (Abel's Test). *Let (f_n) and (g_n) be two sequences of real-valued functions on a set X. Assume that:*
  (i) $\sum_n f_n$ *is uniformly convergent on X.*
  (ii) *There exists $M > 0$ such that $|g_n(x)| \le M$ for all $x \in X$ and $n \in \mathbb{N}$.*
  (iii) $(g_n(x))$ *is monotone for each $x \in X$.*
Then the series $\sum_{n=1}^{\infty} f_n g_n$ is uniformly convergent on X.

Proof. Proof is similar to the last theorem and is left to the reader. □

Consider the series $\sum_{n=1}^{\infty} \frac{\sin nx}{n^p}$ and $\sum_{n=1}^{\infty} \frac{\cos nx}{n^p}$. If $p > 1$, we can apply the M-test to conclude that they are uniformly convergent on $\mathbb{R}$.

When $0 < p \le 1$, we can use Dirichlet's test to show that both series converge uniformly on $[\delta, 2\pi - \delta]$ for $0 < \delta < 2\pi$. See Example 6.1.10.

Example 7.4.11. A Dirichlet series is a series of the form $\sum_n \frac{a_n}{n^x}$ where $x \in \mathbb{R}$. If the series is convergent at $x = \alpha$, then the series $\sum_n \frac{a_n}{n^x}$ is convergent for any $x > \alpha$.

7.5 Power Series

We shall now define a class of functions which are very important in the study of analysis. A discerning reader would have found that so far the only functions which we have introduced rigorously are polynomials, rational functions of the form $P(x)/Q(x)$ where P and Q are polynomials, n-th root functions, and the modulus/absolute value function. The so-called transcendental functions, such as the trigonometric functions and hyperbolic functions, were not defined rigorously and we depended on your encounters with them during your calculus courses. We did make an attempt to define the logarithmic and exponential functions rigorously in Section 6.8.

Assume that we want to find a function f whose derivative is itself. Our forefathers would have started looking for a function of form $a_0 + a_1 x + a_2 x^2 + \cdots + a_n x^n + \cdots$. Computing the derivative formally, we would have arrived at the relations $n a_n = a_{n-1}$ and hence finally $a_n = \frac{a_0}{n!}$. We thus arrive at an expression of the kind $a_0 \left(1 + + \sum_{n=0}^{\infty} \frac{x^n}{n!}\right)$. If we further impose that $f(0) = 1$, we see that $a_0 = 1$. As you may know, this is *the* exponential function.

Exercise 7.5.1. Proceeding as above, try to solve formally the differential equations

$$f'' = -f, \quad \text{with initial conditions } f(0) = 0 \text{ and } f'(0) = 1$$
$$f'' = -f, \quad \text{with initial conditions } f(0) = 1 \text{ and } f'(0) = 0$$

and obtain the following expansions for the sine and cosine functions:

$$\sin x = \sum_{n=0}^{\infty} \frac{(-1)^k}{(2k+1)!} x^{2k+1}$$

$$\cos x = \sum_{n=0}^{\infty} \frac{(-1)^k}{(2k)!} x^{2k}.$$

We take up another example. Assume that $f\colon \mathbb{R} \to \mathbb{R}$ is a C^{∞} function. Let $a \in \mathbb{R}$ be fixed. We then have the Taylor expansion for any $n \in \mathbb{N}$:

$$f(x) = \sum_{k=0}^{n} \frac{f^{(k)}(a)}{k!} (x-a)^k + R_n(x).$$

Since f is infinitely differentiable, it behooves us to consider the series of the type $\sum_{n=0}^{\infty} \frac{f^{(n)}(a)}{n!} (x-a)^n$. This is a series of functions whose n-th term is a power of $(x-a)$.

These considerations lead us to the following.

Definition 7.5.2. A *power series* is an expression of the form $\sum_{k=0}^{\infty} a_k (x-a)^k$ where $a_k, a, x \in \mathbb{R}$. Note that no assumption is made on the convergence of the series.

Note that this is a very special case of the infinite series of functions $\sum_{n=0}^{\infty} f_n(x)$. In the case of a power series, f_n's are of a special form, namely, $f_n(x) = a_n(x-a)^n$.

We are interested in finding $x \in \mathbb{R}$ for which the series $\sum_{n=0}^{\infty} a_n(x-a)^n$ is convergent.

Consider the three power series:
(1) $\sum_{n=1}^{\infty} n^n x^n$,
(2) $\sum_{n=0}^{\infty} x^n$, and
(3) $\sum_{n=0}^{\infty} (x^n/n!)$.

We claim that if $x \neq 0$, then the first series does not converge. For, if $x \neq 0$, choose $N \in \mathbb{N}$ so that $1/N < |x|$. Then for all $n \geq N$, we have $|(nx)^n| > 1$ and hence the series is not convergent. We have already seen that the second series converges absolutely for all x with $|x| < 1$, whereas the third series converges absolutely for all $x \in \mathbb{R}$.

The interesting fact about a power series $\sum_{n=0}^{\infty} a_n(x-a)^n$ is that if it converges at all at x_1, then it will converge at all points x with $|x - a| < |x_1 - a|$. In particular, the set of points x at which $\sum_{n=0}^{\infty} a_n(x-a)^n$ is convergent is an interval centered at a (Theorem 7.5.3). This is not true for an arbitrary series $\sum f_n$ of functions. See Example 6.1.10.

When we say that something is true for all x such that $|x| < R$ for $0 \leq R \leq \infty$, what we mean is this: If $R > 0$, the meaning is clear. If $R = \infty$, this is just a way of saying that something is true for all $x \in \mathbb{R}$. Recall that the meaning of (a, ∞), it is the set of all $x \in \mathbb{R}$ with $x > a$. Think of ∞ as a symbol or a placeholder, and **not** as a number.

Theorem 7.5.3. *Let $\sum_{n=0}^{\infty} a_n(x-a)^n$ be a power series. There is a unique extended real number R, $0 \le R \le \infty$, such that the following hold:*

(i) *For all x with $|x-a| < R$, the series $\sum_{n=0}^{\infty} a_n(x-a)^n$ converges absolutely and uniformly, say, to a function f, on $(-r,r)$ for any $0 < r < R$.*

(ii) *If $0 < R \le \infty$, then f is continuous, differentiable on $(-R, R)$ with derivative $f'(x) = \sum_n na_n x^{n-1}$.*

(iii) *Term-wise integration is also valid: $\int_x^y f(t)\, dt = \sum_n a_n \int_x^y (t-a)^n\, dt$ for $-R < x < y < R$.*

(iv) *For all x with $|x-a| > R$, the series $\sum_{n=0}^{\infty} a_n(x-a)^n$ diverges.*

Strategy: Assume that $a = 0$ for the simplicity of the notation. R is defined to be the least upper bound of $|x|$ for which $\sum_N a_n(x-a)^n$ is convergent. If $|x| < R$, there exists r such that $|x| < r < R$. Hence there exists x_0 such that $|x_0| > r$ and that the power series is convergent at x_0. We now write x^n as $(x^n/x_0^n)x_0^n$ and do a comparison argument.

Proof. Assume $a = 0$. Let $E := \{|z| : \sum_{n=0}^{\infty} a_n z^n \text{ is convergent.}\}$. Note that E is nonempty, as $0 \in E$. The set E may or may not be bounded above. Let $R := \text{lub } E$, if E is bounded above, otherwise $R = \infty$. Then $\sum_{n=0}^{\infty} a_n z^n$ is divergent if $|z| > R$, by very definition. Hence (iv) is proved.

If $R > 0$, choose r such that $0 < r < R$. Since R is the least upper bound for E and $r < R$, there exists $z_0 \in E$ such that $|z_0| > r$ and $\sum a_n z_0^n$ is convergent. Hence $\{a_n z_0^n\}$ is bounded, say, by M:

$$|a_n z_0^n| \le M \text{ for all } n.$$

Now, if $|z| \le r$, then

$$|a_n z^n| \le |a_n| r^n = |a_n z_0^n| (r/|z_0|)^n \le M(r/|z_0|)^n.$$

But the ("essentially geometric") series $M \sum (r/|z_0|)^n$ is convergent. By Weierstrass M-test, the series $\sum_{n=0}^{\infty} a_n z^n$ is uniformly and absolutely convergent to a function f on $|z| \le r$.

We claim that f is continuous at any x with $|x| < R$. For if r is chosen so that $|x| < r < R$, then the series $\sum_n a_n x^n$ is uniformly convergent on $(-r, r)$. This means that the sequence $s_n(x) := \sum_{k=0}^{n} a_k z^k$ is uniformly convergent to f on $(-r, r)$. Since s_n's are polynomials, they are continuous. Hence their uniform limit f is continuous on $(-r, r)$. In particular, f is continuous at x. Since $x \in (-R, R)$ is arbitrary, we conclude that f is continuous on $(-R, R)$.

We claim that f is differentiable. First of all note that the term-wise differentiated series $\sum_n na_n x^{n-1}$ is uniformly convergent on any $(-r, r)$, $0 < r < R$. For, arguing as in the case of uniform convergence, we have an estimate of the form

$$\sum_{k=0}^{\infty} \left| k c_k x^{k-1} \right| \le \sum_{k=0}^{\infty} k \left| c_k \right| r^{k-1}$$

$$= \sum_{k=0}^{\infty} k \left| c_k x_0^k \right| \frac{r^{k-1}}{\left| x_0 \right|^{k-1}} \left| x_0 \right|^{-1}$$

$$\le (M/r) \sum_{k=0}^{\infty} k t^{k-1}, \text{ where } t = (r/\left| x_0 \right|).$$

The series $\sum k t^{k-1}$ is convergent by ratio test. By Weierstrass M-test, the term-wise differentiated series is uniformly convergent, say, to g on $(-r, r)$. The rest of the proof is almost similar to that of the continuity. The partial sums $s_n \to f$ uniformly on $(-r, r)$ and so do s_n' to g on $(-r, r)$. Now we can appeal to Theorem 7.3.15 to conclude that f is differentiable and its derivative is g.

The proof of (iii) uses Item 9 in the set Exercise 7.4.8. and is left to the reader. □

Definition 7.5.4. The extended real number R of Theorem 7.5.3 is called the *radius of convergence* of the power series $\sum_{n=0}^{\infty} a_n (x - a)^n$. The open interval $(-R, R)$ is called the interval of convergence of the power series.

There are explicit formulas for the radius of convergence R, in terms of the coefficients. See Proposition 7.5.11 and Appendix B.12.

Remark 7.5.5. It is important to note that if R is the radius of convergence of a power series $\sum_{n=0}^{\infty} c_n x^n$, it is uniformly convergent *only* on the subintervals $(-r, r)$ for $0 < r < R$. The theorem does not claim that it is uniformly convergent on $(-R, R)$. For example, consider the power series $\sum_{n=0}^{\infty} x^n$. Its radius of convergence is $R = 1$. (Why?) We have seen in Example 7.4.2 that it is not uniformly convergent on $[0, 1)$.

Theorem 7.5.6. *If a power series $\sum_{n=0}^{\infty} a_n (x - a)^n$ has a positive radius of convergence $0 < R \le \infty$, then its sum defines a function, say f, on the interval $(a - R, a + R)$. The function f is infinitely differentiable on this interval. Also, the Taylor series of f in powers of $(x - a)$ is the original power series. We also have $c_n := \frac{f^{(n)}(a)}{n!}$.*

Proof. This is an easy corollary of the earlier results and given as a theorem for ready reference.

By Theorem 7.5.3, the function f is infinitely differentiable on the interval $(a - R, a + R)$. Also, since the term-wise differentiation is valid by the same theorem, we see that $f^{(n)}(a) = n! c_n$. □

Remark 7.5.7. Note that the last theorem says the power series which were obtained by formal manipulations to solve the initial value problems of ordinary

differential equations
   (i) $f' = f$ with $f(0) = 1$,
   (ii) $f'' + f = 0$ with $f(0) = 0$ and $f'(0) = 1$,
   (iii) $f'' + f = 0$ with $f(0) = 1$ and $f'(0) = 0$
are indeed solutions of the initial value problems. We may therefore consider the functions defined by the power series as the exponential, sine, and cosine functions!

It is also important to realize that we can turn the table now. We can define the exponential function by $\exp(x) := \sum_{k=0}^{\infty} \frac{x^k}{k!}$. Theorem 7.5.6 says that $\exp'(x) = \exp(x)$ with $\exp(0) = 1$. That is, exp is the solution of the ordinary differential equation $f' = f$ with the initial condition $f(0) = 1$.

What needs to be established is to prove the periodicity of the sine function with a period denoted by 2π, the identity $\sin^2 x + \cos^2 = 1$, and hence 1 as the bound for sine. This is best done in the context of complex power series (where x is replaced by $z \in \mathbb{C}$). We refer the reader to other books.

Example 7.5.8. Let us exhibit a typical way of using the last result to bring out its significance. Consider the geometric series $\sum_{n=0}^{\infty} x^n$. We know that this series is convergent for $x \in (-1, 1)$ and the sum is $\frac{1}{1-x}$. We also know that the series $\sum_{n=1}^{\infty} nx^{n-1}$, obtained by term-wise differentiation of the geometric series, is convergent to a function g on $(-1, 1)$. Can we identify g? Yes, we can. By the last theorem we know that $f' = g$. Since we know $f'(x) = \frac{1}{(1-x)^2}$, we deduce

$$\sum_{n=1}^{\infty} nx^{n-1} = g(x) = \frac{1}{(1-x)^2}.$$

Exercise 7.5.9. (a) Find an explicit formula for the function represented by the power series $\sum_{k=1}^{\infty} kx^k$ in its interval of convergence.
(b) Find an explicit formula for the function represented by the power series $\sum_{k=1}^{\infty} k^2 x^k$ in its interval of convergence. Use it to find the sum of $\sum_k \frac{k^2}{2^k}$ and $\sum_k \frac{k^2}{3^k}$.

Example 7.5.10. Let us start with the convergent geometric series

$$\frac{1}{1-x} = \sum_{n=0}^{\infty} x^n, \quad |x| < 1.$$

The radius of convergence of this series is 1 and hence we can do term-wise integration on the interval $[0, x]$, (allowed by (iii) of Theorem 7.5.3). We end up with

$$\log(1 - x) = -\sum_{n=1}^{\infty} \frac{x^n}{n}. \tag{7.14}$$

If we can substitute $x = -1$ we shall obtain that the sum of the standard alternating series is $\log 2$. At this point in time, we cannot justify this. We shall return to this in Exercise 7.5.15. See also Example 7.5.14.

The following theorem is quite useful in practice, as it gives two simple formulae to determine the radius of convergence of a power series.

Proposition 7.5.11. *Let $\sum_{n=0}^{\infty} a_n(x-a)^n$ be given. Assume that one of the following limits exists as an extended real number.*
(1) $\lim |a_{n+1}/a_n| = \rho$.
(2) $\lim |a_n|^{1/n} = \rho$.
Then the radius of convergence of the power series is given by $R = \rho^{-1}$.

Proof. (1) follows from the ratio test and (2) from the root test. For instance, if (1) holds and if z is fixed, then

$$\left| \frac{a_{n+1}(z-a)^{n+1}}{a_n(z-a)^n} \right| = |z-a| \left| \frac{a_{n+1}}{a_n} \right| \to |z-a| \rho.$$

By the ratio test, we know that the *numerical* series $\sum_n a_n(z-a)^n$ is convergent if $|z-a| \rho < 1$. Thus the radius of convergence of the series is at least $1/\rho$. By the same ratio test, we know that if $|z-a| \rho > 1$, then the series is divergent. Hence we conclude that the radius of convergence of the given series is $1/\rho$.

The proof of (ii) is similar to that of (i) and is left to the reader. □

Example 7.5.12. We now discuss the convergence of the series $\sum_{n=0}^{\infty}(x^n/n)$. It is easy to see that the radius of convergence is 1. At $x = 1$, we obtain the harmonic series which is divergent. At $x = -1$, we obtain the alternating series which is convergent. Let us see whether we can apply Dirichlet's test to the series on the set $X := [-1, 1-\varepsilon]$. We may take $f_n(x) = x^n$ and $g = 1/n$. We then get

$$|f_n(x)| = \left| \frac{1-x^{n+1}}{1-x} \right| \leq \frac{2}{\varepsilon}.$$

Hence the series is uniformly convergent on X.

Let $\sum a_n x^n$ be a power series whose radius of convergence is $0 < R < \infty$. Then we know that $f(x) := \sum a_n x^n$ defines a C^{∞} function on $(-R, R)$. Assume that the series $\sum a_n R^n$ is convergent. We may then extend f to $(-R, R]$ by setting $f(R) = \sum a_n R^n$. We may also ask whether the series $\sum a_n x^n$ is uniformly convergent on $(-R, R]$. A typical example is the Maclaurin series for $f(x) := \log(1+x)$. Its radius of convergence is 1 and the numerical series at $x = 1$ is the standard alternating series $\sum \frac{(-1)^{n+1}}{n}$, which is convergent. If the Maclaurin series for $\log(1+x)$ is uniformly convergent on $(-1, 1]$, we can then conclude that the sum of the alternating series is $f(1) = \log(2)$. The theorem below says that we can do this. We prove this when $R = 1$. The general case can be reduced to this case by considering the series $\sum a_n R^n x^n$ whose radius of convergence is 1 if that of $\sum a_n x^n$ is R. (This is a hint; the reader is expected to work out the details!)

Theorem 7.5.13 (Abel's Limit Theorem). *If the series $\sum_{n=0}^{\infty} a_n$ converges, then the power series $\sum_{n=0}^{\infty} a_n x^n$ converges uniformly on $[0, 1]$.*

Proof. Given $\varepsilon > 0$, there exists $N \in \mathbb{N}$ such that

$$N \le m \le n \implies \left| \sum_{m+1}^{n} a_k \right| < \varepsilon.$$

If $0 \le x \le 1$, we apply (5.5) on page 163 to $(a_k)_{k=m+1}^{\infty}$ and $(x^k)_{k=m+1}^{\infty}$ to obtain

$$-\varepsilon x^{m+1} < a_{m+1} x^{m+1} + \cdots + a_n x^n < \varepsilon x^{m+1}.$$

That is,

$$\left| \sum_{k=m+1}^{n} a_k x^k \right| < \varepsilon x^{m+1} \le \varepsilon, \text{ for } N \le m \le n \text{ and } x \in [0, 1].$$

If we let $f_n(x) := \sum_{k=0}^{n} a_k x^k$, it follows from the last inequality that the sequence (f_n) is uniformly Cauchy on $[0, 1]$. The conclusion follows. $\square$

Example 7.5.14. The sum of the standard alternating series is $\log 2$. Consider $f(x) = \log(1 + x)$. Then the Taylor series of f at $x = 0$ is given by

$$f(x) = x - \frac{x^2}{2} + \frac{x^3}{3} + \cdots + \frac{(-1)^{n+1} x^n}{n} + \cdots$$

Clearly the power series is convergent at $x = 1$ by the alternating series test. Hence by Abel's limit theorem, we see that the power series is uniformly convergent on $(-1, 1]$ and hence the sum of the series at $x = 1$ is $f(1) = \log 2$. $\square$

Exercise 7.5.15. Go through Exercise 7.5.10. If we take $x = -1$, the infinite series on the right side of (7.14) is the standard alternating series. If we are allowed to put $x = -1$ on the left side, we get $\log(2)$. Can we justify these steps?

Another way of looking at Abel's limit theorem is that it gives us a large supply of power series that are uniformly convergent on $(-R, R]$. We need only start with a convergent series of the form $\sum a_n R^n$.

Example 7.5.16. Find an explicit expression for the function represented by the power series $\sum_{k=0}^{\infty} \frac{x^k}{k+1}$.

Clearly $R = 1$. Also, observe that at $R = -1$, the series is convergent by the alternating series test. By term-wise differentiation, we get

$$(xf(x))' = \sum_{k=0}^{\infty} \left(\frac{x^{k+1}}{k+1} \right)' = \sum_{k=0}^{\infty} x^k = \frac{1}{1-x}, \quad x \in (-1, 1).$$

By the fundamental theorem of calculus (Theorem 6.3.1 on Page 194), we obtain

$$xf(x) = \int_0^x \frac{dt}{1-t} = -\log(1-x), \quad x \in (-1, 1).$$

Note that we can appeal to Abel's theorem (Theorem 7.5.13) to conclude that

$$f(x) = \begin{cases} -\frac{\log(1-x)}{x}, & x \in [-1, 1), x \neq 0 \\ 1, & x = 0. \end{cases}$$

As an application of Abel's Limit Theorem, we give an easy proof of Abel's theorem (Theorem 5.4.6) on Cauchy products of two series.

Theorem 7.5.17 (Abel). *Let $\sum a_n$ and $\sum b_n$ be convergent, say, with sums A and B, respectively. Assume that their Cauchy product $\sum c_n$ is also convergent to C. Then $C = AB$.*

Proof. Consider the power series $\sum_{n=0}^{\infty} a_n x^n$ and $\sum_{n=0}^{\infty} b_n x^n$. From hypothesis, their radii of convergence is at least 1. For $|x| < 1$, both the series are absolutely convergent, and hence by Merten's theorem, we have

$$\sum_{n=0}^{\infty} c_n x^n = \left(\sum_{n=0}^{\infty} a_n x^n \right) \left(\sum_{n=0}^{\infty} b_n x^n \right).$$

In this we let $x \to 1_-$ and apply Abel's theorem to conclude the result.   □

Remark 7.5.18. Two standard applications of the M-test are:
(i) Construction of an everywhere continuous and nowhere differentiable function on $\mathbb{R}$.
(ii) Construction of a space filling curve, that is, a continuous map from the unit interval $[0, 1]$ onto the unit square $[0, 1] \times [0, 1]$.
    We shall not deal with them in our course. You may consult Appendix E in [2].

7.6   Taylor Series of a Smooth Function

In this section we investigate a power series associated with a C^∞ function.
    Let $J \subset \mathbb{R}$ be an interval. Assume that $f: J \to \mathbb{R}$ is infinitely differentiable. Let $a \in J$. Since f is C^{n+1}, we have the n-th Taylor polynomial $T_n(f; a)$ and the corresponding Taylor expansion with a remainder term:

$$f(x) = \sum_{k=0}^{n} \frac{f^{(k)}(a)}{k!} (x-a)^k + \frac{f^{(n+1)}(c)}{(n+1)!} (x-a)^{n+1}. \qquad (7.15)$$

$$= T_n(f; a; x) + R_n(f; a; x), \text{ say}. \qquad (7.16)$$

Since f is C^{n+1} for all n, this leads us to an infinite series of functions, in fact, a power series $\sum_{k=0}^{\infty} \frac{f^{(k)}(a)}{k!} (x-a)^k$. This series is called the Taylor series of f around a and we denote this relation by

$$f(x) \sim \sum_{k=0}^{\infty} \frac{f^{(k)}(a)}{k!} (x-a)^k.$$

If $a = 0$, then the Taylor series of f at 0 is called the Maclaurin series of f.

Example 7.6.1. We list some of the standard Taylor (or Maclaurin) series:

(i) $\exp(x) \sim \sum_{n=0}^{\infty} \frac{x^n}{n!}$.

(ii) $\sin(x) \sim \sum_{n=0}^{\infty} (-1)^n \frac{x^{2n-1}}{(2n-1)!}$.

(iii) $\cos(x) \sim \sum_{n=0}^{\infty} (-1)^n \frac{x^{2n}}{(2n)!}$.

The series listed above are the Maclaurin series of exp, sin, and cos, respectively.

Two natural questions arise now. One is whether the Taylor series converges at points other than a. A better question could be: What is the radius of convergence of the Taylor series? The second question is if the Taylor series converges at x, is the sum equal to $f(x)$?

Let us look at $f : (-1, \infty) \to \mathbb{R}$ defined by $f(x) = \log(1 + x)$. Then f is C^∞ and its Maclaurin series is

$$\log(1 + x) \sim x - \frac{x^2}{2} + \frac{x^3}{3} - \frac{x^4}{4} + \cdots$$

It is easy to see that the radius of convergence of the series is 1. Hence the interval of convergence is $(-1, 1)$. Note that f is C^∞ on $(-1, \infty)$. However, the series is not convergent at $x = 2$.

Let us now revisit the function in Example 4.5.3. The Maclaurin series of the function

$$f(t) = \begin{cases} e^{-1/t}, & t > 0 \\ 0, & t \leq 0 \end{cases}$$

is zero, as $f^{(n)}(0) = 0$ for each $n \geq 0$. Thus the series converges for all $x \in \mathbb{R}$ but its sum is not the value $f(x)$ for $x > 0$. Thus the Taylor series does not represent the function on its interval of convergence.

If the Taylor series of f at $x = a$ converges to f in an interval $(a - R, a + R)$, we then say that the power series $\sum_{n=0}^{\infty} \frac{f^{(n)}(a)}{n!}(x - a)^n$ *represents* f in $(-R, R)$. Note that the interval may be a proper subset of the domain of f. For instance, this happens in the case of $\log(1 + x)$. A similar phenomenon is witnessed in the case of $f(x) = \frac{1}{1-x}$ whose domain of definition is $\mathbb{R} \setminus \{1\}$. Its Maclaurin series is the geometric series $\sum_{n=0}^{\infty} x^n$ and it represents the function only in $(-1, 1)$.

We claim that the Taylor series converges to $f(x)$ iff $R_n(f; a; x) \to 0$. Let us prove this. The Taylor series $\sum_{k=0}^{\infty} \frac{f^{(k)}(a)}{k!}(x - a)^k$ converges to $f(x)$ iff the sequence $s_n(x) := \sum_{k=0}^{n} \frac{f^{(k)}(a)}{k!}(x - a)^k$ converges to $f(x)$, that is, iff $f(x) - s_n(x) \to 0$. In view of (7.16), $f(x) - s_n(x) = R_n(f; a : x)$. The claim follows from this.

In general it may be difficult to verify this. So, we may ask whether there is any easy sufficient condition that will ensure the convergence of Taylor series of f to f in the interval of convergence. Look at the n-th term $f^{(n)}(a)\frac{(x-a)^n}{n}$. From Ratio test, it follows that the series $\sum_{n=0}^{\infty} \frac{(x-a)^n}{n!}$ is convergent for all x. If we are assured that there exists M such that for x in some interval J around a, we have

$\left|f^{(n)}(a)\right| \le M$ for all $n \in \mathbb{Z}_+$, then we conclude that the Taylor series converges to $f(x)$ for $x \in J$.

This is the case for the sine and cosine series. We know that their n-th derivatives are either $\pm \cos x$ or $\pm \sin x$, bounded (in absolute value) by 1 on $J = \mathbb{R}$. Hence we conclude that the Taylor series of the sine (respectively, cosine) function converge to the sine (respectively, to cosine) function on $\mathbb{R}$.

The case of the exponential series is a little more subtle. Let $f(x) = e^x$. Let $x \in \mathbb{R}$ be fixed. Choose $R > 0$ so that $|x| < R$. For any $t \in (-R, R)$, we have $f^{(n)}(t) \le e^R$. Hence the Taylor series of exp converges to e^x for $|x| < R$. Since this is true for any x, we conclude that the Taylor series of the exponential function converges to e^x on $\mathbb{R}$.

Note that since the Taylor series of f is a power series, the points at which it converges form an interval, namely the interval of convergence of the Taylor series.

As an application of Taylor's theorem with Lagrange form of the remainder, we now establish the sum of the standard alternating series is $\log 2$:

$$\log 2 = \sum_{k=1}^{\infty} \frac{(-1)^{k-1}}{k}.$$

Consider the function $f \colon (-1, \infty) \to \mathbb{R}$ defined by $f(x) := \log(1 + x)$. By a simple induction argument, we see that

$$f^{(n)}(x) = (-1)^{n-1}(n-1)!(1+x)^{-n}.$$

Hence the Taylor series of f around 0 is

$$\sum_{n=0}^{\infty} \frac{(-1)^{n-1}}{n} x^n.$$

We now wish to show that the series is convergent at $x = 1$. This means that we need to show that the sum of the series at $x = 1$ is convergent. We therefore take $a = 0$, $b = 1$ in Taylor's theorem and show that the remainder term (in Lagrange's form) $R_n \to 0$. For each $n \in \mathbb{N}$, there exists $c_n \in (0, 1)$ such that

$$R_n := \frac{f^{(n)}(c_n)}{n!} = \frac{(-1)^{n-1}(n-1)!}{n!(1+c_n)^n}.$$

We have an obvious estimate:

$$|R_n| = \left| \frac{(-1)^{n-1}(n-1)!}{n!(1+c_n)^n} \right| \le \left| \frac{1}{n(1+c_n)^n} \right| \le \frac{1}{n}.$$

Hence, $\log 2 = f(1) = \sum_{n=1}^{\infty} \frac{f^{(n)}(0)}{n!} 1^n = \sum_{n=1}^{\infty} \frac{(-1)^{n-1}}{n}.$ $\square$

7.7 Binomial Series

This section is optional and may be omitted on first reading. However, if you are serious about analysis, it is worth going through the proofs, as they teach us good analysis.

We now show how both forms of the remainder are required to prove the convergence of the binomial series.

Theorem 7.7.1 (Binomial Series). *Let $m \in \mathbb{R}$. Define*

$$\binom{m}{0} = 1 \ and \ \binom{m}{k} := \frac{m(m-1)\cdots(m-k+1)}{k!} \ for \ k \in \mathbb{N}.$$

Then

$$(1+x)^m = \sum_{k=0}^{\infty} \binom{m}{k} x^k = 1 + mx + \frac{m(m-1)}{2!} + \cdots, \ for \ |x| < 1.$$

Proof. If $m \in \mathbb{N}$, this is the usual binomial theorem. In this case, the series is finite and there is no restriction on x.

Let $m \notin \mathbb{N}$. Consider $f \colon (-1, \infty) \to \mathbb{R}$ defined by $f(x) = (1+x)^m$. For $x > -1$, we have

$$f'(x) = m(1+x)^{m-1}, \ \ldots, f^{(n)}(x) = m(m-1)\cdots(m-n+1)(1+x)^{m-n}.$$

If $x = 0$, the result is trivial as $1^m = 1$. Now for $x \neq 0$, by Taylor's theorem

$$f(x) = f(0) + xf'(0) + \cdots + R_n = \sum_{k=0}^{n-1} \binom{m}{k} x^k + R_n.$$

Therefore, to prove the theorem, we need to show that, for $|x| < 1$,

$$\left| f(x) - \sum_{k=0}^{n-1} \binom{m}{k} x^k \right| = |R_n| \to 0 \text{ as } n \to \infty.$$

To prove $R_n \to 0$, we use Lagrange's form for the case $0 < x < 1$.

$$|R_n| = \left| \frac{x^n}{n!} f^{(n)}(\theta x) \right| = \left| \binom{m}{n} x^n (1+\theta x)^{m-n} \right| < \left| \binom{m}{n} x^n \right|,$$

if $n > m$, since $0 < \theta < 1$. Letting $a_n := \left| \binom{m}{n} x^n \right|$, we see that $a_{n+1}/a_n = x \left| \frac{m-n}{n+1} \right| \to x$. Since $0 < x < 1$, the ratio test says that the series $\sum_n a_n$ is convergent. In particular, the n-th term $a_n \to 0$. Since $|R_n| < a_n$, it follows that $R_n \to 0$ when $0 < x < 1$.

Let us now attend to the case when $-1 < x < 0$. If we try to use the Lagrange form of the remainder, we obtain the estimate

$$|R_n| = \left|\binom{m}{n} x^n (1 + \theta x)^{m-n}\right| < \left|\binom{m}{n} x^n (1 - \theta)^{m-n}\right|$$

if $n > m$. This is not helpful as $(1 - \theta)^{m-n}$ shoots to infinity if θ goes near 1.

Let us now try Cauchy's form.

$$|R_n| = \left|mx^n \binom{m-1}{n-1} \left(\frac{1-\theta}{1+\theta x}\right)^{n-1} (1 + \theta x)^{m-1}\right|$$

$$\leq \left|mx^n \binom{m-1}{n-1} (1 + \theta x)^{m-1}\right|. \tag{7.17}$$

Now, $0 < 1 + x < 1 + \theta x < 1$. Hence $\left|(1 + \theta x)^{m-1}\right| < C$ for some $C > 0$. Note that C is independent of n but dependent on x.

It follows that $\left|x^n \binom{m-1}{n-1}\right| \to 0$ so that $R_n \to 0$. This completes the proof of the theorem.   □

We now give a second proof of Theorem 7.7.1 on the binomial series. This will use the integral form (as in Theorem 6.5.1) to estimate the remainder term in the Taylor expansion.

Assume that m is not a non-negative integer. Then $a_n := \binom{m}{n} \neq 0$. Since

$$\frac{a_{n+1}}{a_n} = \frac{m-n}{n+1} \to 1, \tag{7.18}$$

the binomial series

$$(1 + x)^m = 1 + \sum_{n=1}^{\infty} \binom{m}{n} \frac{x^n}{n!}$$

has radius of convergence 1. Similarly, the series $\sum_n n\binom{m}{n} x^n$ is convergent for $|x| < 1$. Hence

$$n\binom{m}{n} x^n \to 0 \text{ for } |x| < 1. \tag{7.19}$$

We now estimate the remainder term using (7.19). We have, for $0 < |x| < 1$,

$$R_n = \int_0^x \frac{(x-t)^{n-1}}{(n-1)!} n! \binom{m}{n} (1+t)^{m-n} dt$$

$$= \int_0^x n\binom{m}{n} \left(\frac{x-t}{1+t}\right)^{n-1} (1+t)^{m-1} dt. \tag{7.20}$$

We claim that

$$\left|\frac{x-t}{1+t}\right| \leq |x| \text{ for } -1 < x \leq t \leq 0 \text{ or } 0 \leq t \leq x < 1.$$

Write $t = sx$ for some $0 \le s \le 1$. Then

$$\left| \frac{x-t}{1+t} \right| = \left| \frac{x-sx}{1+st} \right| = |x| \left| \frac{1-s}{1+t} \right| \le |x|.$$

Thus the integrand in (7.20) is bounded by

$$\left| n \binom{m}{n} \left(\frac{x-t}{1+t} \right)^{n-1} (1+t)^{n-1} \right| \le n \left| \binom{m}{n} \right| |x|^{n-1} (1+t)^{m-1}.$$

Therefore, we obtain

$$|R_n(x)| \le n \left| \binom{m}{n} \right| |x|^{n-1} \int_{-|x|}^{|x|} (1+t)^{m-1} \le Cn \left| \binom{m}{n} \right| |x|^{n-1},$$

which goes to 0 in view of (7.19).

This completes the proof of the fact that the binomial series converges to $(1+x)^m$. $\qquad\square$

We now indicate a third way of establishing the convergence of the binomial series of $f_m(x) := (1+x)^m$ to f on $(-1,1)$. This proof is instructive since it will rely heavily on the results from the theory of the power series unlike the earlier two. Keep the notation above. Let $g_m(x) = \sum_{k=0}^{\infty} \binom{m}{k} x^k$. The main steps are the following:

(1) The radius of convergence of the power series g_m is 1.

(2) Note that $f_m(x) = (1+x)^m$ satisfies $(1+x)f'_m(x) = m f_m(x)$ with the initial condition $f_m(0) = 1$. We shall prove that g_m satisfies the similar differential equation

$$(1+x)g'_m(x) = mg_m(x), \text{ with the initial condition } g_m(0) = 1. \qquad (7.21)$$

(3) Using (7.21), we show that g_m/f_m has zero derivative and hence is a constant in $(-1,1)$. Evaluation at 0 yields the constant 1.

Step 1: Since we assume $m \notin \mathbb{Z}_+$, we find that $\binom{m}{n} \ne 0$ for $n \in \mathbb{Z}_+$. We apply the ratio test to the power series g to find that the radius of convergence is 1, see (7.18).

Step 2: The series can be differentiated term by term in $(-1,1)$. We find that

$$g'_m(x) = \sum_{n=1}^{\infty} n \binom{m}{n} x^{n-1} = \sum_{n=0}^{\infty} (n+1) \binom{m}{n+1} x^n.$$

We use the binomial identity $(n+1)\binom{m}{n+1} = m\binom{m-1}{n}$ in the above equation to obtain

$$g'_m(x) = m \sum_{n=0}^{\infty} \binom{m-1}{n} x^n = mg_{m-1}(x). \qquad (7.22)$$

To establish (7.21), we need to figure out $(1+x)g_{m-1}(x)$. We use the binomial identity $\binom{m-1}{n-1} + \binom{m-1}{n} = \binom{m}{n}$ below. We have

$$(1+x)g_{m-1}(x) = (1+x) \sum_{n=0}^{\infty} \binom{m-1}{n} x^n$$

$$= 1 + \sum_{n=1}^{\infty} \binom{m-1}{n} x^n + \sum_{n=1}^{\infty} \binom{m-1}{n} x^{n+1}$$

$$= 1 + \sum_{n=1}^{\infty} \binom{m-1}{n} x^n + \sum_{n=1}^{\infty} \binom{m-1}{n+1} x^n$$

$$= 1 + \sum_{n=1}^{\infty} \left(\binom{m-1}{n} + \binom{m-1}{n-1} \right) x^n$$

$$= \sum_{n=0}^{\infty} \binom{m}{n} x^n = g_m(x). \tag{7.23}$$

From (7.22) and (7.23), it follows that g_m satisfies $(1+x)g_m'(x) = mg_m(x)$ on $(-1, 1)$. Clearly, $g_m(0) = 1$.

Step 3: We differentiate $g_m(x)/(1+x)^m$ and obtain

$$\frac{(1+x)^m g_m'(x) - m(1+x)^{m-1} g_m(x)}{(1+x)^{2m}}$$

$$= (1+x)^{m-1} \frac{(1+x)g_m'(x) - mg_m(x)}{(1+x)^{2m}} = 0.$$

Thus $g_m(x)/(1+x)^m$ is a constant on $(-1, 1)$. Since $g_m(x)/(1+x)^m$ is 1 at $x = 0$, we conclude that $g_m(x) = (1+x)^m$ on $(-1, 1)$.

7.8   Weierstrass Approximation Theorem

The theorem of the title is the following.

Theorem 7.8.1 (Weierstrass Approximation Theorem). *Let $f \colon [a, b] \to \mathbb{R}$ be continuous. Given $\varepsilon > 0$, there exists a real polynomial function $P = P(x)$ such that*

$$|f(x) - P(x)| < \varepsilon \text{ for all } x \in [a, b].$$

In particular, there exists a sequence P_n of polynomial functions such that $P_n \rightrightarrows f$ on $[a, b]$.

What is the significance of this result? It says that the set of polynomials in the set of continuous functions on $[0, 1]$ plays the same role as the set of rational numbers in $\mathbb{R}$. Real numbers are too many and too abstract to deal with. Theorem 1.3.13 on the density of rationals tells us that any real number can be approximated by a rational number to any level of accuracy. For instance, if the ubiquitous π is given and $\varepsilon = 10^{-100}$, the theorem asserts that we can find a rational number r such that $|\pi - r| < 10^{-100}$! Hence for all numerical computations, we may replace π by a rational number.

Assume that we are given a highly complicated continuous function f on $[0, 1]$ and that we are interested in the numerical value of $\int_0^1 f(t)\,dt$. As earlier, if we fix an error tolerance $\varepsilon > 0$, thanks to Weierstrass, there exists a polynomial P such that $|f(t) - P(t)| < \varepsilon$ for $t \in [0, 1]$. Hence we obtain

$$\left| \int_0^1 f(t)\,dt - \int_0^1 P(t)\,dt \right| = \left| \int_0^1 (f(t) - P(t))\,dt \right| \leq \int_0^1 |f(t) - P(t)|\,dt < \varepsilon.$$

Thus if we are interested in the numerical approximation of the integral $\int_0^1 f(t)\,dt$, we may as well deal with the integral $\int_0^1 P(t),dt$, which is easier to compute! Of course, there are far more serious applications of this result. Since our aim is to make you appreciate the result, this will suffice.

The proof, due to Bernstein, uses the definition of *Bernstein polynomials*. We now define Bernstein polynomial B_n (associated with a given f) by

$$B_n(x) := \sum_{k=0}^{n} \binom{n}{k} f\left(\frac{k}{n}\right) x^k (1 - x)^{n-k}.$$

The reason for this definition is given later.

Just to get an idea of what kind of objects they are, let us now find the Bernstein polynomial of the following continuous functions on $[0, 1]$: $f_0(x) = 1$, $f_1(x) = x$ and $f_2(x) = x^2$, $x \in [0, 1]$. Note that these functions are already polynomials and hence their best approximations are themselves!

$$\begin{aligned}
B_n(f_0) &:= \sum_{k=0}^{n} \binom{n}{k} x^k (1 - x)^{n-k} f_0(k/n) \\
&= \sum_{k=0}^{n} \binom{n}{k} x^k (1 - x)^{n-k} \\
&= (x + (1 - x))^n \\
&= 1.
\end{aligned} \tag{7.24}$$

We now compute the Bernstein polynomials of f_1. Recall

$$B_n(f_1) := \sum_{k=0}^{n} \binom{n}{k} x^k (1 - x)^{n-k} (k/n).$$

Now a simple computation shows that $\frac{k}{n}\binom{n}{k} = \binom{n-1}{k-1}$ for $k \geq 1$. Hence,

$$B_n(f_1) \equiv \sum_{k=0}^{n} \frac{k}{n}\binom{n}{k} x^k (1-x)^{n-k}$$

$$= x \sum_{k=1}^{n} \binom{n-1}{k-1} x^{k-1}(1-x)^{n-k}$$

$$= x \left(x + (1-x)\right)^{n-1}$$

$$= x.$$

Hence $B_n(f_1) = f_1$.

We now find $B_n(f_2)$. Our earlier experience with $B_n(f_1)$ shows that we should be ready to simplify the term $\left(\frac{k}{n}\right)^2 \binom{n}{k}$. We start with

$$\frac{k}{n} = \frac{(k-1)+1}{n} = \frac{k-1}{n} + \frac{1}{n} = \frac{k-1}{n-1}\frac{n-1}{n} + \frac{1}{n}, \quad k \geq 1.$$

Note that if $k \geq 2$, we obtain

$$\left(\frac{k}{n}\right)^2 \binom{n}{k} = \frac{k}{n}\frac{k}{n}\binom{n}{k}$$

$$= \frac{k}{n}\binom{n-1}{k-1}, \quad k \geq 1$$

$$= \left(\frac{k-1}{n-1}\frac{n-1}{n} + \frac{1}{n}\right)\binom{n-1}{k-1}$$

$$= \left(1 - \frac{1}{n}\right)\binom{n-2}{k-2} + \frac{1}{n}\binom{n-1}{k-1}, \quad k \geq 2.$$

We are now ready to find $B_n(f_2)$.

$$\sum_{k=0}^{n}\left(\frac{k}{n}\right)^2\binom{n}{k}x^k(1-x)^{n-k} = \left(1 - \frac{1}{n}\right)\sum_{k=2}^{n}\binom{n-2}{k-2}x^k(1-x)^{n-k}$$

$$+ \frac{1}{n}\sum_{k=1}^{n}\binom{n-1}{k-1}x^k(1-x)^{n-k}$$

$$= \left(1 - \frac{1}{n}\right)x^2 + \frac{1}{n}x$$

$$= \left(1 - \frac{1}{n}\right)f_2 + \frac{1}{n}f_1. \tag{7.25}$$

It is clear that $B_n(f_2) \rightrightarrows f_2$ on any bounded subset of $\mathbb{R}$.

Proof. (Weierstrass Theorem) We prove the result when $a = 0$ and $b = 1$. The general case can be deduced from this. See Remark 7.8.2.

We need the identity

$$\frac{x(1-x)}{n} = \sum_{k=0}^{n} x^k (1-x)^{n-k} \left(x - \frac{k}{n} \right)^2. \tag{7.26}$$

Consider

$$1 = (x + (1-x))^n = \sum_{k=0}^{n} \binom{n}{k} x^k (1-x)^{n-k}. \tag{7.27}$$

Differentiate both sides of (7.27) and simplify to obtain

$$0 = \sum_{k=0}^{n} x^{k-1} (1-x)^{n-k-1} (k - nx). \tag{7.28}$$

Multiply both sides of (7.28) by $x(1-x)$, to obtain

$$0 = \sum_{k=0}^{n} x^k (1-x)^{n-k} (k - nx). \tag{7.29}$$

Differentiate both sides of (7.29) and multiply through by $x(1-x)$. On simplification, we get

$$0 = -nx(1-x) + \sum_{k=0}^{n} \binom{n}{k} x^k (1-x)^{n-k} (k - nx)^2. \tag{7.30}$$

Dividing both sides by n^2, we obtain (7.26).

We now define Bernstein polynomial B_n (associated with a given f) by

$$B_n(x) = \sum_{k=0}^{n} \binom{n}{k} f\left(\frac{k}{n}\right) x^k (1-x)^{n-k}.$$

Then by (7.27),

$$B_n(x) - f(x) = \sum_{k=0}^{n} \binom{n}{k} \left(f\left(\frac{k}{n}\right) - f(x) \right) x^k (1-x)^{n-k}. \tag{7.31}$$

Let $\varepsilon > 0$ be given. By uniform continuity of f on $[0,1]$, there exists $\delta > 0$ such that

$$x, y \in [0,1] \text{ and } |x - y| < \delta \implies |f(x) - f(y)| < \varepsilon/4.$$

Let $M > 0$ be such that $|f(x)| \le M$ for $x \in [0,1]$. Choose $N \in \mathbb{N}$ such that $N > \frac{M}{\varepsilon \delta^2}$.

To estimate the term on the right side of (7.31), we observe that $f\left(\frac{k}{n}\right) - f(x)$ is easy to estimate in view of uniform continuity of f provided that k/n is close to x. So we break the sum into two parts and employ the divide and conquer method.

Let $x \in [0,1]$ and $0 \leq k \leq n$. We can write $\{0, 1, 2, \ldots, n\} = A \cup B$ where

$$A := \left\{ k : \left| x - \frac{k}{n} \right| < \delta \right\} \text{ and } B := \left\{ k : \left| x - \frac{k}{n} \right| \geq \delta \right\}.$$

Case 1. $k \in A$. Then we have $|f(x) - f(k/n)| < \varepsilon/4$. Summing over those $k \in A$, we have by (7.27), that

$$\sum \binom{n}{k} \left| f\left(\frac{k}{n}\right) - f(x) \right| x^k (1-x)^{n-k} \leq \frac{\varepsilon}{4}. \tag{7.32}$$

Case 2. $k \in B$. We have, summing over $k \in B$

$$\sum_{k \in B} \binom{n}{k} (|f(k/n)| + |f(x)|) x^k (1-x)^{n-k}$$

$$\leq 2M \sum \binom{n}{k} \left(x - \frac{k}{n} \right)^2 \left(x - \frac{k}{n} \right)^{-2} x^k (1-x)^{n-k}$$

$$\leq 2M\delta^{-2} \sum_{k=0}^{n} \binom{n}{k} x^k (1-x)^{n-k} \left(x - \frac{k}{n} \right)^2$$

$$= 2M\delta^{-2} \frac{x(1-x)}{n}, \text{ by (7.26)},$$

$$\leq 2\varepsilon x(1-x), \text{ since } n > \frac{M}{\varepsilon\delta^2},$$

$$\leq \varepsilon/2, \text{ since } x(1-x) \leq 1/4, \tag{7.33}$$

for, $4x(1-x) - 1 = -(2x-1)^2 \leq 0$ (or by the second derivative test).

It follows from (7.31)–(7.33) that

$$|B_n(x) - f(x)| \leq \frac{3\varepsilon}{4} < \varepsilon,$$

for $x \in [0,1]$ and $n \geq N$. □

Remark 7.8.2. The Weierstrass theorem remains true if $[0,1]$ is replaced by any closed and bounded interval $[a, b]$.

For, consider the map $h \colon [0,1] \to [a,b]$ defined by $h(t) := a + t(b-a)$. Then h is a continuous bijection. Also, $t := h^{-1}(x) = (x-a)/(b-a)$ is continuous. Given a continuous function $g \colon [a,b] \to \mathbb{R}$, the function $f := g \circ h \colon [0,1] \to \mathbb{R}$ is continuous. For $\varepsilon > 0$, let P be a polynomial such that $|f(t) - P(t)| < \varepsilon$ for all $t \in [0,1]$. Then $Q(x) := P \circ h^{-1}(t) = P(\frac{x-a}{b-a})$ is a polynomial in x. Observe that, for all $x = h(t) \in [a,b]$,

$$|g(x) - Q(x)| = \left| f \circ h^{-1}(x) - P \circ h^{-1}(x) \right| = |f(t) - P(t)| < \varepsilon.$$

Remark 7.8.3. The identity (7.26) can also be deduced from (7.24)–(7.25). We shall leave this as an exercise to the reader.

Probabilistic Reason Underlying the Bernstein Polynomial

The proof of the Weierstrass approximation theorem using Bernstein polynomials has its origin in probability. Imagine a loaded or biased coin which turns heads with probability t, $0 \leq t \leq 1$. If a player tosses the coin n times, the probability of getting the heads k times is given by $\binom{n}{k} t^k (1-t)^{n-k}$.

Now suppose that a continuous function f, considered as a payoff, assigns a prize as follows: the player will get $f\left(\frac{k}{n}\right)$ rupees if he gets exactly k heads out of n tosses. Then the expected value E_n (also known as the mean), the player is likely to get out of a game of n tosses is

$$E_n = \sum_{k=0}^{n} f\left(\frac{k}{n}\right) \binom{n}{k} t^k (1-t)^{n-k}.$$

Note that E_n is the n-th Bernstein polynomial of f. It is thus the average/mean value of a game of n tosses.

It is reasonable to expect that if n is very large, the head will turn up approximately nt times. This implies that the prize $f\left(\frac{tn}{n}\right) = f(t)$ and $E_n(t)$ are likely to be very close to each other. That is, we expect that $|f(t) - E_n(t)| \to 0$.

Exercise Set 7.8.4.

(1) (A standard application.) Let $f\colon [0,1] \to \mathbb{R}$ be continuous. Assume that $\int_0^1 f(x) x^n \, dx = 0$ for $n \in \mathbb{Z}_+$. Then $f = 0$.

(2) Show that there exists no sequence of polynomials (p_n) such that $p_n \rightrightarrows f$ on $\mathbb{R}$ where (a) $f(x) = \sin x$, (b) $f(x) = e^x$.

(3) Let $f\colon \mathbb{R} \to \mathbb{R}$ be continuous. The last item says that we may not be able to find a sequence (p_n) of polynomials such that $p_n \rightrightarrows f$ on $\mathbb{R}$. Show that we can still find a sequence (p_n) of polynomials such that $p_n \rightrightarrows f$ on any *bounded* subset of $\mathbb{R}$. (Compare this with Remark 7.5.5.)

(4) Let $f\colon (0,1) \to \mathbb{R}$ be defined by $f(x) := 1/x$. Show that there does not exist a sequence (p_n) of polynomials such that $p_n \rightrightarrows f$ on $(0,1)$.

(5) Keep the hypothesis of the Weierstrass approximation theorem. Can we find a sequence (p_n) of polynomials such that $\sum p_n = f$ on $[0,1]$?

Appendix A

Quantifiers

In this appendix, we shall give a brief and working knowledge of the basics of commonsense logic, the use of quantifiers in mathematical statements, and the kinds of proof.

We believe that readers are already acquainted with the notion of statements and the truth table. In this section, we shall concentrate only on the statement that involves the quantifiers $\forall$, $\exists$ and the connectives *and/or* and the negations of sentences involving these. Most of the statements in mathematics involve these either directly or indirectly. A very large percentage of the proofs are so-called proofs by contradiction or establish the results in contrapositive form. We are sure that after going through this section, you will be more comfortable with these.

Look at the following everyday sentences:

- You can find a page in this book not containing any picture.

- There is a rotten apple in the basket of apples.

- There exists a student in the classroom who is at least 6 feet tall.

Each of these sentences directly or indirectly uses the existential quantifier $\exists$. For example, the first one can be recast as "There exists a page in this book not containing any picture." Another significant observation is that the quantifier depends on a set to make sense. In the first sentence, the set under consideration is the set of pages in this book, in the second it is the set of apples in a basket, and in the third, the set of students in a classroom. What each of the sentences says can be abstracted as follows:

There exists an element in a set X which has some property P.

Now how do we negate each of the sentences above? What does it mean to say that the first sentence is false? How do we prove that the first is false? If we want to prove that the first sentence is false, we need to check each and every page and show that each one of them contains a picture. Another way of saying this is: Given any page of this book, I shall show it contains a picture.

To show that the second sentence is false, given any apple from the basket, we need to show that it is not rotten.

To show that the third is false, we need to show that each student of the classroom is of height less than 6 feet.

You may observe that each of these negations falls in the following pattern:

For each element of the set, we prove the element does not have the property P. In terms of quantifiers, we have the following: The negation of

$$\text{``}\exists x \in X \ (x \text{ has property } P)\text{''}$$

is

$$\text{``}\forall x \in X \ (x \text{ does not have property } P)\text{''}.$$

Now let us look at sentences which use the universal quantifier $\forall$:

- Every page in this book contains at least 500 words.

- Every apple in the basket is ripe.

- Every student in the classroom is at least 5 feet tall.

You would have observed that each of these sentences uses the universal quantifier and involves a set. For the first, the set is the set of pages, for the second it is the set of apples in the basket, and for the third it is the set of students in the classroom.

How do we negate each of these? If the first sentence is false, it means that we can find a page in this book which has less than 500 words in it. That is, we are asserting that there exists a page in this book which has less than 500 words. As for the second, its negation says that there exists at least one apple in the basket which is not ripe. You may wish to say now what the negation of the third is.

You may have observed the following pattern: The negation of the sentence

$$\forall x \in X \ (x \text{ has property } P)$$

is $\qquad \exists x \in X \ (x \text{ does not have property } P).$

Let us consider some statements in mathematics.

(1) There exists a real number x such that $x^2 = 1$.
$\exists x \in \mathbb{R}(x^2 = 1)$.

(2) There exists a rational number r such that $r^2 = 2$.
$\exists r \in \mathbb{Q} \ (r^2 = 2)$.

(3) A real number α is an upper bound of a nonempty set $A \subset \mathbb{R}$ if $\alpha \geq x$ for all $x \in A$.
$\alpha \in \mathbb{R}$ is an upper bound of $\emptyset \neq A \subset \mathbb{R}$ if $(\forall x \in A \ (x \leq \alpha))$.

(4) A nonempty subset $A \subset \mathbb{R}$ is bounded above in $\mathbb{R}$ if there exists $\alpha \in \mathbb{R}$ such that α is an upper bound of A.
A is bounded above in $\mathbb{R}$ if $(\exists \alpha \in \mathbb{R} \ (\alpha \text{ is an upper bound of } A))$.

(5) $A \subset B$ is for each $x \in A$, we have $x \in B$.
$A \subset B$ if $(\forall x \in A(x \in B))$.

Let us look at the statement in Item 3. Note that whether or not α is an upper bound of A depends on whether each element $x \in A$ has a common property, namely of being less than or equal to α. Hence if we want to claim that α is not an upper bound of A, then we have to negate the sentence $\forall x \in A(x \leq \alpha)$. That is, we must show that there exists $x \in A$ such that $x \leq \alpha$ is false. In view of the law of trichotomy, this means that

$$\alpha \in \mathbb{R} \text{ is not an upper bound of } A \text{ if } \exists x \in A \ (x > \alpha).$$

Let us look at the statement in Item 4. This is interesting as it uses two quantifiers to assert something. Let us first negate the sentence as given. The negation is that given any real number $\alpha \in \mathbb{R}$, α is not an upper bound of A. We may write this as

$$A \text{ is not bounded above in } \mathbb{R} \text{ if } \forall \alpha \in \mathbb{R} \ (\alpha \text{ is not an upper bound of } A).$$

Why do we find the statement in Item 4 interesting? Because if we write it in its full glory, it reads as

$$A \text{ is bounded above in } \mathbb{R} \text{ if } (\exists \alpha \in \mathbb{R} \ (\forall x \in \mathbb{R} \ (x \leq \alpha))).$$

Its negation reads as

$$A \text{ is not bounded above in } \mathbb{R} \text{ if } (\forall \alpha \in \mathbb{R} \ (\exists x \in \mathbb{R} \ (x > \alpha))).$$

Thus, if we wish to negate a nested sentence, we do it layer by layer. Another thing to notice is that if a sentence is like

$$\forall x \in X \ (\exists y \in Y \ (y \text{ has some property } P)),$$
the element y may depend on the given x. For example, consider the sentence

$$\forall x \in \mathbb{R} \setminus \{0\} \ (\exists y \in \mathbb{R} \ (xy = 1))).$$

Now, if we take $x = 4$, then $y = 1/4$, whereas for $x = -1$, $y = -1$.

The order in which the quantifiers appear in the statement is important. If we change the order, the truth value of the statement may change. Let us look at an example.

We say that a subset $A \subset \mathbb{R}$ is bounded above (in $\mathbb{R}$) if there exists $\alpha \in \mathbb{R}$ such that for every $x \in A$, we have $x \leq \alpha$, that is, if

$$\exists \alpha \in \mathbb{R}(\forall x \in A(x \leq \alpha)).$$

Let us interchange the quantifiers as follows and define A is bounded above in $\mathbb{R}$ if for every $x \in A$, there exists $\alpha \in \mathbb{R}$ such that $x \leq \alpha$, that is, if

$$\forall x \in A(\exists \alpha \in \mathbb{R}(x \leq \alpha)).$$

(Believe us, this "definition" is given by a few every time this course is taught!) What is wrong with this? If this were the definition, then any subset of $\mathbb{R}$ would be bounded above! For, if $x \in A$ is given, we let $\alpha = x$ (or $\alpha = x + 1$, if we wish to have a strict inequality!). We urge you to think over this again.

As further examples, consider the following statement which assures the existence of an additive identity in $\mathbb{R}$:

$$\exists \theta \in \mathbb{R} \ (\forall x \in \mathbb{R} \ (x + \theta = \theta + x = x))).$$

Consider the sentence which assures the additive inverse in $\mathbb{R}$:

$$\forall x \in \mathbb{R} \ (\exists y \in \mathbb{R} \ (x + y = \theta = y + x)).$$

Note the order in which the quantifiers appear in these two sentences. The first one says that we have the same θ for any $x \in \mathbb{R}$ satisfying the conditions $x + \theta = x = \theta + x$. The second one says if we are given $x \in \mathbb{R}$, there exists y, which may depend on x (in this case, it does!) such that $x + y = \theta$. Of course, this does not mean that y must be different for different x. For example, consider the statement: given $x \in \mathbb{R}$, there exists $y \in \mathbb{R}$ such that $y = x^2$. For $x_1 = 1$ and $x_2 = -1$, we get the same $y = 1$.

Let us now deal with a slightly more complicated sentence. Suppose there is an orchard, full of trees. We make the following statement.

In each tree in the orchard, we can find a branch on which all of the leaves are green.

How do we turn this into a mathematical sentence? Let us fix the notation. We let T denote the set of all trees in the orchard. Let $t \in T$ be a tree. Let B_t denote the set of all branches on the tree t. Let $b \in B_t$ be a branch of the tree t. We let L_b denote the set of all leaves on the branch b. Now we are ready to cast this using quantifiers.

$$\forall t \in T \ (\exists b \in B_t \ (\ \forall \ell \in L_b \ (\ell \text{ is green})))).$$

How do we negate it? As we said earlier, we look at the outermost layer and negate it and move to the next inner layer. Thus, we get: There exists a tree $t \in T$ which does not have the property $(\exists b \in B_t \ (\ \forall \ell \in L_b \ (\ell \text{ is green}))))$. We negate this and so on. Finally we arrive at

$$\exists t \in T \ (\forall b \in B_t \ (\exists \ell \in L_b \ (\ell \text{ is not green})))).$$

Compare these two displayed statements and pay attention to the quantifiers.

Do such complicated sentences occur naturally in mathematics? Yes, when we define the convergence of a sequence of real numbers. Let (x_n) be a sequence of real numbers. We say that (x_n) converges to a real number $x \in \mathbb{R}$ if

$$\forall \varepsilon > 0 \ (\exists N \in \mathbb{N} \ (\forall n \geq N \ (|x_n - x| < \varepsilon))).$$

A more complicated one is when we say that a sequence (x_n) is convergent. We say that (x_n) is convergent if

$$\exists x \in \mathbb{R} \ (\forall \varepsilon > 0 \ (\exists N \in \mathbb{N} \ (\forall n \geq N \ (|x_n - x| < \varepsilon)))).$$

You may try your hand in negating each of these!

We now consider sentences which are combined using the connectives *and* or *or*.

Consider the sentence: A real number $x \geq 0$. This is a combination of two sentences:

The real number $x > 0$ or the real number $x = 0$.

How do we negate such a sentence? Common sense tells us that if the combined sentence is false, then each of the statements is false. That is, the statement "the real number $x > 0$" is false and the statement "$x = 0$" is false. In view of the law of trichotomy, which says that for any real number t exactly one of the following is true (i) $t = 0$, (ii) $t > 0$, (iii) $t < 0$, we conclude that the negation of the combined statement is $x < 0$.

Let us look at one more example. Let A, B be subsets of a set X. When do we say $x \in X$ lies in $A \cup B$?

$$x \in A \cup B \text{ if } x \in A \text{ or } x \in B.$$

This is a combined statement: $x \in A \cup B$ if $x \in A$ or $x \in B$. If this statement is false, then it means that $\in A$ is false **and** $x \in B$ is false.

Thus we arrive at the following: If a statement is of the form P or Q, then its negation is P is false and Q is false, that is, Not P and Not Q.

Suppose you want to buy a mobile phone which has a video camera and Wi-Fi hotspot. If you reject a phone which the shopkeeper shows you, what could be the reason? The handset shown to you either lacked a video camera or Wi-Fi hotspot. Thus among all the handsets in his stock, if V denotes the set of handsets with video facility and W denotes the set of all phones with Wi-Fi hotspot, what you wanted lies in $V \cap W$. If the handset shown to you is not what you wanted, either it does not lie in V or it does not lie in W.

Given two subsets A and B of a set X, we say that $A = B$ if and only if $A \subset B$ and $B \subset A$. This says that for A to equal B, two statements must be simultaneously true, namely, $A \subset B$ and $B \subset A$. Thus

$$A = B \text{ iff } \forall x \in A \ (x \in B) \text{ and } \forall y \in B \ (y \in A).$$

What is the negation of $A = B$? If $A = B$ is false, then it means that one of the two statements is true: (i) there exists $x \in A$ such that $x \notin B$ or (ii) there exists $y \in B$ such that $y \notin A$. Thus we have

$$A \neq B \text{ iff } (\exists x \in A \ (x \notin B)) \text{ or } (\exists y \in B \ (y \notin A)).$$

The upshot of all these is that if we are given a statement like

"every street in the city has at least one house in which we can find a person who is rich and beautiful or highly educated and kind."

you should be able to negate it. We suggest that you write this using quantifiers (with chosen notation) and negate it.

Appendix B

Limit Inferior and Limit Superior

Definition B.1. Given a bounded sequence (a_n) of real numbers, let $A_n := \{x_k : k \geq n\}$. Consider the numbers

$$s_n := \inf\{a_k : k \geq n\} \equiv \inf A_n \text{ and } t_n := \sup\{a_k : k \geq n\} \equiv \sup A_n.$$

If $|x_k| \leq M$ for all n, then $-M \leq s_n \leq t_n \leq M$ for all n. The sequence (s_n) is an increasing sequence of reals bounded above while (t_n) is a decreasing sequence of reals bounded below. Let

$$\liminf a_n := \lim s_n \equiv \text{lub } \{s_n\} \text{ and } \limsup a_n := \lim t_n \equiv \text{glb } \{t_n\}.$$

They are called the limit inferior and limit superior of the bounded sequence (a_n). In case, the sequence (a_n) is not bounded above, then its $\limsup$ is defined to be $+\infty$. Similarly, the $\liminf$ of a sequence not bounded below is defined to be $-\infty$.

Example B.2.

(1) Let (x_n) be the sequence where $x_n = (-1)^{n+1}$. Then $\liminf x_n = -1$ and $\limsup x_n = 1$.

(2) Let (x_n) be the sequence defined by $x_n = (-1)^{n+1} + \frac{(-1)^n}{n}$. If you write down the first few terms of the sequence, you will see the pattern. You will find $\limsup x_n = 1$ and $\liminf x_n = -1$.

(3) Let (x_n) be the sequence defined by $x_n = \frac{n}{5} - \left[\frac{n}{5}\right]$. The hint is the same as in the last example. Show that $\limsup x_n = 4/5$ and $\liminf x_n = 0$.

Proposition B.3. For any bounded sequence (x_n), we have $\liminf x_n \leq \limsup x_n$.

Note that from the very definition, we have $s_n \leq t_n$. Now, $t_n - s_n \geq 0$ and hence $t_n - s_n \to t - s$. Hence $t \geq s$.

Observe that t is the GLB of t_n's. Based on our experience, the best way of exploiting this is to consider $t + \varepsilon$, $\varepsilon > 0$. Since $t + \varepsilon > t$, it is not a lower bound of $\{t_n\}$. Hence there exists N such that $t_N < t + \varepsilon$. What does this mean?

Do you observe that s is the LUB of s_n's? Given $\varepsilon > 0$, what will you do with it? These explorations lead us to the next result.

Theorem B.4. Let (a_n) be a bounded sequence of real numbers with $t := \limsup a_n$. Let $\varepsilon > 0$. Then:

(a) There exists $N \in \mathbb{N}$ such that $a_n < t + \varepsilon$ for $n \geq N$.
(b) $t - \varepsilon < a_n$ for infinitely many n.
(c) In particular, there exists infinitely many $r \in \mathbb{N}$ such that $t - \varepsilon < a_r < t + \varepsilon$.

Proof. Let $A_k := \{x_n : n \geq k\}$.

(a) Note that $\limsup a_n = \inf t_n$ in the notation used above. Since $t + \varepsilon$ is greater than the greatest lower bound of (t_n), $t + \varepsilon$ is not a lower bound for t_n's. Hence there exists $N \in \mathbb{N}$ such that $t + \varepsilon > t_N$. Since t_N is the least upper bound for $\{x_n : n \geq N\}$, it follows that $t + \varepsilon > x_n$ for all $n \geq N$.

(b) $t - \varepsilon$ is less than the greatest lower bound of t_n's and hence is certainly a lower bound for t_n's. Hence, for any $k \in \mathbb{N}$, $t - \varepsilon$ is less than t_k, the least upper bound of $\{a_n : n \geq k\}$. Therefore, $t - \varepsilon$ is not an upper bound for A_k, $k \in \mathbb{N}$. Thus, there exists n_k such that $a_{n_k} > t - \varepsilon$. For $k = 1$, let n_1 be such that $a_{n_1} > t - \varepsilon$. Since $t - \varepsilon$ is not an upper bound of A_{n_1+1}, there exists $n_2 \geq n_1 + 1 > n_1$ such that $t - \varepsilon < a_{n_2}$. Proceeding this way, we get a subsequence (a_{n_k}) such that $t - \varepsilon < a_{n_k}$ for all $k \in \mathbb{N}$. □

We have the following analogous result for $\liminf$.

Theorem B.5. Let (a_n) be a bounded sequence of real numbers with $s := \liminf a_n$. Let $\varepsilon > 0$. Then:

(a) There exists $N \in \mathbb{N}$ such that $a_n > s - \varepsilon$ for $n \geq N$.
(b) $s + \varepsilon > a_n$ for infinitely many n.
(c) In particular, there exists infinitely many $r \in \mathbb{N}$ such that $s - \varepsilon < a_r < s + \varepsilon$.

Proof. We leave this proof as an instructive exercise to the reader. □

Example B.6. We shall illustrate the last two results by considering the sequence with $x_n = (-1)^{n+1}$. We know $\limsup x_n = 1$ and $\liminf x_n = -1$. Let $\varepsilon > 0$. Then if we take $N = 1$, then for all $k \geq 1$ we have $x_n < 1 + \varepsilon$. On the other hand, there exist infinitely many n (namely odd n's) such that $x_n > 1 - \varepsilon = 1/2$.

If we look at $-1 - \varepsilon$ and if we take $N = 1$, then for each $k \geq 1$, we have $x_k > -3/2$, whereas there exist infinitely many k (namely even k's) such that $x_k < -1 + \varepsilon = -1/2$.

Theorem B.7. A sequence (x_n) in $\mathbb{R}$ is convergent iff (i) its bounded and (ii) $\limsup x_n = \liminf x_n$, in which case $\lim x_n = \limsup x_n = \liminf x_n$.

> **Strategy:** Let $x_n \to a$. Let t_n, t, s_n, s be as earlier. To prove $t = s$, we shall prove $t - s = 0$. Hence we need to estimate $|t - s|$. Since $s_n \le s \le t \le t_n$, it is enough to estimate $t_n - s_n$. Since $x_n \to x$, given $\varepsilon > 0$ we end up with
>
> $$x - \varepsilon < x_n \le x + \varepsilon, \text{ for } n \ge N.$$
>
> This leads us to conclude $x_\varepsilon \le s_n \le t_n \le x + \varepsilon$ for $n \gg 0$.
>
> The opposite direction is a simple application of Theorems B.4–B.5.

Proof. Assume that $x_n \to x$. Then (x_n) is bounded. Then $s = \liminf x_n$ and $t = \limsup x_n$ exist. We need to show that $s = t$. Note that $s \le t$. Let $\varepsilon > 0$ be given. Then there exists $N \in \mathbb{N}$ such that

$$n \ge N \implies x - \varepsilon < x_n < x + \varepsilon.$$

In particular, $x - \varepsilon < s_N := \inf\{x_n : n \ge N\}$ and $t_N := \sup\{x_n : n \ge N\} < x + \varepsilon$. But we have

$$s_N \le \liminf x_n \le \limsup x_n \le t_N.$$

Hence it follows that

$$x - \varepsilon < s_N \le s \le t \le t_N < x + \varepsilon.$$

Thus, $|s - t| \le 2\varepsilon$. This being true for all $\varepsilon > 0$, we deduce that $s = t$. Also, $x, s \in (x - \varepsilon, x + \varepsilon)$ for each $\varepsilon > 0$. Hence $x = s = t$.

Let $s = t$ and $\varepsilon > 0$ be given. Using Theorems B.5 and B.4, we see that there exists $N \in \mathbb{N}$ such that

$$n \ge N \implies s - \varepsilon < x_n \text{ and } x_n < s + \varepsilon.$$

$\square$

We give a traditional proof of Cauchy completeness of $\mathbb{R}$.

Theorem B.8. Every Cauchy sequence in $\mathbb{R}$ is convergent.

> **Strategy:** Adapt the proof of Theorem B.7. Since we do not have x in this context, we replace x by x_N where N is such that $|x_n - x_N| < \varepsilon$ for $n \ge N$.

Proof. Let (x_n) be a Cauchy sequence of real numbers. Then it is bounded and hence $s = \liminf x_n$ and $t = \limsup x_n$ exist as real numbers. It suffices to show that $s = t$. It is enough to show $|t - s| < \varepsilon$ for any $\varepsilon > 0$. Since (x_n) is Cauchy, there exists $N \in \mathbb{N}$ such that

$$m, n \ge N \implies |x_n - x_m| < \varepsilon/2, \text{ in particular, } |x_n - x_N| < \varepsilon/2.$$

It follows that for $n \geq N$,

$$x_N - \varepsilon/2 \leq \text{glb } \{x_k : k \geq n\} \leq \text{lub } \{x_k : k \geq n\} \leq x_N + \varepsilon/2.$$

The rest of the proof runs along the lines of the proof of Theorem B.7.

The inequalities displayed above allow us to conclude that

$$x_N - \frac{\varepsilon}{2} < s_n \leq s \leq t \leq t_n \leq x_N + \frac{\varepsilon}{2}, \text{ for } n \geq N.$$

Therefore, $|t - s| < \varepsilon$. □

We now characterize $\limsup x_n$ of a bounded sequence (x_n) as the LUB of the set of limits of convergent subsequences. Let us have a closer look at the players in this game. Given a sequence (x_n), the set $\mathcal{S}$ of all subsequences is a huge set. If you recall our definition of a subsequence, you will see that there is a bijection from the set all infinite subsets of $\mathbb{N}$ onto $\mathcal{S}$. (This says that $\mathcal{S}$ is uncountable. Forget this, if you do not understand.) The set $\mathcal{C}$ of all convergent subsequences is a subset of $\mathcal{S}$. We now collect the limits of elements of $\mathcal{C}$ to form the set S:

$$S := \{x \in \mathbb{R} : \exists (x_{n_k}) \in \mathcal{C} \text{ such that } x_{n_k} \to x\}.$$

What we want to show is that $\limsup x_n = \text{lub } S$ and $\liminf x_n = \text{glb } S$ and that $\limsup x_n, \liminf x_n \in S$. Is it clear why S is nonempty? Is it bounded?

Theorem B.9. Let (x_n) be a bounded sequence of real numbers. Let

$$S := \{x \in \mathbb{R} : x \text{ is the limit of a convergent subsequence of } (x_n)\}.$$

Then

$$\limsup x_n, \liminf x_n \in S \subseteq [\liminf x_n, \limsup x_n].$$

Proof. We prove that S is nonempty. Since (x_n) is bounded, by the Bolzano-Weierstrass theorem there exists a convergent subsequence, (x_{n_k}). If $x_{n_k} \to x$, then $x \in S$.

We claim that S is bounded. Let $x \in S$. Let $M > 0$ be such that $-M \leq x_n \leq M$ for $n \in \mathbb{N}$. Let $x_{n_k} \to x$. Since $-M \leq x_{n_k} \leq M$, it follows that $-M \leq x \leq M$. Hence S is bounded.

We now show that $s = \liminf x_n \in S$. By Item B.5 there exists infinitely many n such that $s - \varepsilon < x_n < s + \varepsilon$. Hence for each $\varepsilon = 1/k$, we can find $n_k > n_{k-1}$ such that $s - \frac{1}{k} < x_{n_k} < s + \frac{1}{k}$. It follows that $x_{n_k} \to s$ and hence $s \in S$. One shows similarly that $t \in S$.

Let $x \in S$. Let $x_{n_k} \to x$. We shall show $x \leq t + \varepsilon$ for any $\varepsilon > 0$. By Item B.4, there exists N such that $n \geq N$ implies $x_n < t + \varepsilon$. Hence there exists k_0 such that if $k \geq k_0$, then $x_{n_k} < t + \varepsilon$. It follows that the limit x of the sequence (x_{n_k}) is at most $t + \varepsilon$. It follows that for any $x \in S$ we have $x \leq t$. Once similarly proves that $s \leq x$ for $x \in S$. Hence $S \subset [s, t]$. □

Note that the last item gives another proof of Cauchy completeness of $\mathbb{R}$. For, $\limsup x_n$ is the limit of a convergent subsequence; see Item 2.7.6! We leave this as an exercise to the reader.

Remark B.10. Note that Item B.9 implies the following.

$$\limsup x_n = \operatorname{lub} S = \max S \quad \text{and} \quad \liminf x_n = \operatorname{glb} S = \min S.$$

In fact, in some treatments, $\limsup x_n$ (respectively, $\liminf x_n$) is defined as $\operatorname{lub} S$ (respectively, $\operatorname{glb} S$). Our experience shows that beginners find our definition (Definition B.1) easy to understand. It allows them to deal with these concepts with more confidence. When one deals with $\limsup$ or $\liminf$, Theorems B.4–B.5 are most useful, especially the first properties stated as (a).

Exercise B.11 (Exercises on limit superior and inferior).

(1) Consider $(x_n) := (\frac{1}{2}, \frac{2}{3}, \frac{1}{3}, \frac{3}{4}, \frac{1}{4}, \frac{4}{5}, \ldots, \frac{1}{n}, \frac{n}{n+1}, \ldots)$. Then $\limsup = 1$ and $\liminf x_n = 0$.

(2) Find the $\limsup$ and $\liminf$ of the sequences whose n-th term is given by:

  (a) $x_n = (-1)^n + 1/n$
  (b) $x_n = 1/n + (-1)^n/n^2$
  (c) $x_n = (1 + 1/n)^n$
  (d) $x_n = \sin(n\pi/2)$

The next result gives a formula for the radius of convergence of a power series $\sum_{n=0}^{\infty} c_n(x-a)^n$ in terms of the coefficients a_n.

Theorem B.12 (Hadamard Formula for the Radius of Convergence). The radius of convergence ρ of $\sum_{n=0}^{\infty} c_n(z-a)^n$ is given by

$$\frac{1}{\rho} = \limsup |c_n|^{1/n} \quad \text{and} \quad \rho = \liminf |c_n|^{-1/n}.$$

Proof. Let $\frac{1}{\beta} := \limsup |c_n|^{1/n}$. We wish to show that $\rho = \beta$.

If z is given such that $|z - a| < \beta$, choose μ such that $|z - a| < \mu < \beta$. Then $\frac{1}{\mu} > \frac{1}{\beta}$ and hence there exists N (by the last lemma) such that $|c_n|^{1/n} < \frac{1}{\mu}$ for all $n \geq N$. It follows that $|c_n| \mu^n < 1$ for $n \geq N$. Hence $(|c_n| \mu^n)$ is bounded, say, by M. Hence, $|c_n| \leq M\mu^{-n}$ for all n. Consequently,

$$|c_n(z-a)^n| \leq M\mu^{-n}|z-a|^n = M\left(\frac{|z-a|}{\mu}\right)^n.$$

Since $\frac{|z-a|}{\mu} < 1$, the convergence of $\sum c_n(z-a)^n$ follows.

Let $|z - a| > \beta$ so that $\frac{1}{|z-a|} < \frac{1}{\beta}$. Then $\frac{1}{|z-a|} < |c_n|^{1/n}$ for infinitely many n. Hence $|c_n| \, |z-a|^n \geq 1$ for infinitely many n so that the series $\sum c_n(z-a)^n$ is divergent. We therefore conclude that $\rho = \beta$.

The other formula for the radius of convergence is proved similarly. $\square$

Exercise B.13. Find the radius of convergence of the power series $\sum_n a_n z^n$, whose n-th coefficient a_n is given below:

(1) $1/(n^2 + 1)$

(2) $2^n - 1$

(3) n^k, k fixed

(4) n^n

(5) $\frac{n^k}{n!}$

(6) $\left(1 + \frac{2}{n^2}\right)^{n^2}$

(7) $(\log n^n)/n!$

(8) $\frac{2^n n!}{(2n)!}$

(9) $\left(\frac{n}{\log n}\right)^n$

(10) $\frac{\sqrt{n+1} - \sqrt{n}}{\sqrt{n^2 + n}}$

(11) $n^2 \frac{1}{2^{2n}}$

(12) $\frac{1}{5} n^4 (n^3 + 1)$

(13) $\frac{(2n)!}{n!}$

(14) $\frac{n!}{(2n)!}$

(15) $\frac{n!}{n^n}$

(16) $\frac{n^{2n}}{(n!)^2}$

(17) $[1 + (-1)^n 3]^n$

(18) $(a^n + b^n)$, $a > b > 0$

(19) $\frac{1}{n^p}$, $p > 0$

(20) $4^{n(-1)^n}$

There are two sharper versions of the ratio and root tests.

Theorem B.14. (1) (Ratio Test) Let $\sum_n c_n$ be a series of nonzero reals. Let

$$\liminf \left| \frac{c_{n+1}}{c_n} \right| = r \text{ and } \limsup \left| \frac{c_{n+1}}{c_n} \right| = R.$$

Then the series $\sum_n c_n$ is
(i) absolutely convergent if $0 \le R < 1$,
(ii) divergent if $r > 1$.
(iii) The test is inconclusive if $r \le 1 \le R$.

(2) (Root Test) Let $\sum_n a_n$ be a series of reals. Let $\limsup |a_n|^{1/n} = R$. Then the series $\sum_n a_n$ is absolutely convergent if $0 \le R < 1$, divergent if $R > 1$, and the test is inconclusive $a = 1$.

Proof. Go through the proofs of Theorem 5.1.22 (on the ratio test) and those of Theorem B.12 and Theorem 5.1.22. You should be able to write down the proofs on your own. □

Appendix C

Topics for Student Seminars

The topics listed here are selected with the aim of training students to be comfortable with so-called hard analysis and to reinforce their understanding by offering different proofs or perspectives.

1. The field $\mathbb{Q}$ of rational numbers does not enjoy the LUB property. See Theorem 1.3.18 and Remark 1.3.20.

2. Euler's Constant. See Ex. 2.8.4.

3. Euler's number e. See Section 2.3 on page 44.

4. Abel's summation by parts. See Section 5.2.

5. Abel's, Dirichlet's, and Dedekind's tests. See Section 5.2.

6. Convex Functions. See Section 4.6. See also the article (g) in the list of expository articles by S. Kumaresan [8].

7. Classical Inequalities: Arithmetic-Geometric, Holder's, Minkowski, Cauchy-Schwarz, and Bernoulli's inequality. Same reference as in the last item.

8. Mean value theorems for integrals. See Section 6.4.

9. Cauchy's form of the remainder in Taylor expansion. See Section 4.7.

10. Integral form of the remainder in Taylor expansion. See Section 6.5.

11. Binomial series with non-integral index. See Section 7.7.

12. Results that use the curry-leaves trick.

13. Results that use the divide and conquer trick.

14. Logarithmic and exponential functions. See Section 6.8.

15. Nested interval theorem and its applications. See the article (b) in the list of expository articles by S. Kumaresan [8].

16. The role of LUB property in real analysis. See the article (a) in the list of expository articles by S. Kumaresan [8].

17. Abel's theorem on Cauchy product of series. See Theorem 5.4.6.

18. Decimal expansion and other expansion to the base of a positive integer $a \geq 2$. See the article (f) in the list of expository articles by S. Kumaresan [8].

19. Extension theorem for uniformly continuous functions; application to the definition of x^a. See Section 3.8.

20. Monotone functions; inverse function theorem; its application to $x \mapsto x^{1/n}$. See Section 3.5.

21. Cauchy's generalized mean value theorem and its applications to L'Hospital's rules. See Section 4.3.

22. Existence of C^∞ functions with compact support. See Example 4.5.3 and Exercise 4.5.4.

23. Riemann's theorem on rearrangement of series. Refer to [1] Theorem 8.33 in [1] or Theorem 3.54 in [7].

24. Weierstrass approximation theorem: Bernstein's proof. See Section 7.8. For a proof using the so-called Landau kernels, refer to the article (h) on "Approximate Identities" in the list of expository articles by S. Kumaresan [8].

25. Thomae's function; discussion on points of continuity and integrability on $[0,1]$. See page 76 and page 183.

26. Sharp forms of ratio and root tests in terms of lim sup and lim inf. See Theorem B.14.

27. Hadamard's formula for the radius of convergence of a power series. See Theorem B.12.

28. Cantor's construction of real numbers using Cauchy sequences. See an article to be found in the list of expository articles by S. Kumaresan [8].

29. Dedekind's construction of real numbers via cuts. See [7].

30. Functions of bounded variation and rectifiability. See [1].

31. Peano curves; main purpose here is to make the students see applications of p-adic (binary, ternary) expansions. Refer to [2].

32. Sets of measure zero and Lebesgue's characterization of Riemann integrability. Refer to Theorem 7.3.12 in [2].

33. Oscillation of a function at a point, points of discontinuity, F_σ sets. Refer to Section 5.6 in [4].

34. Construction of a function which is continuous everywhere but differentiable nowhere. Refer to Appendix E in [2].

Appendix D

Hints for Selected Exercises

Contents

D.1 Chapter 1

Exercise Set 1.1.3:

Ex. 3: Observe that $y^{n+1} - x^{n+1} = y(y^n - x^n) + (y - x)x^n$ and use induction.

Ex. 1.2.10: If a lower bound α of A, belongs to A, then $\alpha = \text{glb } A$.

Exercise Set 1.3.8:

Ex. 4: $J_n := \left(-\frac{1}{n}, 1 + \frac{1}{n}\right)$.

Exercise Set 1.3.15:

Ex. 1: Yes! Show that if $a \neq b$, then $C_a \neq C_b$.

Ex. 3: Choose $r \in \mathbb{Q}$ such that $\frac{a}{t} < r < \frac{b}{t}$.

Exercise Set 1.3.26:

Ex. 3: Note that $t_1 m + \cdots + t_n m \leq t_1 x_1 + \ldots + t_n x_n \leq t_1 M + \ldots + t_n M$.

Exercise Set 1.3.27:

Ex. 1: A is a singleton.

Ex. 3: Show that glb B is a lower bound of A and lub B is an upper bound of A.

Ex. 6: $\max\{\text{lub } A, \text{lub } B\} = \text{lub } (A \cup B)$.

Ex. 8: GLB property of $\mathbb{R}$: Let B be a nonempty subset of $\mathbb{R}$ and bounded below, then there exist $\beta \in \mathbb{R}$ such that $\beta = \text{glb } B$.

Ex. 10: (b) $\mathbb{N} \setminus \{1\}$, (c) $\mathbb{Z}$, (e) $\mathbb{R}$.

Ex. 11: $b + a$.

Ex. 13: glb $(bA) = ba$.

Ex. 15: lub $(B) = \frac{1}{\text{glb } (A)}$.

Ex. 19: Note that $x^2 - 5x + 6 = (x - 2)(x - 3)$. Hence $x \in A$ iff $x \in (2, 3)$.

Ex. 20: glb $\left(\{x \in \mathbb{R} : x + \frac{1}{x} : x > 0\}\right) = 2$ and the set is not bounded above.

Ex. 22: The set can be written as $\{\frac{1}{n} + \frac{1}{m} : m, n \in \mathbb{N}\}$.

Ex. 25: (1) shows that the hypothesis of intervals being closed is necessary. (2) shows that the hypothesis of the intervals being bounded is necessary.

Exercise Set 1.4.8:

Ex. 1: $(2/7, 6)$.

Ex. 3: $(-3, \infty)$.

Ex. 5: $(-2, -1) \cup (1, 2)$.

Ex. 7: $(-\infty, 0) \cup (1, \infty)$.

Ex. 9: $|x| < 2$.

Ex. 11: $(-3/2, 1/2)$.

Ex. 13: $(-3, 0)$.

D.2   Chapter 2

Ex. 2.1.6: $\forall x \in \mathbb{R} \, (\exists \varepsilon > 0 \, (\forall k \in \mathbb{N} \, (\exists n_k \geq k \, (|x - x_{n_k}| \geq \varepsilon))))$.

Exercise Set 2.1.28:

Ex. 1: (a) 0, (b) 7 and (c) 5.

Ex. 5: Use the formula in Item 9 of Theorem 1.4.2, Proposition 2.1.18, and Theorem 2.1.26 on algebra of convergent sequences.

Ex. 7: False.

Ex. 9: True.

Ex. 11: If $a_n \to 0$, then $a_n^n \to 0$.

Ex. 12: $z_n = \begin{cases} x_k & \text{if } n = 2k - 1 \\ y_k & \text{if } n = 2k. \end{cases}$

Ex. 15: Find $x_1, x_2, \ldots x_n$, explicitly and observe their pattern.

Ex. 16: For each $n \in \mathbb{N}$, choose $x_n \in \left(a - \frac{1}{n}, a + \frac{1}{n}\right) \cap \mathbb{Q}$. Similarly choose $y_n \in \left(a - \frac{1}{n}, a + \frac{1}{n}\right) \cap \mathbb{R} \setminus \mathbb{Q}$.

Exercise Set 2.2.5:

Ex. 3: (x_n) is eventually constant.

Ex. 5: Show that $|x_m - x_n| \le c^{n-1} \frac{1}{1-c} |x_2 - x_1|$.

Exercise Set 2.3.4:

Ex. 3: Show that (a_{n+1}/a_n) converges to a limit less than 1.

Ex. 4: Consider $b_n := a_n - 2^{1-n}$. Is it monotone?

Exercise Set 2.4.3:

Ex. 4: Observe that $(b^n)^{1/n} \le (a^n + b^n)^{1/n} \le (2b^n)^{1/n}$. Use Item 4 of Theorem 2.5.1.

Ex. 2.5.2: Write $a = 1 + h$ for some $h > 0$ and use $a^n \ge \frac{h^2}{2} \times n(n-1)$.

Exercise Set 2.6.5:

Ex. 2: Show that $x_{2^n} \ge n/2$.

Ex. 3: Note that for any k, $x_k + \frac{1}{x_k} \ge 2$.

Ex. 5: For $c > 0$, there exists $N \in \mathbb{N}$ such that $c + d/n < \frac{3c}{2}$ for all $n \ge N$. This implies $\frac{an^2+d}{cn+d} > \frac{an+d/n}{2/(3c)}$.

Ex. 7: Define $a_n := \frac{n^n}{(n!)^{1/n}}$.

Ex. 8: The limit is 1, 0, or -1 depending on whether $a > b$, $a = b$, or $a < b$.

Exercise Set 2.7.10:

Ex. 2: True! It is bounded above.

Ex. 4: Just negate the definition of a bounded sequence.

Ex. 7: Suppose (r_n) converges to $r \in [0, 1]$. Choose ε so that $(r - \varepsilon, r + \varepsilon)$ does not contain one of the endpoints. Does this contradict density of rationals?

Exercise Set 2.8.3:

Ex. 1: The limit exists and is a root of $(x^2 - 2)^2 - x = 0$.

Ex. 2: The limit is $\ell = 2$.

Ex. 3: The limit is $\sqrt{a}$.

Ex. 4: Show that $|x_n - x_{n-1}| \le \frac{b}{a+b} |x_{n-2} - x_{n-1}|$.

Ex. 6: The limit is 2.

Exercise Set 2.8.4:

Ex. 1: $\gamma_n - \gamma_{n+1} = \int_n^{n+1} t^{-1} - \frac{1}{n+1} > 0$.

Ex. 2: $\gamma_n \le \gamma_1$ for all n. Also, $\gamma_n > \sum_{k=1}^n \left[\frac{1}{k} - \int_k^{k+1} t^{-1} dt \right] > 0$.

Exercise Set 2.8.5:

Ex. 5: The limit of γ_{2n} and γ_{2n+1} satisfies the equation $\ell^2 - \ell - 1 = 0$.

Ex. 6: Show that $|\gamma_n - \ell| \le \frac{1}{\ell} |\gamma_{n-1} - \ell|$.

Exercise Set 2.8.7:

Ex. 2: The limit of the given sequence is 0.

Ex. 4: Draw pictures.

Ex. 5: Note that $x_{n+1} = \sqrt{2 + x_n}$ for $n \geq 1$. Show that $\lim x_n = 2$.

Ex. 6: If $(\sin n)$ converges, $\sin(n+2) - \sin n \to 0$. Use the trigonometric identities:

$$\sin A - \sin B = 2 \sin \frac{A - B}{2} \cos \frac{A + B}{2}$$
$$\cos A - \cos B = 2 \sin \frac{A + B}{2} \sin \frac{B - A}{2}.$$

D.3   Chapter 3

Exercise Set 3.1.4:

Ex. 2: f is continuous only at $x = 4$.

Ex. 6: Show (i) $f(n) = nf(1)$ for $n \in \mathbb{N}$, (ii) $f(-x) = f(x)$ for any $x \in \mathbb{R}$, (iii) $f(n) = nf(1)$ for $n \in \mathbb{Z}$, (iv) $f(1/q) = \frac{1}{q}f(1)$ for $q \in \mathbb{N}$, and (v) $f(x) = xf(1)$ for $x \in \mathbb{Q}$. Now define $g(x) := f(1)x$.

Ex. 8: Draw a picture. The function is continuous at all $x \notin \mathbb{Z}$.

Ex. 9: f is continuous on $\mathbb{R}$.

Ex. 3.2.4: For $\varepsilon > 0$, choose $\delta := \varepsilon / L$.

Ex. 3.2.6: For $f(c) > 0$, define $\varepsilon := f(c)/2$ and apply the definition of continuity of f at c.

For the general setup, observe that

$$x \in J \ \& \ |x - c| < \delta \implies |f(c)| \leq |f(c) - f(x)| + |f(x)|.$$

Exercise Set 3.3.4:

Ex. 3: Let $f(x) \neq f(y)$. Then $[f(x), f(y)] \subset f([a, b])$ or $[f(y), f(x)] \subset f([a, b])$.

Exercise Set 3.3.10:

Ex. 1: Apply IVT to $x - \cos x$ on $[0, \pi/2]$.

Ex. 3: If such an f exists, it is not a constant. Its image will be countable. Use Ex. 3 on page 81.

Ex. 5: Consider $g \colon [0, \pi] \to \mathbb{R}$ given by $g(x) = f(x) - f(x + \pi)$.

Ex. 6: Given $c \in \mathbb{R}$, consider the polynomial $q(X) := p(X) - c$.

Ex. 8: What are its values, say, at $\pm R$ where $R \gg 0$ and at 0?

Exercise Set 3.4.7:

Ex. 1: $c = \min f$ and $d = \max f$.

Ex. 2: No, otherwise it contradicts the Weierstrass theorem.

Ex. 3: No, otherwise it contradicts the extreme value theorem.

Ex. 5: What is the equation of the line joining the points (a, c) and (b, d) in $\mathbb{R}^2$?

Ex. 6: Split the interval: $(0, 1) = (0, 1/3) \cup [1/3, 2/3] \cup (2/3, 1)$ and construct an onto map from $[1/3, 2/3]$ to $[0, 1]$.

Ex. 8: Similar to Ex. 4 on page 84.

Exercise Set 3.7.8:

Ex. 2: Consider $f(x) = x^2$ on $\mathbb{R}$. Recall that any Cauchy sequence is bounded.

Exercise Set 3.7.12:

Ex. 1: For $p > 0$, we have that $\mathbb{R} = \cup_{k \in \mathbb{Z}}[kp, (k+1)p)$ is a disjoint union. Use uniform continuity of f on $[0, p]$.

Ex. 2: Is the derivative bounded on $[0, \infty)$?

Ex. 4: Note that f is uniformly continuous on $[0, R]$.

Ex. 6: Use $\lim_{x \to 0} \frac{\sin x}{x} = 1$ and define $f(x) = -\frac{\sin x}{x}$ on $[-1, 0)$ and $f(0) = -1$. Then f is uniformly continuous on $[-1, 0)$ and hence on $(-1, 0)$. Similarly, the given function is uniformly continuous on $(0, 1)$.

Ex. 8: If false, there exist x_n such that $|f(x_n)| \geq n$ for each $n \in \mathbb{N}$. Let (x_{n_k}) be a convergent subsequence by Bolzano-Weierstrass. Note that $|f(x_{n_r})| \leq 1 + |f(x_{n_N})|$, since (x_{n_k}) is Cauchy.

Ex. 10: Can there exist an $L > 0$ such that $|f(x)| \leq L|x|$ for all $x \in [0, 1]$?

Ex. 11: (a) Not uniformly continuous. (b) Uniformly continuous.

Ex. 3.8.8: Let $x_n, y_n \in \mathbb{Q}$ with $x_n \to x$ and $y_n \to y$. Then $a^{x_n + y_n} \to a^{x+y}$.

Ex. 3.8.9: Observe that $f(1) = f(\frac{1}{2} + \frac{1}{2}) \geq 0$ and $f(n) = [f(1)]^n$ for $n \in \mathbb{N}$.

Ex. 3.8.10: Formally manipulate $\exp(x)\exp(y)$. Then justify the steps. This exercise is best done at the end of Section 7.5.

D.4 Chapter 4

Ex. 4.1.5: Define $f_1(x) := \begin{cases} x^{n-1} & x \geq 0 \\ x^{m-1} & x < 0. \end{cases}$

When is f_1 continuous at $x = 0$?

Exercise Set 4.1.10:

Ex. 3: If it were, then $f_1(x) = x^{-2/3}$ for $x \neq 0$. Can one define $f_1(0)$ suitably so that f_1 is continuous at 0?

Ex. 5: $a^n f'(a) - f(a)na^{n-1}$.

Ex. 6: Let $f, g: \mathbb{R} \to \mathbb{R}$ be differentiable functions. Find the limit of $\lim_{x \to a} \frac{g(a)f(x) - f(a)g(x)}{x-a}$. What is the answer?

Ex. 8: Differentiate the identity, multiply the result by x, and add 1 to it.
Answer: $(n+1)\frac{x^n}{x-1} - x\frac{x^{n+1}-1}{(x-1)^2} + 1$.

Exercise Set 4.1.14:

Ex. 2: $r > 1$. You may use the auxiliary function f_1 to arrive at this:

$$f(x) = x^r \sin(1/x) = f(0) + x^{r-1} \sin(1/x)(x - 0), \quad (x \neq 0).$$

Exercise Set 4.2.11:

(2) Consider $f(t) := \log(1 + t)$ on $[0, x]$.
(4) Show that $\sin x - x$ is decreasing. Could you have used $x - \sin x$?
(6) You may need to compute the derivative of $\sin x/x$ and again the derivative of its numerator.

Exercise Set 4.2.16:

Ex. 1: Between any two roots of P we have a root of P'.

Ex. 3: Use mean value theorem to conclude that f is Lipschitz. For the second part, argue by contradiction.

Ex. 6: (a) Consider $f(x) = x^{1/n} - (x - a)^{1/n}$. It is decreasing; evaluate f at 1 and b/a.

Ex. 7: Apply the mean value theorem to the numerator of the difference quotient $\frac{f(c+h)-f(c)}{h}$ and use the hypothesis.

Ex. 9: If it has two distinct zeros, then $f'(c) = 0$ for some $c \in (0, 1)$.

Ex. 11: $\lim_{x \to} \frac{f(x)-f(0)}{x} = \lim_{x \to 0} (1 + 2x \sin(1/x)) = 1$. In any interval around 0, the points of the form $x = 2/(n\pi)$ lie on both sides of 0.

Ex. 12: Consider $h(x) := f(x)e^{g(x)}$.

Ex. 16: False. Observe that $3 = f(5) - f(2) = f'(c)3 = 3(f(x)^2 + \pi) > 3$.

Ex. 19: Mean value theorem applied to $f(1/n) - f(1/m)$ shows that the sequence (a_n) is Cauchy.

Ex. 20: Consider $e^{-\lambda x} f(x)$.

Ex. 23: Answer: 7. Can you justify this?

Ex. 28: The derivative is positive at all nonzero reals.

Ex. 29: If it has at least three zeros, its derivative must vanish at two distinct points. You may need the last exercise.

Ex. 31: If false, there exists a sequence (t_n) of distinct points in $[a, b]$ such that $f(t_n) = 0$. Apply Bolzano-Weierstrass theorem to find a c such that $f(c) = 0 = f'(c)$.

Ex. 32: If f has two distinct fixed points, say, $x \neq y$, what is $f(x) - f(y)$? Apply the mean value theorem to this.

Exercise Set 4.3.7:

Ex. 1: By Darboux theorem, we conclude that f' is a (rational) constant, that is, $f'(x) = r \in \mathbb{Q}$. Hence $f(x) = rx + s$ where $r \in Q$ and $s \in \mathbb{R}$.

Exercise Set 4.6.11:

Ex. 1: The precise formulation is as follows: f convex iff

$$c, x, y \in (a, b) \text{ with } x < c < y \implies \frac{f(x) - f(c)}{x - c} \le \frac{f(d) - f(x)}{d - x}.$$

Draw pictures.

Ex. 3: Remember that e^x is also convex!

Ex. 6: $\max\{f, g\}$ is convex and $\min\{f, g\}$ may not be convex.

Ex. 8: Fix $c \in \mathbb{R}$. Assume $f'(c) > 0$. From the last exercise, we obtain $f(x) \ge f(c) + f'(c)(x - c)$, for $x \in \mathbb{R}$.

D.5   Chapter 5

Ex. 5.1.14: Let t_n denote the n-th partial sum of $\sum c_n$. If $n \ge N$, observe that $t_n = s_n - s_N + b$.

Exercise Set 5.1.24:

Ex. 2: Observe that the n-th term is $7(\frac{7}{9})^n$.

Ex. 4: Apply the ratio test to the series $\sum \frac{n!}{n^n}$.

Ex. 7: Observe that $2^n - n \ge 2^n - 2^{n-1} = 2^{n-1}$.

Ex. 9: Observe that $1 + a_n \ge 1/2$ if $n \gg 0$. Hence $\left|\frac{a_n}{a_n + 1}\right| \le 2 |a_n|$.

Ex. 5.1.26: Consider (1) $f(x) = \frac{1}{x^p}$, (2) $f(x) = \frac{1}{x \log x}$, and (3) $f(x) = \frac{\log x}{x^p}$, respectively.

Exercise Set 5.3.9:

Ex. 2: Observe that $\frac{\ell}{2} \le \frac{a_n}{b_n} \le \frac{3\ell}{2}$ for $n \gg 0$.

Ex. 6: Note that $a_n < 1$ for $n \gg 0$. Hence $a_n^2 \le a_n$ for $n \gg 0$. We may use the comparison test.

Ex. 8: Convergent except at $\frac{\pi}{2}$ and $x = \frac{3\pi}{2}$.

Ex. 10: $\sum \frac{(-1)^n}{\sqrt{n}}$.

Ex. 13: If and only if $a = 0 = b$.

Ex. 18: The hypothesis allows us to obtain an estimate of the form $|a_n| \le Cn^{-2}$.

Ex. 19: Observe that $|a_n| < 1$ for $n \gg 0$. For such n, we have $|a_n^3| \le a_n^2$.

Ex. 22: True.

Ex. 25: Apply the root test. Recall $n^{1/n} \to 1$.

Ex. 28: Keep the notation in the proof of the root test. From $a_k \le a_N R^{-N} r^k$, deduce an estimate for $a_k^{1/k}$ for $k \ge N$.

Ex. 32: Go through Example 5.3.8.

Ex. 35: Observe that $(n!)^{1/n} \ge 1$. Use this to estimate the n-th term.

D.6 Chapter 6

Ex. 6.1.17: Divide and conquer strategy; see Item 2 in Example 6.1.14.

Ex. 6.2.6: (iii) Note that $fg = 4^{-1}\left((f+g)^2 - (f-g)^2\right)$.

Ex. 6.2.8: (a) Since $1 + 4x^{90} \leq 4(1 + x^{90})$, it follows that $\int_0^1 \frac{x^4}{\sqrt{1+4x^{90}}} \geq \frac{1}{2\sqrt{2}}\int_0^1 x^4\, dx$.

(b) Observe that the minimum of $x^4(x-4)$ is attained at $x = 3$.

Exercise Set 6.2.12:

Ex. 4: If false, then using the extreme value theorem, either $f(x) \leq M < 0$ or $f(x) > m > 0$ for $x \in [a, b]$. Use the monotonicity of the integral.

Ex. 6: Take $g = f$.

Ex. 7: The upper bound for the limit is M. We need to show that the limit is at least $M - \varepsilon$ for any $\varepsilon > 0$. Let $c \in [a, b]$ be such that $f(c) = M$. Let $\delta > 0$ be such that for $x \in [c - \delta, c + \delta]$, we have $f(x) > M - \varepsilon$. Observe that

$$\left(\int_a^b f^n(x)\, dx\right)^{1/n} \geq \left(\int_{c-\delta}^{c+\delta} f^n(x)\, dx\right)^{1/n} \geq (2\delta)^{1/n}(M - \varepsilon).$$

Ex. 9: Let m and M be the minimum and maximum of f on $[a, b]$. Use monotonicity of the integral to $m \leq f(x) \leq M$ to conclude that $\frac{1}{b-a}\int_a^b f(x)\, dx$ lies between m and M. Use the intermediate value theorem.

Ex. 14: Take $x = \frac{|a_i|}{\|a\|_p}$ and $y = \frac{|b_i|}{\|b\|_q}$ in Young's inequality (6.18) and sum over i.

Exercise Set 6.3.10:

Ex. 1: Apply the first fundamental theorem of calculus to the equation.

Ex. 2: Ans: $g(x)^2 g'(x)$.

Ex. 5: Observe that $g'(x) \geq g(x)^{1/2}$ so that $g'(x)g(x)^{-1/2} \geq 1$. Integrate both sides of this inequality and use integration by parts.

Exercise Set 6.4.6: Ex. 1: Observe that $\left|\frac{c^n}{1+c_n}\right| \leq \frac{|a|^n}{1+a}$.

Exercise Set 6.4.10: All the problems are easy applications of the integral test, Theorem 6.4.8. For example, the obvious choice in (1) is $f(x) = \frac{1}{1+x^2}$.

Exercise Set 6.7.2: We shall do only one in this set, the rest are similar.

Ex. 1: $\displaystyle\lim \frac{r}{r^2 + n^2} = \lim \frac{1}{n}\sum_{r=1}^{n} \frac{\frac{r}{n}}{1 + (\frac{r}{n})^2} = \int_0^1 \frac{x}{1 + x^2}\, dx$.

Exercise Set 6.7.3:

Ex. 2: If you know the sum $\sum_{r=1}^{n} r = \frac{n(n+1)}{2}$, how will you find the sum

$$-1 + 2 - 3 + 4 - \cdots + (-1)^n n?$$

Or, how do you find the sum of $\sum_{k=1}^{n} 2k + 1$? Ans: 0.

A smarter solution is based on the observation that the given sequence looks like the difference $s_n - s_m$ of partial sums of the standard alternating series. Hence the limit is 0.

Exercise Set 6.7.4:

Ex. 1: Let $M = \int_a^b f$, then there exists a partition P such that $L(f, P) > M/2(b-a)$. If $\{J_k : 1 \leq k \leq n\}$ are the partitioning subintervals, can $f(x) < M/2(b-a)$ for all $x \in J_k$ and for all k?

Ex. 4: If $0 < a < b$, what is the relation between $1/a$ and $1/b$? What is the relation between $m_i(f)$, $m_i(g)$, $M_i(f)$, and $M_i(g)$?

Ex. 6: If f is not zero, then there exists $c \in [a, b]$ such that $m := f(c) > 0$. There exists $\delta > 0$ such that $f(x) > m/2$ for $(c - \delta, c + \delta) \cap [a, b]$.

Ex. 8: Divide and conquer method. Go through the integrability of Thomae's function on page 183.

Ex. 12: Our intuition should say that we should exploit the fact that $x^n \to 0$ for $0 \leq x < 1$. We may employ the divide and conquer method near $x = 1$.

Ex. 13: Same as the last one.

Ex. 14: Use integration by parts to show that

$$\left| \int_a^b f(t) \sin(nt) \, dt \right| \leq \frac{3M}{n},$$

where $|f(t)| \leq M$ and $|f'(t)| \leq M$ for $t \in [a, b]$.

Exercise Set 6.9.20:

Ex. 1: Answers (a) $p > 1$, (b) $p < 1$, (c) $p > 1$, (d) $p > 1$.

Ex. 4: Yes, it is $\frac{\pi}{2}$.

Ex. 6: Answers: (a) diverges, (b) diverges, (c) converges, (d) converges.

Ex. 7: If $L \neq 0$, there exists $R > 0$ such that for $x > R$, we have $f(x) > L/2$. For $x > R$, we obtain $\int_R^x ft) \, dt > L(x - R)/2$.

D.7 Chapter 7

Ex. 7.3.9: Uniformly convergent, of course!

Exercise Set 7.3.10:

Ex. 2: f_n takes a maximum value at $n/(n + 1)$.

Ex. 5: Observe that $f_n(\frac{1}{n}) = 1/2$. Use Proposition 7.3.6.

Ex. 7: For $x \in [0, 1]$, observe that $f_n(x) \leq 1/n$. For $x > 1$, observe that $f_n(x) \to 1$ pointwise.

Ex. 9: f_n attains the maximum at $n/(n + 1)$.

Ex. 12: The sequence converges pointwise to $f(x) = x^2$. It is not uniform.

Ex. 14: Note that $\lim_{x \to 1_-} f_n(x) = 1/2$. Hence $f_n(x_n) > 1/4$ at some x_n.

Ex. 16: The sequence converges pointwise to 0. It is uniform iff $p > 2$.

Ex. 19: Show that g_n has a maximum $\sqrt{\frac{n}{2e}}$ at $\sqrt{\frac{1}{2n}}$.

Ex. 23: The sequence converges pointwise at $x \neq 0$.

Ex. 27: The result remains valid if we take $\sin nt$ in place of $\cos nt$.

Exercise Set 7.3.20:

Ex. 3: $1/2$ and 0.

Ex. 4: The pointwise limit 0.

Ex. 7: The pointwise limit is $f(x) = 1$ for $x \in [0, 1]$ and $f(x) = x$ for $x \in [1, 2]$.

Ex. 9: Darboux theorem 4.3.3.

Exercise Set 7.3.21:

Ex. 1: Modify the proof of Theorem 7.3.1.

Ex. 4: The domain is not bounded.

Ex. 6: Use the uniform continuity of f: Given ε, choose $\delta > 0$, and then N, so that $1/N < \delta$.

Ex. 7: Given $\varepsilon > 0$, let $\delta > 0$ correspond to the uniform continuity of g on $[-M, M]$. Let $N \in \mathbb{N}$ correspond to the uniform convergence of (f_n). Now, for $n \geq N$ and $x \in X$, we have the estimate:

$$|g(f_n(x)) - g(f(x))| < \varepsilon.$$

Ex. 9: Note that (f_n) is uniformly Cauchy on X. Adapt the proof of any Cauchy sequence of real numbers is bounded.

Ex. 10: We need to find a uniform estimate of $\left| \frac{g(x + \frac{1}{n}) - g(x)}{\frac{1}{n}} - g'(x) \right|$.

Ex. 12: Use the estimate $|f_n(x_n) - f(x)| \leq |f_n(x_n) - f(x_n)| + |f(x_n) - f(x)|$.

Exercise Set. 7.4.8:

Ex. 1-5: All are amenable to M-test. One may use calculus to find M_n, if required.

Ex. 6: Note that $\frac{1}{1+x^n} \leq \frac{1}{x^n - 1} \leq \frac{2}{x^n}$, for $n \gg 0$.

Ex. 10: Use Ex. 8b.

Exercise Set 7.8.4:

Ex. 1: Figure this out: $0 = \int_0^1 f(x) P_n(x)\, dx \to \int_0^1 f^2(x)\, dx$!

Ex. 2: (a) How does $p_n(x)$ behave if $x \gg 0$? (b) If $N > \deg p_n$, then observe $e^x > \frac{x^N}{N!}$. Go through the strategy of Theorem 3.3.7 and the estimate (3.4).

Ex. 4: Observe that p_n are uniformly continuous on $(0, 1)$. You may find Ex. 7.3.4 to be of use.

Ex. 5: If $q_n \rightrightarrows f$, use telescopic sum trick to get p_n's.

Bibliography

[1] Tom M. Apostol, *Mathematical Analysis*. Addison-Wesley Publishing Company, Inc., 1974.

[2] Robert G. Bartle and Donald R. Sherbert, *Introduction to Real Analysis*, John Wiley & Sons Inc. New York 1972.

[3] Edward D. Gaughan, *Introduction to Analysis*: Fifth Edition, Brooks/Cole Publishing Company, 1998.

[4] Richard R. Goldberg, *Methods of Real Analysis*, Blaisdell Publishing Company, 1964.

[5] M.H. Protter and C.B. Morrey, *A First Course in Real Analysis*, Springer-Verlag, 1991.

[6] Kenneth A. Ross, *Elementary Real Analysis*, Springer, 1980.

[7] Walter Rudin, *Principles of Mathematical Analysis*, Third Edition, McGraw Hill Inc., 1976.

The books [2], [3], [4], and [6] are excellent as collateral reading material.

[8] We also refer the reader to the following articles written by S. Kumaresan. They are available for download from http://main.mtts.org.in/expository-articles.

   (a) The Role of LUB Axiom in Real Analysis
   (b) Nested Interval Theorem and Its Applications
   (c) An Outline of Real Analysis
   (d) Construction of Real Numbers
   (e) Properties Equivalent to the LUB Property
   (f) Decimal, Binary and Ternary Expansions of Real Numbers
   (g) Convex Functions on $\mathbb{R}$ and Inequalities
   (h) Approximate Identities

Index

For Product Safety Concerns and Information please contact our EU
representative GPSR@taylorandfrancis.com
Taylor & Francis Verlag GmbH, Kaufingerstraße 24, 80331 München, Germany